Implications, Policies, and Frameworks of Intelligent Computing for a Sustainable Future

Anchit Bijalwan
https://orcid.org/0000-0002-9612-9073
British University Vietnam, Vietnam

Ali Al-Dulaimi
British University Vietnam, Vietnam

IGI Global
Scientific Publishing
Publishing Tomorrow's Research Today

Vice President of Editorial	Melissa Wagner
Director of Acquisitions	Mikaela Felty
Director of Book Development	Jocelynn Hessler
Production Manager	Mike Brehm
Cover Design	Jose Rosado

Published in the United States of America by
IGI Global Scientific Publishing
701 East Chocolate Avenue
Hershey, PA, 17033, USA
Tel: 717-533-8845 | Fax: 717-533-7115
Website: https://www.igi-global.com E-mail: cust@igi-global.com

Library of Congress Cataloging-in-Publication Data

LCCN: 2026015332 (CIP Data Pending)
ISBN13: 9798337395494
Isbn13Softcover: 9798337395500
EISBN13: 9798337395517
British Cataloguing in Publication Data
A Cataloguing in Publication record for this book is available from the British Library.

Table of Contents

Detailed Table of Contents

Artificial Intelligence (AI) is reshaping the face of digital connectivity and cybersecurity. Increasing dependence on highly integrated systems, thereby enhancing applications of AI for accurate detection of threats to optimize network performance and to support intelligent decision-making. The study applies a mixed methodology comprising bibliometrics and case studies to understand secure and ethical AI strategies across critical infrastructures. Study proposes a five-dimensional (security, transparency, interoperability, resilience, and e-governance) model for organizations and policymakers for the developing a responsible AI system towards secure connected digital futures. Study also has implications for strategic policy formulation for governments and organizations related to infrastructure design and ethical oversight in driving digital resilience for trustworthy connected futures.

In the fourth industrial revolution (4IR) context, innovations and changes in the construction industries worldwide require all nations to adopt innovations and technologies to improve productivity and achieve sustainable development goals (SDGs). Artificial intelligence technology (AI) has been developed and implemented in several industries in Vietnam. However, the application of AI is not as widespread among construction companies in the Vietnamese construction industry, and there still seems to be some uncertainty about the benefits and barriers of AI. This paper identified the application of AI technology in construction, its benefits and obstacles in the Vietnamese context. The research results support enhancing the implementation of AI to improve the productivity of the construction industry not only in more advanced countries but also in developing countries toward sustainable development goals.

This research examines the critical role of community-based financial mobilization in supporting innovative startup activities in Vietnam. Since its emergence in 2012, crowdfunding has served as an alternative financing solution, enabling entrepreneurs to raise capital from the public. Although this model has achieved notable successes, with platforms like FirstStep establishing market credibility, crowdfunding development in Vietnam faces significant challenges, particularly the absence of clear legal frameworks. It should be noted that academic research on Vietnamese startup crowdfunding remains notably scarce, making this study among the pioneers in this field. The research evaluates the current state of crowdfunding and analyzes success factors within Vietnam's innovation ecosystem, ultimately proposing regulatory frameworks to foster creativity and development in the startup landscape.

The European Union's General Data Protection Regulation (GDPR) mandates users' "Right to be Forgotten," yet in recommendation systems, merely deleting user data from storage is insufficient as information remains encoded in model parameters. This paper presents the first comprehensive evaluation framework for machine unlearning in collaborative filtering-based recommendation systems, systematically comparing state-of-the-art strategies across privacy guarantee verification, model utility preservation, and computational efficiency. We propose Adaptive SISA (A-SISA), enhancing the standard SISA framework with intelligent user clustering and dynamic shard sizing optimized for recommendation contexts. Extensive experiments on Movie Lens 25M and Amazon Product Reviews datasets demonstrate that A-SISA achieves 94.2% privacy protection with only 12% computational cost of retraining while maintaining 89.7% utility retention. Our findings provide practical guidelines for deploying GDPR-compliant machine unlearning in production recommendation systems.

Convolutional Neural Networks (CNNs) have become the cornerstone of modern computer vision as they can learn complex hierarchical feature representations and have provided state-of-the-art performance in medical imaging, Robustness is improved with adversarial training but at the cost of a careful tuning of its parameters, a higher computational burden, and a degradation of accuracy on clean data. Standard regularization methods such as weight decay and dropout are not tightly connected with robustness to strong attacks. In this work we consider spectral regularization as a computationally efficient off-the-shelf approach to improve adversarial robustness with no or small accuracy sacrifices. Spectral norm constraints on CNN weight matrices are computationally efficient and decrease sensitivity to perturbations, yet have been under-exploited in the literature. Results on CIFAR-10 and ImageNet demonstrates the enhanced robustness with ResNets at the baseline performance.

Mobile Edge Computing environments present challenges for learning agents, including delayed feedback and dynamic task arrivals. This paper introduces TARA (Trace-Aware Reinforcement Architecture), a deep reinforcement learning framework designed to improve temporal credit assignment and policy stability in volatile edge networks. TARA integrates eligibility traces into a dual-head neural architecture that separates state-value and action-value estimation. Simulations demonstrate improved learning responsiveness and delay-energy tradeoffs, supporting efficient task offloading in real-time Mobile Edge Computing deployments.

The explosive popularity of cloud computing has enhanced the vulnerability of enterprises to high-tech threats, one of the most virulent of which is fileless malware. On a Microsoft Security Incident dataset, which combines real system logs and network traffic, the model reaches a classification accuracy of 96%, thus beating baseline settings that do not include MHA. The results of training are stable: 97.2 proportions of correct answers and 95.5 proportions on the validation dataset, as well as the loss decrease are less than 0.01. The strength of the framework comes out of sophisticated pre-processing, an intelligent choice of feature engineering, and context-sensitive intra pattern analysis. Since it enables scalable, real-time anomaly detection and constant monitoring, the system can be used to shore-up cloud infrastructure against persistent threats which reside in memory.

In the digital age, data is becoming a valuable asset, and its protection is paramount. Phishing attacks pose a significant threat to data security, often leading to unauthorised access to sensitive information. This paper presents a novel approach for phishing detection that preserves user privacy. Utilizing federated learning, we propose a system that allows for collaborative model training without compromising individual data privacy. Our work involves distributing a global model across clients, who then gives back the model updates with their local data. These updates are aggregated to enhance the global model while ensuring the credibility of client contributions. FedCred is novel aggregation algorithm framed to ensure the credibility of the clients. The effectiveness of our approach is demonstrated through the use of comprehensive dataset, feature extraction techniques, and a robust model design. This research contributes to the cybersecurity field by offering a privacy-preserving solution to combat phishing threats.

This article surveys the current state-of-the-art in AI and NLP, and describes a Chatbox Tutor to learn STEM from chatbots, showing that current algorithms are ready to teach students in schools. The system uses principles from the learning sciences, where it provides adaptive hints, reflective questions, and metacognitive prompts to help students solve difficult maths and physics problems. A quasi-experimental study with 120 undergraduate engineering students compared an experimental group that used the Chatbox Tutor with a control group that used standard course materials. Compared to the control group, subjects who were exposed to the Chatbox Tutor learnt more (28% vs 14%), retained more (82% vs 65%), and were more engaged. Surveys of students in the study revealed that students found it helpful for getting quick questions answered and solving problems, but difficult to trust and understand vague questions.

This study addresses the early detection of propeller blade faults in multirotor unmanned aerial vehicles through vibration signal analysis. Triaxial acceleration data were collected under controlled hover conditions to capture representative blade states, including healthy, damaged, and unbalanced configurations. Diagnostic features were extracted from three complementary domains: statistical measures in the time domain, harmonic content in the frequency domain, and transient signatures via wavelet-based time–frequency analysis. Results show that damaged blades induce broadband vibration increases, while unbalanced blades generate distinct narrowband peaks at harmonic frequencies. These findings highlight axis-specific sensitivities—particularly along the Y-axis—and identify robust feature sets for distinguishing fault types. The proposed framework provides a validated signal processing basis for reliable fault detection, offering a foundation for future integration with machine learning and real-time UAV health monitoring systems.

Since 2020, Artificial Intelligence (AI) has emerged as a transformative force across multiple domains, with especially strong effects in the creative industries. Generative

AI (GenAI) is increasingly employed not only to streamline production but also to expand possibilities for collaboration and co-authorship, where AI contributes to creative decision-making and idea generation. This paper examines an experimental framework integrating GenAI into creative practice with a focus on reusing recycled materials at the intersection of new media. The central proposition is that discarded resources can be reconceived through GenAI to gain renewed significance and function. Initial experiments transformed textile waste from the fashion industry into digital creative code, generating novel patterns and outcomes. Progressing from discovery to development, these explorations culminated in the Tech:rial collection, illustrating how GenAI-assisted methods can support experimentation with repurposed materials and shape emerging trajectories of creativity.

This study investigates how spatial visual hierarchy shapes user experience (UX) in augmented reality (AR) storytelling applied to printed merchandise. While AR packaging and campaigns are increasingly used in creative industries, little research addresses how layout and visual sequencing affect comprehension and immersion of this technological format. Using a mixed-method design, the project analyzed pioneering AR campaigns from major advertising companies through structured observation and focus groups, identifying six recurrent spatial patterns. These informed the development of two prototypes tested with users in a controlled A/B experiment. Results showed that high-hierarchy designs significantly improved narrative clarity, emotional immersion, and perceived flow compared to low-hierarchy versions. The findings highlight spatial hierarchy as a key mechanism linking design structure with user engagement in hybrid media, combining physical and digital realms. This work contributes to design research by bridging theory and practice in AR-enhanced storytelling.

This pilot study explores how exposure to negative news about Vietnamese crime in Japan affects migrant workers' daily life, workplace experiences, and intention to

Preface

INTRODUCTION

The rapid advancement of intelligent computing has fundamentally transformed the ways in which individuals, organizations, and governments address complex technological and societal challenges. Artificial intelligence, machine learning, cloud computing, edge intelligence, cybersecurity, and data-driven decision-making continue to reshape industries while creating new opportunities for innovation, sustainability, and economic growth. At the same time, these technological developments present important questions regarding privacy, governance, security, ethics, and long-term resilience. As intelligent systems become increasingly integrated into critical infrastructures and everyday life, interdisciplinary research that bridges technical innovation with practical implementation and policy development has become more essential than ever.

Implications, Policies, and Frameworks of Intelligent Computing for a Sustainable Future bring together a diverse collection of scholarly research that examines the evolving landscape of intelligent computing across numerous disciplines and application areas. The contributions presented throughout this volume highlight innovative computational methods, emerging technological frameworks, and practical solutions designed to address contemporary challenges in cybersecurity, education, finance, engineering, cloud computing, creative design, and digital transformation. Collectively, the chapters emphasize the importance of responsible innovation while demonstrating how intelligent technologies can contribute to secure, sustainable, and human-centered solutions.

Beyond presenting technical advancements, this volume also underscores the importance of ethical governance, regulatory compliance, interdisciplinary collaboration, and societal impact. Readers will encounter research that explores both foundational algorithmic developments and practical applications capable of improving organizational performance, protecting digital infrastructures, enhancing educational outcomes, supporting sustainable development, and fostering creative innovation. The breadth of perspectives represented throughout the book reflects the increasingly interdisciplinary nature of intelligent computing and its growing influence across virtually every sector of society.

CHAPTER OVERVIEW

Chapter 1 examines the expanding role of artificial intelligence in strengthening cybersecurity and digital connectivity while proposing a multidimensional framework that integrates security, transparency, interoperability, resilience, and governance to support trustworthy digital ecosystems.

Chapter 2 investigates the adoption of artificial intelligence within Vietnam's construction industry, discussing both the opportunities for improving productivity and the technological, organizational, and regulatory barriers that continue to influence implementation.

Chapter 3 explores the development of crowdfunding as an alternative financing mechanism for entrepreneurial ventures, evaluating the current landscape in Vietnam while proposing regulatory strategies that encourage innovation and sustainable economic development.

Chapter 4 evaluates modern machine unlearning techniques designed to satisfy privacy regulations while maintaining recommendation system performance, introducing an adaptive framework that balances privacy protection, computational efficiency, and model utility.

Chapter 5 presents a spectral regularization framework that enhances the robustness of convolutional neural networks against adversarial attacks while maintaining predictive accuracy through computationally efficient optimization techniques.

Chapter 6 introduces a reinforcement learning architecture designed to optimize task offloading within mobile edge computing environments, improving learning responsiveness, policy stability, and resource efficiency.

Chapter 7 proposes an adaptive business intelligence dashboard framework that combines real-time analytics, user profiling, and intelligent visualization techniques to improve decision-making and user engagement.

Chapter 8 presents a deep learning framework for anomaly detection and performance optimization within edge security environments by integrating transformer models, graph neural networks, and reinforcement learning to create adaptive cybersecurity solutions.

Chapter 9 examines the application of deep learning techniques for detecting fileless cyberattacks within cloud computing environments, demonstrating highly accurate and scalable approaches for real-time threat detection.

Chapter 10 introduces a federated learning framework for phishing detection that preserves user privacy while improving collaborative model training through a credibility-based aggregation mechanism.

Chapter 11 investigates the use of artificial intelligence-powered conversational tutoring systems to improve STEM education through adaptive guidance, personalized learning support, and enhanced student engagement.

Chapter 12 presents a multi-domain vibration analysis framework for the early diagnosis of propeller faults in unmanned aerial vehicles by combining statistical, frequency-based, and wavelet-based analytical techniques.

Chapter 13 explores the application of generative artificial intelligence within sustainable creative practice, demonstrating how AI-assisted design can transform recycled materials into innovative interdisciplinary works.

Chapter 14 examines the influence of visual hierarchy on user experience within augmented reality storytelling, illustrating how interface design contributes to immersion, comprehension, and user engagement.

Chapter 15 investigates the relationship between negative media exposure, perceived discrimination, and migration intentions among Vietnamese workers in Japan, highlighting the broader social implications of technology, communication, and workforce mobility.

Chapter 16 concludes the volume with a comprehensive review of artificial intelligence-driven firewall technologies that integrate machine learning, anomaly detection, reinforcement learning, and real-time threat intelligence to strengthen enterprise cybersecurity while improving operational efficiency.

CONCLUSION

We hope this collection serves as a valuable resource for researchers, practitioners, policymakers, educators, and graduate students interested in the continued advancement of intelligent computing. By bringing together diverse perspectives spanning technical innovation, policy development, sustainability, and real-world implementation, this volume contributes to the ongoing dialogue surrounding the responsible development of intelligent technologies. It is our hope that the ideas presented throughout these chapters will encourage future research, interdisciplinary collaboration, and innovative solutions that help shape a more secure, ethical, and sustainable technological future.

Anchit Bijalwan
British University Vietnam, Vietnam

Ali Al-Dulaimi
British University Vietnam, Vietnam

Chapter 1
Emerging Trends in AI Applications for a Secure and Connected Future

Harish Kumar
http://orcid.org/0000-0002-7204-7321
Indian Institute of Management, Kashipur, India

ABSTRACT

Artificial Intelligence (AI) is reshaping the face of digital connectivity and cybersecurity. Increasing dependence on highly integrated systems, thereby enhancing applications of AI for accurate detection of threats to optimize network performance and to support intelligent decision-making. The study applies a mixed methodology comprising bibliometrics and case studies to understand secure and ethical AI strategies across critical infrastructures. Study proposes a five-dimensional (security, transparency, interoperability, resilience, and e-governance) model for organizations and policymakers for the developing a responsible AI system towards secure connected digital futures. Study also has implications for strategic policy formulation for governments and organizations related to infrastructure design and ethical oversight in driving digital resilience for trustworthy connected futures.

1. INTRODUCTION

Artificial Intelligence (AI), cloud computing, Internet of Things (IoT), and emerging communication technologies have made societies learn, communicate, work and keep digital things safe in new ways. This digital culture brings various issues in the fields of cybersecurity, ethical domain and governance. AI is stepping into the spotlight as a major player in creating safer and smarter digital environments.

DOI: 10.4018/979-8-3373-9549-4.ch001

AI is helping us to improve tasks, operational efficiencies, uncovering odd bits of information, and making decisions on autopilot across different sectors (Achuthan et al., 2024). To foster a digitally secure and interconnected future, it is essential to strike a thoughtful equilibrium between technological advancement and the imperatives of safety, privacy, and ethical accountability (Bahangulu and Owusu-Berko, 2025; Zou, 2024).

AI has transformative potential in managing data flows, intelligent automation, securing communication networks, IoT applications and cloud-based systems (Das et al., 2025; Siraparapu, and Azad, 2024). AI possesses transformative capabilities in the management of data flows, intelligent automation, protecting communication networks, Internet of Things applications, and cloud-based systems. AI can use machine learning (ML) and deep learning (DL) models to create a strong and adaptable infrastructure that can deal with threats that change all the time. Quantum computing and federated learning can create a strong cryptographic base for AI-driven cybersecurity (Bijalwan and Bijalwan, 2025; Achuthan et al., 2024; Sharma and Kumar, 2023).

The convergence of AI with critical systems and advanced networks have widely opened new opportunities for a more intelligent, automatic and efficient systems. Using generative AI can make cybersecurity and digital resilience better. "Generative Adversarial Networks (GANs)" and "Variational Autoencoders (VAEs)" make things like images, text, and audio for data augmentation and simulations. But it also makes new cyber threats possible, like fake news, deepfakes, and making content without permission (Fu, Hadid, and Damer, 2025; Ashutosh et al., 2024). Therefore, strategic frameworks are required to deal with new cyber threats and moral issues in places where technology is advanced.

This research is trying to fill in those gaps by putting together learning from recent studies and real-world case examples. The aim is to support the development of responsible and secure AI-powered applications, that are crucial in building a digitally safe and connected future. The research sets following objective -

To identify strategic approaches for organizations and governments to implement secure AI solutions for digital resilience and a connected future.

The results revolved around five interconnected aspects e.g. security, transparency, interoperability, resilience and governance to AI adoption towards digital connected environment. Study shows few empirical evidence and technology exemplars that gives an actionable roadmap. This cross-business approach is key to being able to spot systematic gaps and possible areas of convergence.

2. REVIEW OF LITERATURE

AI has established itself as a driving force to combat cyber security and ethical oversight challenges by securing, managing and upgrading the intelligent digital infrastructure (Ashutosh et al., 2024). Pioneering AI solutions provide dynamic threat detection and response, by analysing user behaviour and looking for anomalies in different phases. Zou (2024) emphasises the role of ethical governance in AI development, including ensuring fairness-aware algorithms, regular audits for bias detection and stakeholder inclusivity. Bahangulu and Owusu-Berko (2025) note the importance of data ethics and governance in ensuring that AI systems. In literature review, study discusses the most recent development in AI based applications for secure and connected systems.

2.1 AI-Enabled IoT and Infrastructure Security

The fusion of AI with IoT and cloud-based technology is generally acknowledged as a transformative enabler to intelligent automation, predictive analytics and real-time decision support. But integration also adds some core vulnerabilities, especially in multi-cloud and hybrid cloud settings. Ali et al. (2025) present an extensive taxonomy of security and privacy issues in these architectures, including threats from cross cloud attacks, side channel leaks and insecure APIs. In their review, they highlight the role of cryptographic primitives, zero-trust architectures and AI enabled threat detection solutions that are necessary to enable secure data sharing between institutions while ensuring regulatory compliance. They suggested a layered security framework for trusted infrastructure, end-to-end encryption, multi-factor authentication and continuous enterprise monitoring.

Ransomwares are also attacking IoT devices, as these devices have less computational power and non-uniform security protocols (Yan and Khoei, 2025). During COVID-19, attacks like Ryuk, Maze and BlackCat have exposed the dependency from connected systems especially in healthcare or municipality infrastructure. According to Siraparapu and Azad (2024) to cope with such threats, the secure system architecture (SSA) needs to incorporate encryption data, access control protocol intrusion detection mechanism and anonymization.

AI is an integral component of 6G advance network for intelligent resource assignment, anomaly detection and predictive maintenance (Das et al., 2025). Alotaibi and Barnawi (2023) reviewed AI and IoT deployments in 6G networks for understanding the state of the art in IDSes which revealed drawbacks of traditional techniques and proposed intelligent architectures, by combining ML, DL and SDN which could be applied to detect and prevent attacks. The authors propose the application of edge intelligence and network function virtualization (NFV) for scalability and agility enhancement.

Alotaibi and Barnawi (2023) investigate how AI can be used to ensure large-scale IoT deployments within secure 6G networks. Their study emphasizes the insufficiency of conventional intrusion detection systems (IDS), and suggests intelligent infrastructures that leverage ML, DL and SDN to discover and prevent cyber-attacks. The authors suggest combining edge intelligence with NFV to increase scalability and responsiveness. Methods like deep learning (DL), reinforcement learning (RL) and federated learning (FL) are being incorporated into network optimization and cyber security schemes. These initiatives highlight the necessity for both governments and corporations to integrate AI solutions that are not only secure but also adaptable in dynamic in contexts.

2.2 Privacy-Preserving AI in IoT Ecosystems and Encryption Strategies

Tsouplaki, Fung, and Kalloniatis (2025) suggested a layered framework of AI and IoT infrastructure. Filaly et al. (2025) used hybrid techniques for an extensive analysis of encryption methodologies within Hadoop-based big data ecosystems and introduced a blockchain based framework for asymmetric encryption to enhance data security in decentralized systems. This approach is consistent with strategic objectives of permeating with each other applications as well as robustness against failures and paves a way for decentralized key management and strong data integrity. The research looks at newer trends like post-quantum encryption and blockchain integration, showing possibility of making "future-proof" AI systems against new threats.

Rustemi and Dalipi (2025) examine the use of blockchain, AI, and IoT in smart agriculture and emphasize on how joint deployment has the potential to improve transparency, traceability and automation. Their study recognizes the applications of block-chain in securing supply chains, authenticating data provenance and smart contracts for autonomous behaviour. The authors suggest a strategy using blockchain for decentralized key management and AI for intelligent automation. Fu et al. (2025) examine the utilization of generative AI technologies within healthcare sectors, including diagnostic support, caregiving robotics, cognitive therapy and rehabilitation instruments. While these applications provide substantial advantages in personalization and efficiency, but also present challenges concerning explainability, bias, and trust.

2.3 AI for Critical Infrastructure Protection

Raval et al. (2024) investigate how AI contributes to security frameworks across sectors like energy, finance, agriculture, and healthcare. They delineate a structured classification of AI-enhanced safeguards, including biometric authentication, anomaly detection, and access control based on risk assessment. The authors put forward an AI-driven security architecture for smart grids, which was evaluated based on accuracy, training time, and ROC curves. This model illustrates the strategic use of AI to increase service continuity, real-time threat detection and adaptive governance. The study also addresses open challenges, like algorithmic discrimination, information exchange between AI and human, and societal support for the collaboration between human and AI.

Increasing array of IoT and ransomware threats coupled with the potential shifts towards 6G networks further complicate that landscape requiring proactive planning by both organizations and governments. Further, many devices do not have OTA update functionalities which make them vulnerable consistently to the known exploits (Siraparapu, and Azad, 2024). This is indeed distressing in part of the critical infrastructure segments. For example, ransomware targeting hospital networks have disrupted emergency care's services, encrypted patient histories and led to the diversion of immediate patient care (Yan, and Khoei, 2025).

The emergence of Explainable AI (XAI) platforms is a ray of hope to facilitate the transparency and trust. XAI seeks to make AI decision-making more interpretable and understandable to the user, enabling them to make informed decisions, while minimizing liability. AI workload is prevailing, with the increasingly popular applications of AI, the demand on both scalable and secure infrastructure becomes urgent. As cloud architectures move from centralised to decentralised edge and fog computing, there are requirements for modular and energy-efficient data centres (Agarwal et al., 2025).

The literature surveyed in this work provides a strong base to derive strategic directions for the deployment of AI with security, transparency, interoperability and long-term resilience as main concerns. By aggregating the lessons from across domains and technologies, this review provides a roadmap for organizations and governments to traverse the vast AI applications. In an increasingly digital ecosystem, strategic convergence with resilience and connectivity objectives will be key to the responsible harnessing of AI's transformative power. The research reinforces the need for sector cross-collaboration and regulatory intervention to help scale AI solutions responsibly.

Based on retrieved literature, this work realizes few research gaps for example lack of scalability, need of domain-specific XAI models with the trade-offs described above in cybersecurity. Additional research is needed on how to implement privacy-preserving AI operationally at scale across domains. Additionally, the development of standardized methods and standard architectures that support secure and interoperable AI across different digital infrastructures is required to improve cybersecurity, data privacy, and interoperability between digitally connected ecosystems.

3. RESEARCH METHODOLOGY

This study employs a mixed-methods approach combining systematic literature review (SLR), comparative analysis, and conceptual synthesis. Over 24 peer-reviewed journal articles published between 2020 and 2025, including surveys and case studies were analysed to perform this research. Study adopts comparative framework analysis approach to evaluate technical architectures, security protocols, and governance models across sectors. The results are also supported by real-world case studies to contextualize the strategic recommendations. This mixed methodology facilitates a holistic comprehension of the technical and policy aspects of secure AI implementation.

A comparative framework analysis offers a systematic approach for evaluating and differentiating diverse frameworks relevant to a particular discipline. This process might include looking at policy frameworks, governance structures, technological infrastructures, or strategic systems. Scholars in disciplines like public policy, governance, education, and technology often examine the practical efficacy, coherence, or synergy of these frameworks (Ragin, 2014; Srivastava and Thomson, 2009). The procedure consists of seven simple phases.

a). Define the Comparative Objective

The first step in making a useful comparison. This means making clear what the study is trying to find out and what kinds of frameworks are being looked at (Ragin, 2014). The primary objective of this research is to identify strategies for public and private entities to leverage AI technology to enhance the resilience of the digital realm and foster a more interconnected future. The analysis centres on contrasting AI implementation models across key sectors namely healthcare, energy, finance, and smart infrastructure to evaluate how well they support resilience goals and uphold ethical principles.

b). Select Frameworks for Comparison

This step involves identifying and selecting the frameworks to be analysed. Frameworks may be drawn from academic literature, policy documents, organizational strategies, or international standards. The study uses 24 peer-reviewed journal articles published between 2020 and 2025. The study looked at different frameworks for comparison, such as AI governance models, cybersecurity architectures, privacy-preserving AI systems, and ethical oversight mechanisms.

c). Develop a Comparative Matrix

A matrix needs to be constructed to categorize the frameworks according to designated dimensions or metrics, facilitating systematic comparison. These criteria may derive from theoretical frameworks, policy objectives, or empirical practices (Srivastava, and Thomson, 2009). The study applies a range of analytical dimensions such as approaches to threat mitigation, methods for ensuring explainability, interoperability across platforms, resilience to system faults, and adherence to ethical standards to evaluate and contrast the frameworks systematically.

d). Data Collection and Coding

The article's comparative framework analysis relies on data from various qualitative sources. Study uses 24 journal articles that were published between 2020 and 2025 related with AI frameworks that incorporate fairness-aware algorithms and bias audits (Zou, 2024) to analyse how various policy models address ethical alignment, accountability, and regulatory compliance in the formulation of rules, standards, and governance frameworks for AI, cybersecurity, and digital infrastructure.

To understand standardization specifications of detail descriptions about the system architecture, protocol, algorithm, and security mechanisms utilized by AI-based infrastructures, study relies upon multiple documents to compare which technical solutions are efficient in terms of solution/application space and scalability feasible for different sectors or platforms. Study uses and analyses the cryptographic procedures and zero-trust architecture for multi-cloud environment (Ali, 2025), AI-enabled intrusion detection system (IDS) with deep learning and federated learning (Bijalwan and Bijalwan, 2025; Alotaibi, and Barnawi, 2023), and Blockchain consensus models such as PoA and hybrid algorithms for IoT security (Sun, 2025).

Strategic alignment was examined by luminating long-term goals around AI adoption, digital resilience and promotion of modernised infrastructure. Study integrates smart grid security frameworks with AI anomaly detection and predictive maintenance (Raval et al., 2024). Study cites few real-world applications and evaluations of AI systems in particular fields to reinforce theoretical explanations like Ransomware attacks on hospital networks influence at emergency services (Yan and Khoei, 2025), Blockchain-AI integrated solution in smart agriculture for traceability and automation (Rustemi and Dalipi, 2025) and Generative AI applications to medical diagnostics and robotic caregiving (Fu, Hadid, and Damer, 2025).

e). Comparative Analysis

After coding the data, the study conducts a comparative analysis to identify recurring themes, significant differences, and actionable insights. This process includes surfacing effective practices such as the use of federated learning to safeguard privacy, spotting shortcomings like the absence of scalable explainable AI models, and tracing areas where frameworks either align or diverge, particularly in sector-specific security implementations.

f). Synthesis and Interpretation

Insights obtained from the comparative analysis are amalgamated to formulate strategic guidance, advance theoretical development, and impact relevant policy discussions. This step connects empirical observations to the broader research objective (Yin 2018). Study proposes a five-dimensional strategic framework for responsible AI deployment (section 4 and section 5). Study also provides actionable recommendations (section 6) for policymakers and organizations, including unified standards, ethical governance models, and resilient infrastructure design for building secure and connected digital ecosystems.

g). Validation and Reflexivity

To ensure rigor, study validates its findings through triangulation, peer review, case studies, and thematic synthesis (Section 4 and section 5).

4. RESULTS & ANALYSIS

Five interdependent dimensions (figure 1) like security, transparency, interoperability, resilience and governance to AI application for digital resilience are identified through the study. Each dimension covers a very strategically relevant aspect related to how AI systems will be aligned with long-term objectives in the creation of secure, connected and trusted digital ecosystems.

4.1 Security and Threat Mitigation

Safety continues to be the priority when it comes to deploying AI in critical infrastructure. With the increasing deployment of AI as smart grid control, healthcare decision support, financial risk assessment and IoT network management component, intelligent systems face sophisticated cyber threats such as adversarial attacks, data poisoning and model inversion. AI-based intrusion detection systems (IDS), particularly those utilizing deep learning (DL) frameworks, have demonstrated efficacy in identifying botnets and DDoS attacks within extensive IoT networks (Alotaibi and Barnawi, 2023).

Scalability, reactiveness are still important challenges. IDSs that are based on DL, often demand significant computational resources and training data, which may not be possible in edge or resource-limited environments. Additionally, the adaptive models are required to handle cyber threats that keep changing over time. Newer paradigms like federated learning and edge intelligence provide some answers to this problem, as the model training is decentralized, so anything that would interfere with threats could be detected locally (Agarwal et al., 2025; Raval et al., 2024).

Figure 1. Strategic Framework for AI deployment for a secure and connected future

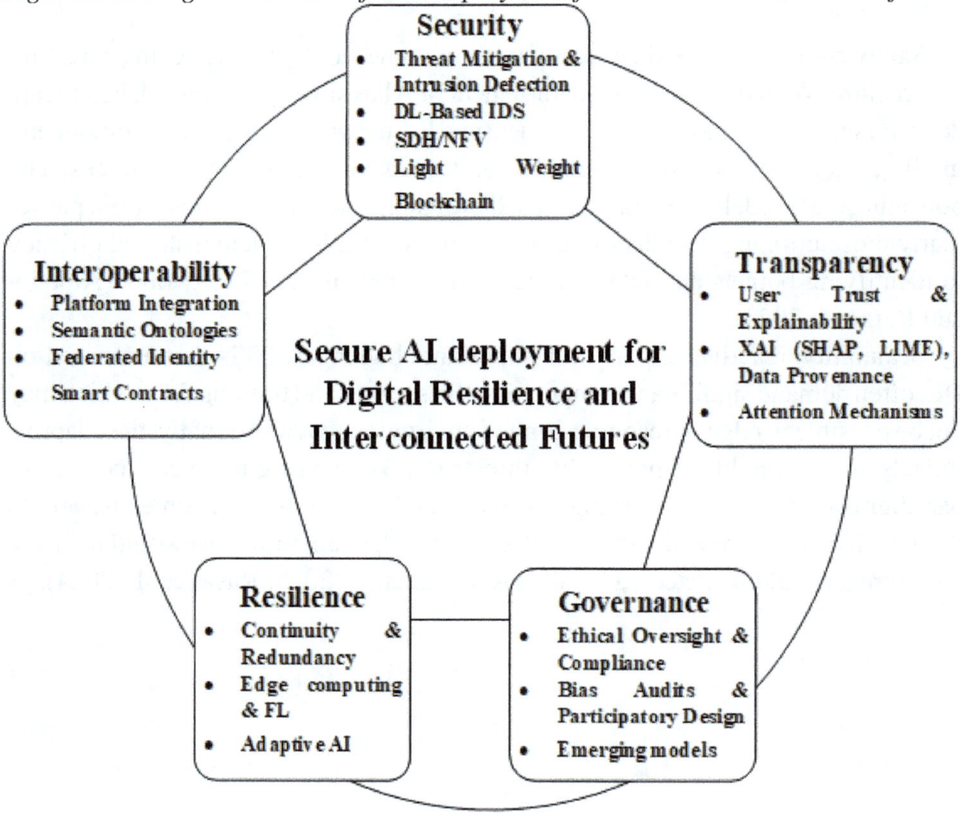

The AI consideration along with SDN and NFV enhances the security through programmable and flexible network architectures. SDN decouples control from data and AI gives the ability for reconfiguration of network flows based on detected threat (Alotaibi and Barnawi, 2023). NFV enables security functions to be deployed in fixed or flexible policy-driven slices on a virtual infrastructure. Sun et al. (2025) emphasize on lightweight approaches such as proof-of-authority (PoA) or hybrid approaches to manage trade-offs between IoT universe security and efficiency.

4.2 Transparency and Explainability

In critical domains such as healthcare, border control and law enforcement, trust is crucial for AI systems. Adil et al. (2025) emphasize the growing relevance of explainable AI (XAI) in biometric technologies, particularly those involving cognitive and behavioural identification. For systems like facial recognition, iris scans, and gait analysis, it is essential that both users and regulatory bodies can comprehend the rationale behind authentication outcomes. However, the absence of a standardized metric for what constitutes a valid "explanation" poses challenges and making it difficult for AI systems to develop trustworthiness and equally difficult for legal frameworks to enforce accountability.

Recent developments in XAI techniques are SHAP "(SHapley Additive exPlanations)", LIME "(Local Interpretable Model-agnostic Explanations)" and attention mechanisms within neural networks. Such techniques provide visibility into which features are contributing most greatly to decisions. It also applies in terms of data lineage and model training transparency.

4.3 Interoperability Across Platforms

Scalable and fault tolerant AI deployment requires interoperability. With multi-cloud and hybrid cloud deployments, AI systems need to be able seamlessly run on different platforms with diverse data formats and APIs. Without interoperability, data silos form, workflows slowdown in fragmented situations, and security is at risk. Ali et al. (2025) highlight these problems in multi-cloud environments including a weak Identity and Access Management (IAM) strategy, inconsistent encryption standards, and different compliance frameworks.

Federated identity management and standardized policy frameworks are two good ways to bring access control together. Interoperability in blockchain-based IoT infrastructures is even harder because there aren't enough resources and there are so many different protocols. Sun et al. (2025) point out that there is a place for an architecture supporting lightweight and decentralized orchestration, while allowing cross-chain communication, centralized consensus and simple design. Solutions such as sidechains, interoperability bridges, and smart contract standardization are currently being considered to solve these problems. Integration can be made semantically, when ontologies, metadata standards and schema-matching tools bridge systems or files, so that AI models are able to understand and work with the data. This is especially relevant in health care and smart city applications, where the data comes from sensors, wearables and legacy systems.

4.4 Resilience and Redundancy

The digital resilience describes the capability of an organization to continue operation under, or quickly recover from, a disruptive event. AI can be used to improve resiliency with predictive maintenance, anomaly detection, and adaptive control. Mitra et al. (2024) discuss a Kalman filter technique for identifying anomalous consumption patterns while preserving user privacy. To keep grid systems stable, even when there are cyber threats or data breaches, it's important to find a balance between privacy and security. To keep operations running smoothly, it's important to build in redundancy, like backup systems and failover mechanisms. Artificial intelligence can help fine-tune redundancy by identifying vulnerable points, forecasting potential breakdowns, and managing resources dynamically. Intelligent technologies may swiftly transfer workloads, alter network traffic routes, and reinstate services to minimize downtime in cloud-based systems. Edge computing enhances system stability by relocating data processing nearer to the source. This indicates a reduced dependence on centralized infrastructure. Despite network issues or latency surges, enterprises utilizing edge AI models can continue to operate effectively (Agarwal et al., 2025). Federated learning enhances robustness by enabling simultaneous model training across all nodes (Raval et al., 2024).

4.5 Governance and Ethical Alignment

For AI to be used responsibly, there must be strong ethical standards and good governance. Raval et al. (2024) caution against algorithmic bias in critical infrastructure, resulting in discriminatory outcomes and diminishing public trust. Values alignment requires that AI systems exercise human rights and refrain from causing harm, while promoting inclusiveness. This necessitates various types of training data, de-biasing approaches, and participatory design methods. Governance also includes responsible banking, auditable and accountable practices, and stakeholder inclusiveness. Governance frameworks of AI need to be established to make sure that the AI models running on IoT will be explainable, auditable and comply with ethical principles. This includes requiring XAI in high-stakes applications such as healthcare, autonomous vehicles or criminal sentencing.

The organizations should create clear roles and responsibilities related to the supervision of AI systems, along with right of appeal and redress. Transparency in model design, training data and performance metrics is significantly important for compliance as well public review. It is important to have cross-sectoral collaboration on standardizing governance. Public-private partnerships can help speed the adoption of responsible AI practices and guarantee they meet societal norms.

5. DISCUSSIONS

The emergent dimensions in section 4 (figure 1) intersect across sector and technologies, requiring a multifaceted approach that brings together technical innovation, policy reform, and cross-sector cooperation.

5.1 Sectoral Convergence Requires Unified Standards

The absence of common rules that different sectors can adopt remains a key obstacle for resilience. Silos are imposed by sector-specific protocols, data formats and governance models that globally slow down the uptake and expansion of AI solutions. For example, smart meters used in power networks and blockchain systems used in supply chains are often built on platforms that don't work with each other. The lack of interoperability makes it harder to share and coordinate data, especially when it comes to cybersecurity threats (Sun et al., 2025; Mitra, Chakraborty, and Mitra, 2024).

Unified standards are needed to make AI systems work together across different fields to solve these problems. This includes shared ontologies of data, APIs that interoperate and security protocols that are standardised. International agencies such as the ISO, IEEE and NIST have already began to tackle these challenges; however, a multidisciplinary synthesis is mandatory. To start blockchain-based IoT systems, protocols can be made to allow smart contracts to be sent and received across chains. These mechanisms help make different decentralized networks work together and make systems run more smoothly (Sun et al., 2025).

5.2 Edge Intelligence and Federated Learning

Federated learning and edge computing are becoming important technologies for protecting users' privacy (Raval et al., 2024). In federated learning, data stays on devices and updates are communicated to a central server. These can make systems more resilient, with computation distributed throughout the grid rather than concentrated in central assets. Due to the edge computing, even when there're network failures or being attacked from outside, the edge nodes can still work independently as they do conventionally. Adopting edge intelligence and federated learning needs to take the resources constraints, communication overheads, and model convergence into account. The design of scalable, robust, lightweight architectures, efficient aggregation algorithms and secure communication protocols are critical for large scale deployment. Alotaibi and Barnawi (2023) work on efficient networks capable of supporting large scale IoT deployment in 6G, within which edge computing can benefit from SDN and NFV to generate intelligent reconfigurable network infrastructure.

5.3 Explainability Must Be Embedded in Design

Explainability is a fundamental requirement for trust in AI system, especially where the consequences of decisions are ethically or legally consequential or have safety-critical implications. Post-hoc explanation approaches are useful but for mission-critical systems that need real-time interpretability and auditability, these are not sufficient. Rather, explainability needs to be built into those AI models and interfaces. The AI Act also wants to make sure that people can see and control high-risk AI. In addition, explainability supports stakeholder involvement and contribute to participatory design. Communicating AI decisions to non-experts could allow organizations to engage users, regulators and civil society in the development and governance of these systems.

5.4 Lightweight Architectures of Blockchain to enhances Trust

Utilizing blockchain technology has the potential of building trust, transparency and decentralization in AI applications. This aspect is particularly useful in a multi-stakeholder environment, e.g. in supply chains or smart cities and in collaborative research network (Sun et al., 2025).

But classical blockchain framework like PoW, are resource extensive and it is not efficient and limited for the IoT environment. This is a problem that has attracted the attention of researchers, who are investigating lightweight consensus mechanisms including PoA, PoS and hybrids to achieve both security and efficiency. Sun et al. (2025) Suggest a classification of fault tolerance and lightweight blockchain designs that can be profiled with IoT to facilitate secure data exchange and implementation of decentralised AI. In multi-cloud environments, blockchain can behave as a reliable intermediary for managing access controls and identities (Ali et al., 2025). In federated learning, blockchain makes it safe to combine data. Furthermore, privacy-preserved mechanisms like homomorphic encryption can be leveraged for enhancing the trust of blockchain-supported AI systems.

5.5 Governance and Ethical Alignment

To safely use AI, we need to have good morals and good government. Technology changes how we make decisions about banking, health care, energy, and public safety. Raval et al. (2024) assert that algorithmic bias in critical systems may erode trust and yield inequitable outcomes. Governance needs openness, responsibility, and participation. Businesses need to hire people to oversee AI and let workers question and fix mistakes. For compliance and disclosure, keep records of the design of the model, the training data, and the results of the performance evaluation. To keep governance norms, people from different fields need to work together. Working together between the public and private sectors can make AI more accountable and better for society. We need to figure out how long it will take to make AI better in a way that is both ethical and long-lasting.

6. POLICY, CONTRIBUTIONS, AND STRATEGIC IMPLICATIONS

The strategic use of artificial intelligence (AI) in essential systems and services requires a strong policy framework and coordinated action among all stakeholders. The results of this study (figure 1), which revolved around five aspects of secure, transparent, interoperable and resilient AI technologies, reinforce the pressing need to bring technological innovation in line with governance, ethics and long-term digital resilience. Combining perspectives from energy systems (Mitra, Chakraborty, and Mitra, 2024), biometric authentication (Adil et al., 2025), multi-cloud deployments (Ali et al., 2025), blockchain-driven IoT (Sun et al., 2025) as well as 6G for massive IoT coverage (Alotaibi, and Barnawi, 2023), this research provides a holistic synthesis of AI deployment challenges and strategic facilitators cross-industry. This crossing of domains is necessary to see systemic vulnerabilities and convergence opportunities.

The research proposes a five-dimensional framework (figure 1) security, transparency, interoperability, resilience and governance that could help organizations and governments assess the quality of their AI systems in relation to DR goals. Every aspect is substantiated with data and examples, thus providing a practical pathway for implementation. The study emphasizes edge intelligence, federated learning, explainable AI and lightweight blockchain architectures and standardised policy engines as the emerging enablers for secure and scalable deployment of AI applications. AI systems need to be co-designed with the communities, particularly in cases of public services and surveillance. Participatory Design is a good way to make sure that we build systems which are inclusive, context-aware, and ethically grounded.

Enterprises need to create governance structures that can focus on risk assessment, ethical compliance, and stakeholder engagement. These models should be consistent with sectoral policies and international norms. Organizations must prioritize infrastructure that promotes secure data exchange, cross-platform interoperability and real-time threat detection. It also means adding SDN/NFV, blockchain, and federated learning to the architectures of businesses (Alotaibi, and Barnawi, 2023). AI systems should be built to do more than just work well. They should also be able to handle problems and be long-lasting, flexible, and able to reinstate quickly. Redundant system architecture, distributed control structures, and adaptive learning mechanisms are essential components (Mitra, Chakraborty, and Mitra, 2024). Interoperability is that which allows for collaboration, scalability and innovation. It is critical to build AI across industries, platforms and borders. Investments in interoperability pay long-term dividends in terms of efficiency and resilience.

7. CONCLUSIONS

This research highlights the need for a multi-dimensional approach that governments and organizations should consider in adopting AI systems that are secure, transparent and interoperable while being resilient in an increasingly digitally connected future. The five crucial dimensions of security and threat mitigation, transparency and explainability, interoperability between systems, resilience and redundancy, and governance with ethical alignment demonstrate that the deployment of AI is not simply a technical exercise but a systemic challenge that requires coordinated public-policy responses across sectors. Incorporating edge intelligence and federated learning provides a possible way forward to improve privacy and scalability, as does embedding explainability in the design of AI which is critical for ensuring accountability in high stakes domains. While blockchain is good in providing trust, it needs to go toward lighter architectures in resource-constrained environments. Furthermore, strategic use of AI that steps to the long-term goals of digital resilient can protect critical infrastructure and public services with trust, innovation and benefits for society in a world increasingly driven by intelligence and connectivity.

Though the research learns from across energy, healthcare, finance and IoT sectors, it is predominantly literature based lacking empirical testing through field deployment or longitudinal case studies. This restriction makes it hard to evaluate how well a given AI framework can deal with real-world limitations like latency, data sparsity or adversarial problems. Several of the strategic advice could need to be altered continuously to keep pace with emerging regulatory and technology dynamics. The study does not deal in depth in the socio-cultural and geopolitical implications around how AI is deployed like digital sovereignty, indigenous data rights or regional imbalances of infrastructure. Future research may focus on empirical validation of AI resilience models. Studies should investigate the combination of AI and next generation technologies as quantum-computer, neuromorphic systems, holographic communications to improve security, scalability and explainability.

REFERENCES

Achuthan, K., Ramanathan, S., Srinivas, S., & Raman, R. (2024). Advancing cybersecurity and privacy with artificial intelligence: Current trends and future research directions. *Frontiers in Big Data*, *7*, 1497535. DOI: 10.3389/fdata.2024.1497535 PMID: 39703783

Adil, M., Farouk, A., Ali, A., Song, H., & Jin, Z. (2025). Securing tomorrow of next-generation technologies with biometrics, state-of-the-art techniques, open challenges, and future research directions. *Computer Science Review*, *57*, 100750. DOI: 10.1016/j.cosrev.2025.100750

Agarwal, P., Bijalwan, A., Patel, A., Deshpande, A. A., & Panchal, V. (2025). Edge AI demystified: From its origins to future frontiers. In *Edge artificial intelligence: Foundations, techniques, and applications* (pp. 1–38). Wiley., DOI: 10.1002/9781394355037.ch1

Ali, S., Talpur, D. B., Abro, A., Alshudukhi, K. S. S., Alwakid, G. N., Humayun, M., Bashir, F., Wadho, S. A., & Shah, A. (2025). Security and privacy in multi-cloud and hybrid cloud environments: Challenges, strategies, and future directions. *Computers & Security*, *104599*, 104599. Advance online publication. DOI: 10.1016/j.cose.2025.104599

Alotaibi, A., & Barnawi, A. (2023). Securing massive IoT in 6G: Recent solutions, architectures, future directions. *Internet of Things : Engineering Cyber Physical Human Systems*, *22*, 100715. DOI: 10.1016/j.iot.2023.100715

Bahangulu, J. K., & Owusu-Berko, L. (2025). Algorithmic bias, data ethics, and governance: Ensuring fairness, transparency and compliance in AI-powered business analytics applications. *World Journal of Advanced Research and Reviews*, *25*(2), 1746–1763. DOI: 10.30574/wjarr.2025.25.2.0571

Bijalwan, A., & Bijalwan, J. G. (2025). An efficient framework to detect sexual predators early by using federated learning approach. In *2025 International Conference on Next Generation of Green Information and Emerging Technologies (GIET)* (pp. 1–6). IEEE. https://doi.org/DOI: 10.1109/GIET65294.2025.11234823

Das, B. R., Hasan, S. R., Sabuj, S. R., Hossain, M. A., & Ray, S. K. (2025). A comprehensive survey on emerging AI technologies for 6G communications: Research direction, trends, challenges, and opportunities. *International Journal of Intelligent Networks*, *6*, 113–150. DOI: 10.1016/j.ijin.2025.06.001

Filaly, Y., Berros, N., & Younes El Bouzekri, E. L. (2025). A comprehensive survey on big data privacy and Hadoop security: Insights into encryption mechanisms and emerging trends. *Results in Engineering*, *106203*, 106203. Advance online publication. DOI: 10.1016/j.rineng.2025.106203

Fu, B., Hadid, A., & Damer, N. (2025). Generative AI in the context of assistive technologies: Trends, limitations and future directions. *Image and Vision Computing*, *154*, 105347. DOI: 10.1016/j.imavis.2024.105347

Mitra, S., Chakraborty, B., & Mitra, P. (2024). Smart meter data analytics applications for secure, reliable and robust grid system: Survey and future directions. *Energy*, *289*, 129920. DOI: 10.1016/j.energy.2023.129920

Ragin, C. C. (2014). *The comparative method: Moving beyond qualitative and quantitative strategies*. Univ of California Press. DOI: 10.1525/9780520957350

Raval, K. J., Jadav, N. K., Rathod, T., Tanwar, S., Vimal, V., & Yamsani, N. (2024). A survey on safeguarding critical infrastructures: Attacks, AI security, and future directions. *International Journal of Critical Infrastructure Protection*, *44*, 100647. DOI: 10.1016/j.ijcip.2023.100647

Rustemi, A., & Dalipi, F. (2025). Synergizing IoT, AI, and blockchain for smart agriculture: Challenges, opportunities, and future directions. *Internet of Things : Engineering Cyber Physical Human Systems*, *101778*, 101778. Advance online publication. DOI: 10.1016/j.iot.2025.101778

Sharma, V. K., & Kumar, H. (2023). Enablers driving success of artificial intelligence in business performance: A TISM-MICMAC approach. *IEEE Transactions on Engineering Management*, *71*, 4665–4675. DOI: 10.1109/TEM.2023.3236768

Siraparapu, S. R., & Azad, S. M. A. K. (2024). Securing the IoT landscape: A comprehensive review of secure systems in the digital era. *e-Prime-Advances in Electrical Engineering. Electronics and Energy*, *10*, 100798. DOI: 10.1016/j.prime.2024.100798

Srivastava, A., & Thomson, S. B. (2009). *Framework Analysis: A Qualitative Methodology for Applied Policy Research (*SSRN *Scholarly Paper No. ID 2760705). Rochester, NY: Social Science Research Network.*

Sun, X., Yu, X., Huang, Q., Wang, Z., Guo, J., Huang, Z., & Xie, F. (2025). Reliability techniques and architectures for blockchain-enabled internet of things: Current applications, systematic review, and future trends. *Computer Standards & Interfaces*, *104068*. Advance online publication. DOI: 10.1016/j.csi.2025.104068

Tsouplaki, A., Fung, C., & Kalloniatis, C. (2025). Enhancing IoT privacy with artificial intelligence: Recent advances and future directions. *Internet of Things : Engineering Cyber Physical Human Systems, 101752*, 101752. Advance online publication. DOI: 10.1016/j.iot.2025.101752

Yan, P., & Khoei, T. T. (2025). Securing the internet of things: A comprehensive review of ransomware attacks, detection, countermeasures, and future prospects. *Franklin Open, 11*, 100256. DOI: 10.1016/j.fraope.2025.100256

Yin, R. K. (2018). *Case study research and applications* (Vol. 6). Sage.

Zou, R. (2024, September). Research on ethical issues, data privacy protection, algorithmic bias, and regulatory policy of Artificial Intelligence Technology in Digital Transformation. In *International Conference on Economic Management and Green Development* (pp. 234-242). Singapore: Springer Nature Singapore. https://doi.org/DOI: 10.1007/978-981-96-3236-7_21

Chapter 2

Artificial Intelligence Adoption in the Vietnamese Construction Industry: Benefits and Barriers

Paul Harbottle
Qatar Energy LNG, Qatar

An Le Thi Hoai
Massey University, New Zealand

Quynh To Thi Huong
http://orcid.org/0009-0007-4314-0969
Hanoi University of Civil Engineering, Vietnam

ABSTRACT

In the fourth industrial revolution (4IR) context, innovations and changes in the construction industries worldwide require all nations to adopt innovations and technologies to improve productivity and achieve sustainable development goals (SDGs). Artificial intelligence technology (AI) has been developed and implemented in several industries in Vietnam. However, the application of AI is not as widespread among construction companies in the Vietnamese construction industry, and there still seems to be some uncertainty about the benefits and barriers of AI. This paper identified the application of AI technology in construction, its benefits and obstacles in the Vietnamese context. The research results support enhancing

DOI: 10.4018/979-8-3373-9549-4.ch002

the implementation of AI to improve the productivity of the construction industry not only in more advanced countries but also in developing countries toward sustainable development goals.

1. INTRODUCTION

The term Artificial Intelligence (AI) has been used since around the 1950s, since the beginning of information technologies, and with the rise of computer power, its usefulness has been used even further to the forefront, especially now with the availability of the Internet and powerful new algorithms. It is now gaining even more attention for its real-world use and is being widely recognised for its immense potential and that it could make a significant impact on the world in general. There is an issue however in that there is a significant difference in the way countries view the use of AI especially between that of developed and developing countries. Their national strategies can differ immensely in the use of AI. Developed countries normally have a strong technological infrastructure with strong economies and they want to use AI to drive their economic growth further, whilst developing countries do not normally have these attributes and are still trying to work their way through the AI processes and bring themselves up to speed with it, but with varying degrees of success (Demaidi, 2025). These developing countries are all required to first try and attain a set of Sustainable Development Goals (SDGs), such as poverty reduction, improving infrastructure, industry, and innovation (Santos & Jocson, 2024; Regona et al., 2022a; Abioye et al., 2021).

In the past, construction projects in developed countries still had to rely on human analysis alone for their data information collation and that was limited by the extent of knowledge that was available to their project team personnel at that moment in time. Nowadays, greater digital access and the use of AI software and technologies have given projects the ability to analyse vast amounts of data more accurately and much quicker than ever before, this can allow projects and project managers to make more informed decisions on how a project should move forward (Regona et al., 2022a). However, the use of AI in the construction industry in developing countries is still uncertain, this is generally down to the fact that the digital infrastructure in those countries is still developing and for them to achieve nationwide access to AI is a major challenge to most of them (Santos & Jocson, 2024). In Vietnam, the application of AI has been popular in many different areas such as agriculture, environment, business and health, attracting the research interest of many scientists (Pham et al., 2024). AI has also brought many benefits in supporting teaching, energy management and use, or government operations (Figure 1). Although its benefits have been proven, the implementation of AI in the Vietnamese construction industry is still extremely limited. The above status quo implies that to promote AI adoption in the construction industry, there is a need to study its application and advantages, parallel to anticipate and overcome the difficulties and challenges in applying AI in the Vietnamese construction industry context.

Figure 1. AI applications in Vietnam

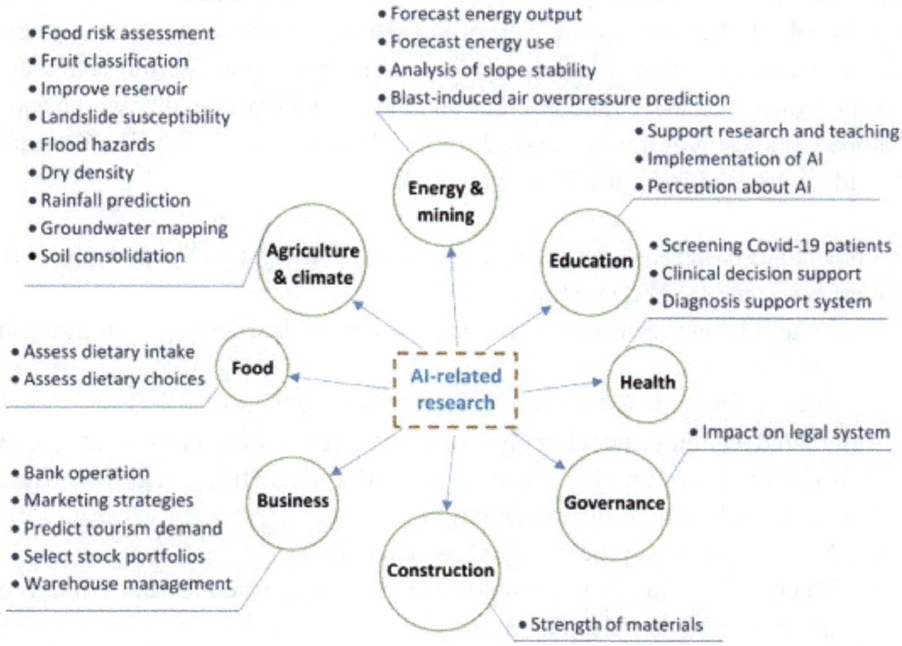

(Source: (Pham et al., 2024)

2. AI APPLICATIONS IN THE CONSTRUCTION INDUSTRY

2.1 Overview of AI Applications

Artificial Intelligence (AI) is the study of how to make machines do things. There are three main types of AI, and those are Artificial Narrow Intelligence (ANI), Artificial General Intelligence (AGI) and Artificial superintelligence (ASI) (Abioye et al., 2021). In which, ANI is sometimes referred to as weak intelligence and is concerned with a machine learning form of AI, where the machine specializes in just one area and solves one problem at a time. AGI referred to as strong AI, is more concerned with making machines operate at the very least at the same level as human beings, machine intelligence is where a machine such as a computer can be as just as smart if not smarter than a normal human being. ASI concerns itself with building machines that can exceed the capabilities of human beings across many different domains and can control itself autonomously, this is known as machine consciousness. The general theory of artificial intelligence includes the study of neural-like and multidimensional neural-like elemental growing networks, temporary and long-term memory study of the functional organization of the brain of the artificial intelligence system, also of the sensor systems, modulating systems, motor systems, conditioned and unconditioned reflexes, purposeful behaviour of thinking "consciousness", subconscious and the artificial personality which can be developed as a result of training and education (Yashchenko, 2014). The 7 main AI Subfields illustrated in Figure 2 below include:

1) Knowledge-based systems. (Expert systems, intelligence agents, case-based reasoning, and Linked systems).
2) Computer vision (Scene reconstruction, motion analysis, image restoration and recognition).
3) Robotics (Climbing, actuation, locomotion, and sensing).
4) Automated planning and scheduling (automated planning and automated scheduling). 5) Optimization (Evolutionary algorithms, genetic algorithms, different evolution, and particle swarm evolution).
    6) Natural language processing (speech and Text).
    7) Machine learning (Supervised learning, unsupervised learning, reinforcement learning and deep learning).

Figure 2. Artificial Intelligence and the Subfields

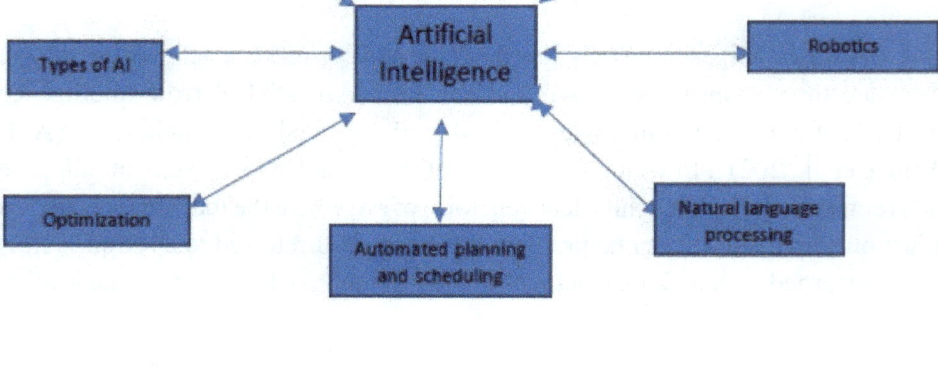

2.2 AI in the Construction Industry

AI in the construction industry still has knowledge gaps that need to be addressed there is a shortage of studies on the deployment experience of AI in large-scale projects. While there are research papers describing the potential of AI and surveys highlighting the challenges of AI deployment in the industry, there is still a need for more detailed studies from the experiences of companies involved in projects with high safety standards. There has been very little analysis conducted on the development paths for AI technologies in the construction industry, indicating a gap in understanding the evolution and potential applications of AI in architectural design, engineering design and construction services. There has also been a limited number of studies that have reviewed the potential of AI in the construction industry such as modular construction and robotics. There remain grey areas towards future barriers and opportunities that technologies and robotics may bring to a construction site. In addition, existing literature does not provide a comprehensive understanding of how AI tools and technologies are applied and used in the entire construction project lifecycle, i.e., planning, design, and construction stages (Li et al., 2025; Patil et al., 2024). AI may be an emerging subject in the construction industry, but it still lacks applied knowledge and most of the literature focuses on one algorithm throughout the construction process, so the use of AI remains uncertain and there is still that lack of understanding due to the limited research and development in AI (Yaseen et al., 2020).

There has been a lot of money spent on the research and development of AI in the construction industry over the past few years, especially from 2014 to 2019 when the global construction industry invested billions of dollars in both construction and engineering technologies and a lot of those billions has included AI technologies (Regona et al., 2022b). Despite this increase in expenditure, construction processes have not changed much over the last four decades, there is still a lack of knowledge of AI in the construction industry, hence its application in the industry remains costly, error-prone, and time-consuming which has led to its slow adoption in construction projects (Na et al., 2022). Although the industry is slowly changing to automated and digitalized processes which has allowed it to help improve quality, productivity, and safety it is gaining more and more attention from businesses who are beginning to realize its potential (Shukla et al., 2019). The adoption of AI software and technologies in construction has already helped to enhance construction operations as well as increase industry productivity. It has done this by providing better competitive advantages against the more conventional approaches that were used in the past. There are distinct types of subfields of AI which have different applications in construction (Abioye et al., 2021), as can be seen in Table 1 below.

Once these subfields are applied, they can tackle all types of complex problems which can support project management decisionmaking for real-time problems.

Table 1. AI Applications in the Construction Industry

AI SUBFIELDS	APPLICATION IN CONSTRUCTION
Machine learning	Risk Management, Real-Time Monitoring
Knowledge-Based Systems	Project Management, Design, Cost Planning
Computer Vision	Asset Management, and Maintenance Automated Quality control
Robotics	Autonomous vehicles, and Machinery
Natural Language Processing	Voice-operated GPS systems, Digital assistants, Speech to text digital software
Automated Planning and Scheduling	Construction Estimating and Bidding Construction Project Management
Optimization	Supply Chain, Design,| safety monitoring, Quality control

3. ARTIFICIAL INTELLIGENCE ADOPTION IN THE VIETNAMESE CONSTRUCTION INDUSTRY

3.1. The Vietnamese Construction Industry

In recent years, the average growth of Vietnam's construction industry has been about 7.6% per year; in 2022 alone, it increased by 8.5%. According to statistics, in the 2018-2022 period, the construction industry's contribution to GDP growth was about 6% on average; in 2022, it was 6.1%. In the first quarter of 2023, the industrial and construction sector increased by 1.95%, contributing 0.12 percentage points to the 3.32% increase in the overall GDP; down 0.4% compared to the same period last year.

The Government has set a general goal by 2030 to improve the capacity of the construction industry to undertake all stages of management, design, procurement, and construction of modern, complex, largescale construction projects and gradually compete and expand the market to operate abroad. Improving the technology content and application of technology in construction investment management, design, construction, and construction quality management has been considered the leading solution to overcome difficulties and promote the development of the industry in the 4.0 era.

3.2. Artificial Intelligence Adoption

AI is already being used in various applications on construction projects in Vietnam, such as self-driving trucks, drones, and automated equipment, and there are already some machines that can perform tasks such as laying bricks and performing land surveys (Pham et al., 2024; Van Tam et al., 2024). AI has also been integrated into project management and design where with machine learning, AI systems can analyze vast amounts of data from past projects to predict outcomes and make recommendations to help with planning and decision making. It can also allow team managers to see real-time data of the workforce on the project to assess obstacles with labour productivity, it can enhance safety and track equipment productivity. The benefits of this could be less waste, fewer delays a safer more productive site with better cost control. There can be some challenges and limitations however when trying to implement AI systems on construction projects, especially in developing countries, where the use of AI can sometimes still be in its infancy. AI systems are particularly vulnerable to cyber-attacks and any small mistakes in data interpretation can lead to costly or critical errors on a project, that could also endanger the life and health of the building occupants or before that, those of the construction workers. Construction companies are limited to the use of AI, based on the availability of electricity, computing power, and connectivity. Since some projects can take place in areas that are not well connected it is sometimes difficult to access the software needed to perform AI functions. These tools require electricity and a connection to the cloud to access shared data and perform their calculations (Pham et al., 2024; Van Tam et al., 2024).

3.3. Major Perceived Benefits of Using AI in the Construction Industry

A construction project involves organizing and directing every part of a construction project's lifecycle, to deliver it on time and within budget, construction projects can certainly benefit from the implementation of AI. However, different to any other form of technology, AI has its limitations as well (Pham et al., 2024; Regona et al., 2022b). One of the obvious benefits of using AI in construction is that it allows participants to focus their efforts more on the project itself and more innovative tasks while the AI can look after the repetitive more mundane tasks such as generating reports and importing and exporting data. Predictive analytics can make predictions about future outcomes of projects using historical data combined with statistical modelling, data mining techniques and machine learning. Project participants can make informed decisions, forecast costs, manage risk, and optimize resources which will allow companies to have valuable insight into the future of their projects. Quality control can help identify problems and correct issues before releasing them to the end user, it can save time and resources while ensuring high-quality products are always available and can be expedited at any time when required. Resource optimization allows a project to allocate available resources most efficiently to achieve their desired goals which will help to reduce the cost of labour and other expenses by delivering quality products on pre-determined timelines..

3.4. The Main Barriers of AI Implementation for the Construction Industry Participant

According to the International Monetary Fund (IMF), the rise of AI could lead to a widening gap between the richer, more developed nations and the poorer, less developed ones. This study found that the area where AI is expected to have the most effect was in the Job market. The manpower of developing countries has long been used by the more developed countries for outsourcing purposes, and one of those industries mentioned was the construction industry, where it is believed that the skills required in some of the construction industries' discipline trades could soon be replicated by AI (robotics for instance), resulting in large scale job losses in these countries (Santos & Jocson, 2024). Besides, the shortage of skilled qualified labour also is one of the major challenges in the Vietnamese construction industry which poses a risk to the overall construction quality (Van Tam et al., 2024). These smaller-sized construction companies also struggle to sustain their businesses and compete against the larger construction companies, the smaller companies can often lack the capital and the heavy equipment required to take on large-scale projects. There needs to be close collaboration between the government, the private sector, and foreign investors to help maximize the construction industry's potential.

There is a lack of awareness and resistance to change by companies when they are trying to implement AI into their business model, and from a construction company perspective, the new challenge of implementing green construction means there is a need for a technology adaption, but also there is the lack of a clear set of regulations to follow set out by the government, combine that with the education of society, and it is making it very difficult for companies to transition from normal construction designs to newer greener friendly ones (Van Tam et al., 2024). All these factors can make it difficult for construction companies to meet the required goals of what the government are setting out, about reducing carbon emissions. The adoption of digital technologies such as BIM and construction management software can improve efficiency for construction companies. Nevertheless, the investment capital can be expensive to implement, which may not be a problem for the larger businesses, but for smaller construction businesses the cost of trying to implement these technologies will probably outweigh the benefits of using them.

4. CONCLUSIONS

AI is making a tremendous impact on the way things are done in construction projects. Its innovative approach has helped to improve efficiency, and productivity and has allowed project teams to solve the many challenges that were both time-consuming and costly in the past. AI can help improve communication and collaboration among project stakeholders by providing real-time alerts and updates. The study talks about the different subfield types and components of AI and the way they can work in construction projects. It shows the advantages, benefits, challenges, and limitations of the use of AI in the construction industry. Therefore, several conclusions can be drawn from the research: 1) By using AI in construction projects there is the potential for it to help upscale project management, where performing core tasks such as developing schedules, assessing risk, and estimating cost can be done at a faster pace. It will improve accuracy; AI can avoid human errors and help to deliver a higher degree of accuracy with things like scheduling and drive operational efficiency by freeing the project team from mundane tasks allowing them to spend more time on more complex work and drive innovation within the team. The project managers need to seriously consider how they use AI correctly to achieve their required goals. 2) The use of AI in the construction industry can have a social impact as well, especially in developing countries, with such things as the transparency of AI and its algorithmic bias. There are also ethical impacts that can arise from AIdriven decisions. The management of these needs to be considered and be part of a project's team final AI strategy. 3) Developing countries that are just embarking on or are in the early stages of implementing AI as part of their economic growth require proper regulation and management in the use of AI throughout every sector in which it is used. If there are clear regulations and guidelines in place, then the use of AI can be done both ethically and safely. 4) Despite the benefits and challenges that AI can bring with its implementation, the overall optimism around AI suggests a brighter future where the technology will significantly narrow the gap not only in the construction industry but also in the health, education, environmental and help with the economic development between those of the developing countries and the more developed ones.

REFERENCES

Abioye, S. O., Oyedele, L. O., Akanbi, L., Ajayi, A., Delgado, J. M. D., Bilal, M., & Ahmed, A. (2021). Artificial intelligence in the construction industry: A review of present status, opportunities and future challenges. *Journal of Building Engineering*, *44*, 103299. DOI: 10.1016/j.jobe.2021.103299

Demaidi, M. N. (2025). Artificial intelligence national strategy in a developing country. *AI & Society*, *40*(2), 423–435. DOI: 10.1007/s00146-023-01779-x

Li, H., Zhang, Y., Cao, Y., Zhao, J., & Zhao, Z. (2025). Applications of artificial intelligence in the AEC industry: A review and future outlook. *Journal of Asian Architecture and Building Engineering*, *24*(3), 1672–1688. DOI: 10.1080/13467581.2024.2343800

Na, S., Heo, S., Han, S., Shin, Y., & Roh, Y. (2022). Acceptance model of artificial intelligence (AI)-based technologies in construction firms: Applying the Technology Acceptance Model (TAM) in combination with the Technology–Organisation–Environment (TOE) framework. *Buildings (Basel, Switzerland)*, *12*(2), 90. DOI: 10.3390/buildings12020090

Patil, D., Rane, N. L., Desai, P., & Rane, J. (Eds.). (2024). *Trustworthy artificial intelligence in industry and society*. Deep Science Publishing. DOI: 10.70593/978-81-981367-4-9

Pham, H. T., Nong, D., Simshauser, P., Nguyen, G. H., & Duong, K. T. (2024). Artificial intelligence (AI) development in the Vietnam's energy and economic systems: A critical review. *Journal of Cleaner Production*, *438*, 140692. DOI: 10.1016/j.jclepro.2024.140692

Regona, M., Yigitcanlar, T., Xia, B., & Li, R. Y. M. (2022a). Artificial intelligent technologies for the construction industry: How are they perceived and utilized in Australia? *Journal of Open Innovation*, *8*(1), 16. DOI: 10.3390/joitmc8010016

Regona, M., Yigitcanlar, T., Xia, B., & Li, R. Y. M. (2022b). Opportunities and adoption challenges of AI in the construction industry: A PRISMA review. *Journal of Open Innovation*, *8*(1), 45. DOI: 10.3390/joitmc8010045

Santos, J. V., & Jocson, J. (2024). Adoption of artificial intelligence technologies in the Philippine construction industry: A review of literature. *Journal of Interdisciplinary Perspectives*, *2*(8), 461–471. DOI: 10.69569/jip.2024.0304

Shukla, A. K., Janmaijaya, M., Abraham, A., & Muhuri, P. K. (2019). Engineering applications of artificial intelligence: A bibliometric analysis of 30 years (1988–2018). *Engineering Applications of Artificial Intelligence*, *85*, 517–532. DOI: 10.1016/j.engappai.2019.06.010

Van Tam, N., Toan, N. Q., & Van Phong, V. (2024). Investigating potential barriers to construction digitalization in emerging economies: A study in Vietnam. *International Journal of Information Management Data Insights*, *4*(1), 100226. DOI: 10.1016/j.jjimei.2024.100226

Yaseen, Z. M., Ali, Z. H., Salih, S. Q., & Al-Ansari, N. (2020). Prediction of risk delay in construction projects using a hybrid artificial intelligence model. *Sustainability (Basel)*, *12*(4), 1514. DOI: 10.3390/su12041514

Yashchenko, V. (2014, August). *Artificial intelligence theory (Basic concepts). In 2014 Science and Information Conference*. IEEE.

Chapter 3
Beyond Traditional Finance:
Mapping Vietnam's Crowdfunding Landscape in the Digital Age

Anh D. Pham
http://orcid.org/0000-0002-5199-2682
Research Insitute for Banking, Banking Academy of Vietnam, Hanoi, Vietnam

Men T. Bui
Faculty of Finance, Banking Academy of Vietnam, Hanoi, Vietnam

ABSTRACT

This research examines the critical role of community-based financial mobilization in supporting innovative startup activities in Vietnam. Since its emergence in 2012, crowdfunding has served as an alternative financing solution, enabling entrepreneurs to raise capital from the public. Although this model has achieved notable successes, with platforms like FirstStep establishing market credibility, crowdfunding development in Vietnam faces significant challenges, particularly the absence of clear legal frameworks. It should be noted that academic research on Vietnamese startup crowdfunding remains notably scarce, making this study among the pioneers in this field. The research evaluates the current state of crowdfunding and analyzes success factors within Vietnam's innovation ecosystem, ultimately proposing regulatory frameworks to foster creativity and development in the startup landscape.

DOI: 10.4018/979-8-3373-9549-4.ch003

1. INTRODUCTION

Capital is broadly acknowledged as a fundamental prerequisite for the sustainable development of any new business venture. Hence, securing adequate funding is one of the key challenges entrepreneurs face in transforming their innovative concepts into successful enterprises. Empirical evidence suggests that nascent companies frequently encounter significant barriers when accessing traditional financing channels, such as bank loans or bond issuances. This financial constraint has compelled enterprises to explore alternative funding mechanisms actively. Among these alternatives, crowd-funding has emerged as a distinctive form of alternative finance, where entrepreneurs solicit financial support, typically in small denominations, from the general public or 'the crowd,' rather than relying on a limited pool of professional investors such as banks or venture capitalists to fund their business initiatives. This democratized approach to capital raising represents a paradigm shift in entrepreneurial finance.

In Vietnam, crowdfunding platforms officially emerged in late 2012. Several prominent crowdfunding websites operate in the market, including IG9.vn, FirstStep, Comicola, Fundstart, Charity Map, and FundingVN. Among these, FirstStep has distinguished itself by supporting diverse entrepreneurial sectors, encompassing agriculture, software development, entertainment, gaming, and charitable initiatives. The growing share of software and gaming ventures among these projects reflects the rapid technological progress and the rising computational and data-processing demands that increasingly shape these emerging fields (Bijalwan, 2025). The platform's success can be attributed to its innovative features, including integrated e-wallet payment systems, automated and flexible refund mechanisms, transparent operations, efficient contributor management, and notably high project success rates, which have significantly enhanced market confidence in Vietnam's crowdfunding ecosystem.

Crowdfunding provides startups with three essential tools: market research capabilities, capital acquisition, and marketing opportunities. The success of numerous small and medium-sized projects demonstrates the huge potential of crowdfunding in Vietnam. However, the development of crowdfunding channels in the country faces multiple challenges. In Asia broadly and Vietnam specifically, the distinction between innovation and improvement remains ambiguous. This conceptual divergence is exemplified in international patent practices: while the United States requires demonstrable novelty for patent issuance, Japan's system accommodates incremental improvements. This East-West cultural dichotomy in understanding innovation creates complexities in legal frameworks, technological intellectual property rights, and idea protection. Project initiators in crowdfunding face a particular dilemma: while transparency is essential for fundraising, there exists legitimate concern about intellectual property protection when publicly disclosing innovative concepts.

Up until the present time (December 2024), Vietnam lacks comprehensive legal frameworks and systematic regulations governing crowdfunding activities, with no specific provisions defining participating stakeholders' qualifications, rights, and responsibilities. Existing regulations are insufficient for effective management, monitoring, and supervision of crowdfunding operations, hampering efforts to maximize benefits while minimizing potential risks. Therefore, this study provides a comprehensive analysis of the current crowdfunding landscape for Vietnamese startups and relevant policy recommendations for development. Our in-depth analysis examines capital mobilization from both investors and crowdfunding platforms, identifying key trends, challenges, and opportunities. Based on these findings, policy implications and recommendations are put forward to foster the development of crowdfunding for Vietnamese startups, advocating for appropriate regulatory frameworks, identifying critical success factors, and exploring the potential impact of state recognition and clear legal provisions.

In brief, this study contributes to the understanding and advancement of crowdfunding as a viable financing option for Vietnamese startups, supporting the growth of the Vietnamese startup ecosystem and unlocking the potential of crowdfunding as a catalyst for innovation and entrepreneurship in the country.

2. THEORETICAL BACKGROUND ON CROWDFUNDING

Crowdfunding encompasses diverse financing methods that operate both within and beyond traditional financial frameworks. In its broad context, it includes informal channels that bypass traditional financial intermediaries, such as borrowing from family and friends (characterized by rapid access and favorable interest rates), as well as seeking investment partners or utilizing internet-based crowdfunding platforms. More specifically, it represents the process of raising capital from the public to fund specific projects or ventures as an alternative to professional financiers such as banks or venture capitalists. This conceptualization is supported by Belleflamme et al. (2014) and Kleemann et al. (2008), who define crowdfunding as a public call, predominantly through internet platforms, for financial resources through donations or exchanges for future products or rewards to support specific initiatives. However, it is important to note that this conceptualization excludes internet-based peer-to-peer lending and fan-initiated fundraising initiatives (Mollick, 2014).

Crowdfunding typically facilitates project financing through relatively small contributions from a large number of individuals via Internet platforms, eliminating the need for financial intermediaries. The crowdfunding ecosystem encompasses four primary models, differentiated by the benefits offered to funders. The equity-based model enables investors to receive company shares in exchange for their capital contribution, effectively becoming shareholders in the venture. In the lending-based approach, contributors act as lenders, a model particularly suitable for small enterprises lacking traditional collateral for bank loans. The reward-based model involves project initiators offering predetermined benefits (reward packages) commensurate with contribution levels, independent of project profitability. Finally, the donation-based model is typically utilized by social organizations, charities, and NGOs for non-profit initiatives such as disaster relief or humanitarian assistance, with no expected return for contributors.

Unlike traditional financial intermediation where banks mediate between depositors and borrowers, crowdfunding facilitates direct investment relationships between entrepreneurs and funders through online platforms (Schwienbacher & Larralde, 2010). Prominent platforms include Kickstarter, Indiegogo, Fundable, and Kiva. This mechanism particularly benefits startups with limited internal capital and difficulties accessing external funding due to insufficient collateral.

The advantages of crowdfunding extend beyond mere capital acquisition. Through customer self-service mechanisms, it enables profit enhancement, cost reduction, and productivity improvements. Contribution amounts can range from modest sums for specific projects to seed funding for early-stage ventures, particularly beneficial for small enterprises (Schwienbacher & Larralde, 2010). Moreover, crowdfunding transcends pure financing: backers often participate in product development, testing, marketing, and strategic decision-making processes (Belleflamme et al., 2014).

However, this funding channel might have several limitations and challenges. First, projects face the fundamental risk of failure due to insufficient funding mobilization, often compounded by entrepreneurs' concerns about intellectual property protection when publicly sharing innovative concepts. The public nature of crowdfunding campaigns can also create psychological barriers, as founders fear the stigma associated with public failure, which could potentially damage their professional reputation and future opportunities with investors and customers. Aside from that, some entrepreneurs avoid crowdfunding since they perceive a misalignment between their product market and the crowdfunding community or they feel unable to offer appropriate rewards. The resource-intensive nature of campaign management presents another significant challenge, as founders must commit substantial time and effort to create compelling project descriptions, manage regular backer updates, and maintain consistent communication with a large pool of potential investors (Gleasure, 2015; Hui et al., 2014). Meeting these communication and relationship-building demands calls for strong networking and communication skills, which have been shown to be key drivers of entrepreneurial feasibility and new venture creation in the Vietnamese context (Pham et al., 2020).

Crowdfunding remains an emerging research field, with recent studies predominantly employing case study methodologies (Bi et al., 2017; Schwienbacher &

Larralde, 2010; Gleasure, 2015). Researchers focus on understanding founder and backer motivations, identifying success factors, and analyzing project outcomes within specific contexts. This study contributes to this growing body of knowledge by examining success and failure factors within Vietnam's innovation ecosystem, specifically investigating geographical influences, stakeholder motivations and barriers, and critical success factors in crowdfunding initiatives.

3. RESEARCH METHODS

To comprehensively address the established research objectives, this study employed a dual-targeted survey methodology encompassing two stakeholder groups: Startup enterprises and Financial solution providers. The survey instruments were specifically tailored for each respondent category, with questionnaires designed to capture relevant data aligned with their respective roles in the crowdfunding ecosystem. The data collection was conducted during the first quarter of 2021, focusing primarily on participants in Hanoi and Ho Chi Minh City.

3.1. Survey of Startup Enterprises

The study employed a stratified random sampling approach, drawing from a comprehensive database of startups compiled from multiple authoritative sources: (i) local business registration authorities (Provincial Departments of Planning and Investment); and (ii) prominent startup incubation and acceleration programs, including National Innovation Startup Festival (Techfest Vietnam), Global Startup Competition (VietChallenge), Kawai Startup Fair (hosted by Foreign Trade University), and B.A Startup Competition (hosted by the Banking Academy of Vietnam). This consolidated database utilized computer-assisted random sampling to identify the final survey participants.

The survey implementation followed a two-phase approach:

- Phase 1 involved a pilot study with 30 initial respondents to validate questionnaire clarity and eliminate potential ambiguities or misinterpretations in the survey instrument. Based on this preliminary feedback, the questionnaire was refined before proceeding to the main data collection phase.
- Phase 2 comprised the distribution of the finalized questionnaire to the remaining selected respondents through both online survey links and physical questionnaires delivered to specific addresses.

Regarding the response rate and sampling characteristics:

Of the 1,200 questionnaires initially distributed, 662 responses were received (representing a 55% response rate), with 605 valid responses providing complete and reliable data suitable for analysis. The sample encompasses startups from diverse industries, predominantly technology (60%), e-commerce (23%), communication (19%), education (18.2%), and finance (16%). Other fundamental characteristics of the surveyed startups are presented in **Figure 1**. The relatively low response rate can be attributed to several factors: the comprehensive nature of the questionnaire requiring extensive information, the complexity of questions that may have deterred potential respondents, and the presence of incomplete or inconsistent responses deemed unreliable. Moreover, as the target respondents were company CEOs with significant time constraints, their willingness to complete the lengthy survey was limited.

Figure 1. Main characteristics of startup enterprises in the survey sample (% respondents)

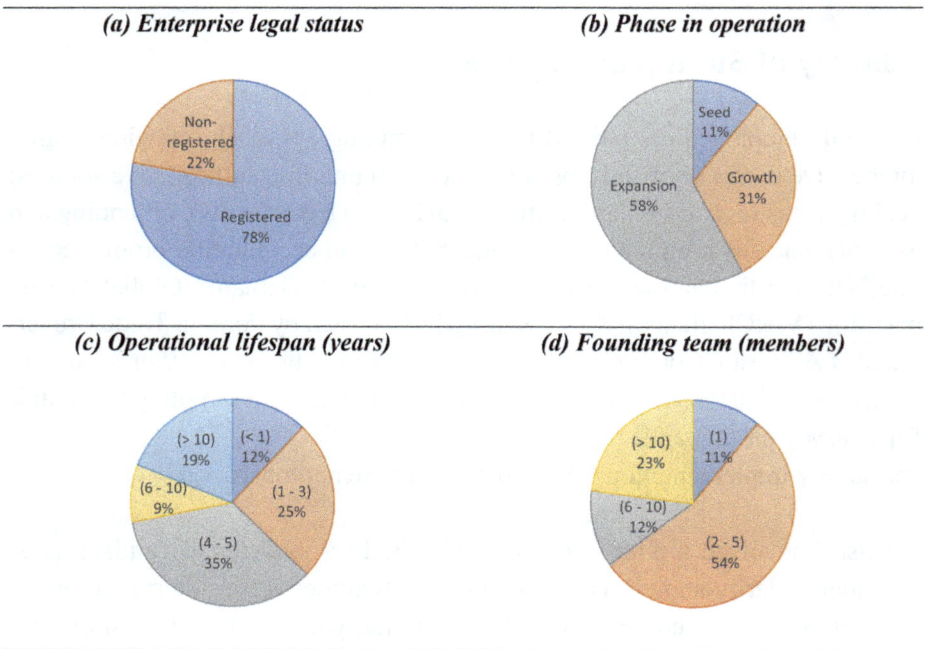

Source: Author's survey results

3.2. Survey of Startup Financial Solution Providers

The survey development started with a pilot phase involving 20 initial investors (both management-level executives and specialists) selected from the master list. These participants were engaged in a comprehensive questionnaire evaluation process via email, with the option for direct telephone interviews. They were specifically tasked with assessing multiple dimensions of the survey instrument, including content appropriateness, linguistic clarity, difficulty level, structural coherence, and optimal length. Participants could provide feedback either through email responses or scheduled telephone interviews, according to their preference. This preliminary testing phase was crucial for gathering detailed evaluations on the questionnaire structure, identifying potential linguistic ambiguities, and assessing response difficulties. The accumulated feedback was systematically analyzed to refine the final survey instrument, which was subsequently converted into an online format.

The main study employed a stratified random sampling method, with the sampling frame derived from the database of potential investors maintained by the National Agency for Technology Entrepreneurship and Commercialization Development (2075.com.vn). Computer-assisted random sampling was utilized to generate the final list of survey participants from this comprehensive database.

Regarding the Response Rate and Sampling Characteristics

From an initial distribution of 500 questionnaires, 336 responses were received, with 324 meeting the quality criteria for analysis. The investor survey achieved a notably higher response rate of 67%, significantly exceeding the startup enterprise survey response rate (55%). This enhanced response rate can be attributed to the more concise and focused design of the investor questionnaire, which prioritized directness and efficiency in data collection. The sampling structure for startup financial solution providers is reported in **Figure 2**.

Figure 2. Composition of startup financial solution providers in the survey sample (% respondents)

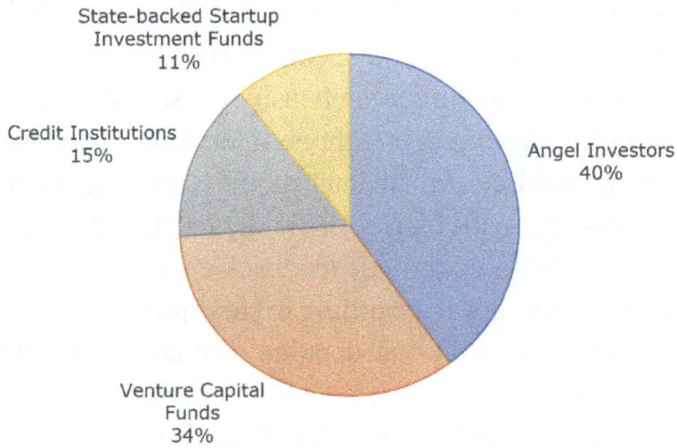

Source: Author's survey results

4. COMMUNITY FINANCE MOBILIZATION FOR STARTUPS IN VIETNAM: A CURRENT STATE ANALYSIS

4.1. Capital Mobilization From Investors

In Vietnam, support for startups has been increasingly emphasized with the emergence of startup support centers such as co-working spaces, incubators, and accelerators. Alongside the development of the startup ecosystem, a network of angel investors has also been established, with prominent names including iAngel Network and Vietnam Angel Network (VAN). The core members of these alliances are domestic individual investors, and their activities primarily focus on sharing angel investment experiences, connecting investors with startup projects, and managing investments in these projects. Notably, the establishment of the Mekong Angel Investors Network (MAIN) in 2016, a collaborative effort between the Mekong Business Initiative (MBI) and Lotus Fund, marked a promising new turning point in the linkage between international angel investors (mainly from the US, the EU, and Australia) and domestic investors in nurturing the development of startup projects in the sub-region (Tran et al., 2019). Although investor networks have emerged, the angel investment model is still quite novel, and the duration and experience in operating the network model are not yet extensive. The investment effectiveness and spillover effects for startup projects and enterprises remain unclear. Moreover, investors within the networks still lack close sharing of investment tools and experiences or tight cooperation in co-investing. The only bright spot from these networks lies in their stronger voice in policy recommendations and proposals to the government regarding startup investment policies.

Observing global practices reveals that, by general standards, angel investors are individuals with education, investment and business experience, and certain financial resources used to invest in startups in their early stages. For example, in the United States, angel investors typically hold a master's degree, have an average age of 56, with at least 8 years of investment experience and 14 years of business experience (Tia Sang, 2021). However, in Vietnam, most angel investors are current or former entrepreneurs and managers, and many are retired. Due to the lack of professional “hard” standards for knowledge, experience, and finance applied to angel investors, assessing the credibility of individual investors in Vietnam remains a significant concern for startups whenever they can connect and cooperate. This significantly limits the opportunities for domestic investors to connect and collaborate with international angel investors due to substantial differences in standardized knowledge and competence. It is also considered an indirect barrier for Vietnamese startups in the process of expanding connections and developing business on a global scale. It is noteworthy that according to AngelList statistics, over 5,400 foreign angel investors are paying special attention to Vietnam (Dang, 2020), and to date, only a few prominent Vietnamese startups, such as Vntrip, Logivan, Bizzi, etc., have received investment from foreign angel investors.

Figure 3. Extent of informal funding for startups (% population)

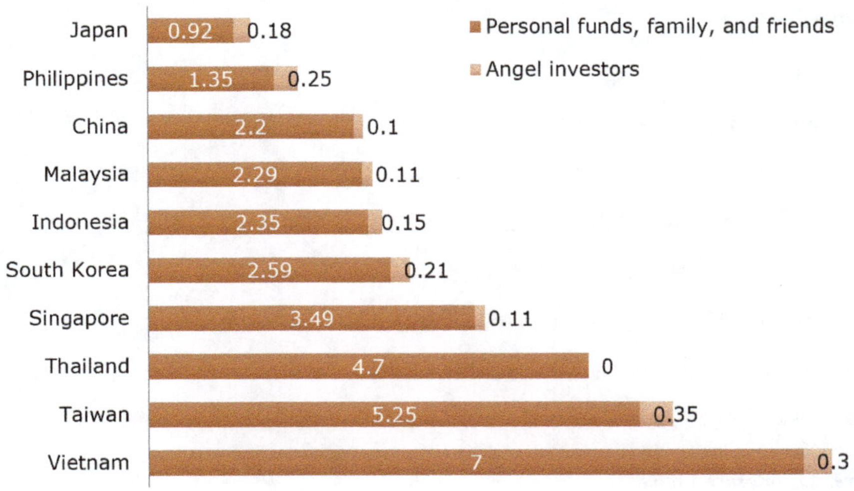

Source: Au et al. (2016)

The scale of angel investment in Vietnam remains modest as this investment model is still relatively new in the context of the recently formed startup ecosystem. Au et al. (2016) pointed out that the angel investment component accounts for less than 5% of the total informal investment for startups. This is true for most developing Asian countries (including Vietnam) (Figure 3).

Figure 4. Angel investment reception across development stages of Vietnamese startups (% respondents)

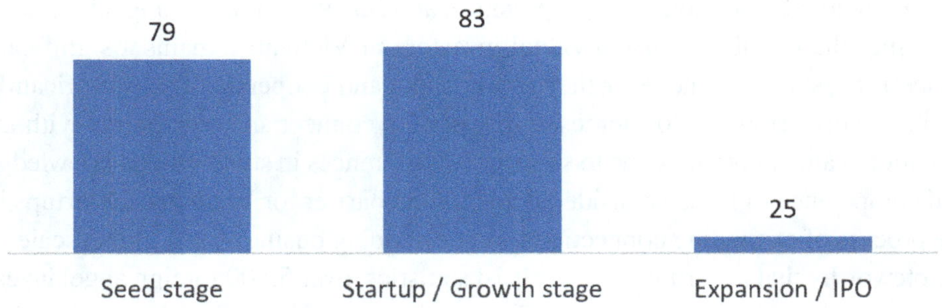

Source: Author's survey results

Figure 5. Primary startup sectors attracting angel investment in Vietnam (% respondents)

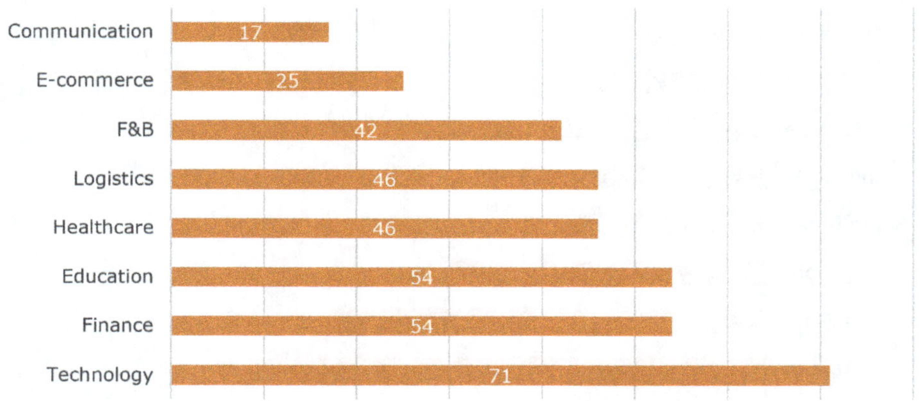

Source: Author's survey results

Figure 6. Per-deal value of angel investments in Vietnam (% respondents)

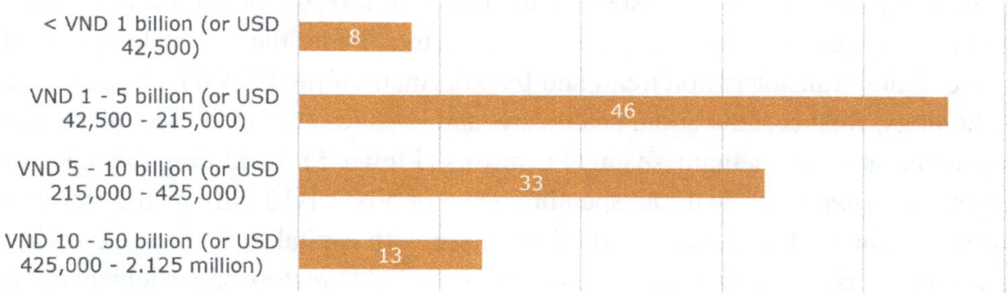

Source: Author's survey results

Figure 7. The Role of Angel Investors in Investee Companies (% respondents)

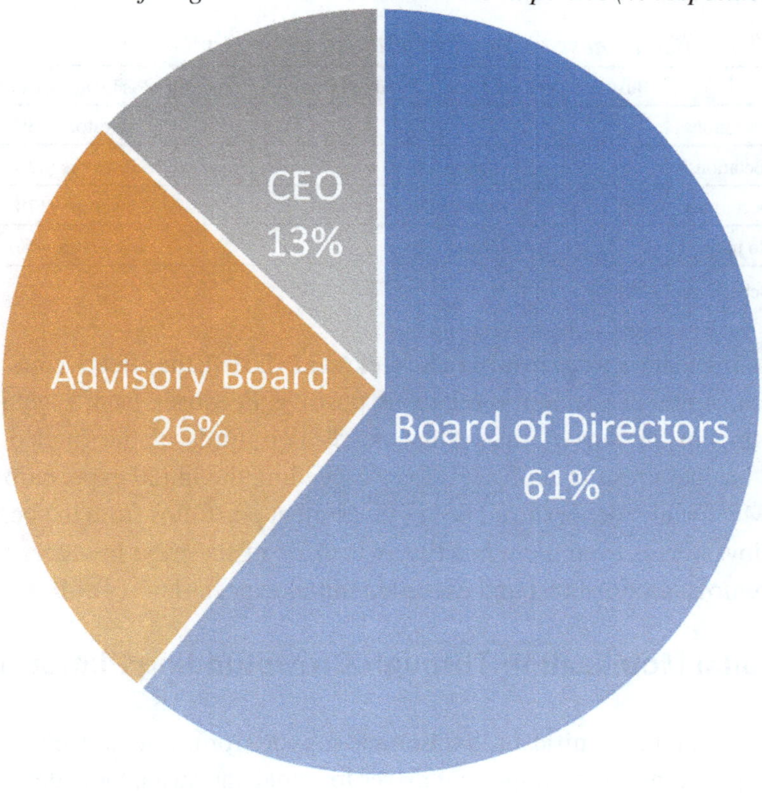

Source: Author's survey results

Our survey results show that among the development stages of startups, angel investors still prioritize their financial resources in the seed stage (nurturing ideas and testing the market) to create a solid "launchpad" to help startups develop and confidently move towards the stage of approaching investment from venture capital funds (**Figure 4**). Regarding investment sectors, under the overarching impact of the digital transformation trend and the aftermath of the COVID-19 pandemic, technology, finance, education, healthcare, and logistics are the top sectors receiving attention from Vietnamese angel investors (**Figure 5**). Regarding the value of angel investments per deal, the spending level of VND 1-10 billion still accounts for the majority (**Figure 6**). In addition, along with capital investment activities, angel investors also require a position on the Board of Directors or participate in the Advisory Board (**Figure 7**) to facilitate support in development orientation, expand connections with investment and business partners, as well as equip founders with additional knowledge, skills, and vision to ensure the effectiveness of their capital contribution.

Table 1. The Effectiveness of Angel Investment in Vietnam

Evaluation	Business Performance of Funded Startups	Portfolio Performance of Investors
Above expectations	22 startups (27%)	4 investors (33%)
Meets expectations	32 startups (39%)	5 investors (42%)
Below expectations	21 startups (25%)	2 investors (17%)
Too early to judge	8 startups (10%)	1 investors (8%)

Source: Scheela et al. (2018)

According to the survey results of Scheela et al. (2018), the effectiveness of angel investment capital in Vietnam is relatively positive. Feedback from angel investors participating in the survey shows that nearly two-thirds of the funded startups achieved business results equivalent to or exceeding the initial expectations of investors. Meanwhile, the performance of investment portfolios from the perspective of angel investors is even more positive, with 75% of surveyed investors reporting that the returns obtained met and exceeded initial expectations (**Table 1**).

4.2. Capital Mobilization Through Crowdfunding Platforms

Despite receiving significant attention and support from various levels and sectors and being positioned as an important driver for rapid and sustainable development, startups in Vietnam still face numerous difficulties in accessing capital compared to traditional enterprises due to their small scale, low level of diversification, and inability to meet collateral requirements. Particularly, since the global financial crisis of 2008-2009 and the COVID-19 pandemic in 2020, lending activities by credit institutions and capital funding from investment funds for businesses have been tightened, with all disbursement decisions being considered with a high degree of prudence to minimize risks for capital providers in the context of a slow economic recovery. This financial bottleneck is consistent with macro-level evidence showing that domestic private investment is highly sensitive to monetary policy, since credit tightening tends to restrict capital access for non-state enterprises more sharply than for others (Dang et al., 2020). Consequently, given these systemic constraints within traditional banking channels, online fundraising has emerged as a viable and accessible solution for startups. Globally, the development of crowdfunding platforms has been creating new competition for traditional banks and investment funds, not only promoting the development of financial markets and enhancing competitiveness but also helping to increase profits for businesses (Fenwick et al., 2017). Furthermore, the emergence of this new lending model contributes to reducing systemic risk and increasing the diversification of funding products.

First, regarding the *participants*:

The participants involved in crowdfunding activities include backers (contributors), crowdfunding project owners (fundraisers), and crowdfunding service providers. Among them, backers can be any organizations or individuals with suitable financial capabilities and the desire to help crowdfunding projects become a reality. Crowdfunding project owners (fundraisers) are organizations or individuals with ideas for a business or charitable project who need community support to gather sufficient funds to implement their projects. Crowdfunding service providers are business organizations that establish an environment for crowdfunding project owners to introduce their projects and seek help from the community. Currently, the legal status of these service providers has not been clearly defined by Vietnamese law. However, considering their main functions, it can be seen that they play the role of intermediaries, bridging the gap between fundraisers and backers. Additionally, in some cases where requested by fundraisers, these organizations can act as consultants, providing support for certain tasks as required.

Second, regarding the *purposes of participating in crowdfunding*:

When participating in crowdfunding activities, each participant has their own objectives. First, for crowdfunding project owners, the purpose can be charitable fundraising or an initial step to realize a business startup project. For business fundraising projects, since crowdfunding essentially involves seeking initial help and support from society, crowdfunding project owners are usually individual entrepreneurs or business organizations with limited capital. Through crowdfunding, fundraisers promote their projects and market their products before launch. Besides, when conducting crowdfunding, project owners also have the opportunity to survey market demand, gather feedback, and assess the project's success. For backers, contributing to a crowdfunding project primarily demonstrates their appreciation for the project's idea, supports startup entrepreneurs, and encourages creativity or charitable activities in the community. Moreover, backers may also aim for certain benefits if the project succeeds, such as rewards, shares, or loan interest. Thus, individuals who contribute to crowdfunding activities for business projects also become investors. Finally, crowdfunding service providers are business organizations that act as intermediaries connecting crowdfunding project owners with individuals who have the desire to contribute and support those projects. While the relationship between fundraisers and backers in a crowdfunding project may not always be commercial (in the case of charitable projects), the relationship between fundraisers and crowdfunding service providers is always a commercial one. Crowdfunding service providers benefit from establishing an environment (website) for crowdfunding project owners to post their project content, and therefore, they will receive a certain percentage of the funds as specified in predefined rules or agreements with the crowdfunding project owners.

Third, regarding *crowdfunding models*:

In Vietnam, professional platforms mainly conduct crowdfunding based on the reward model. According to Luong (2019), this model is favored because it essentially functions as pre-sales (products are sold before being manufactured), and backers feel more assured when the money they contribute will be reciprocated with a reward. Other crowdfunding models, especially equity-based and lending-based, have not even been implemented in Vietnam.

Fourth, regarding the *legal basis*:

Current Vietnamese law does not have specific provisions on each party's status, powers, and responsibilities in crowdfunding relationships. However, based on general provisions on civil, commercial, or e-commerce activities, the legal basis for defining the rights and responsibilities of the parties are as follows: the crowdfunding service use agreement (between the crowdfunding project owner and the crowdfunding service provider); the agreement between the crowdfunding project owner and the backers (expressed through the introduction, images, and commitments that the project owner publishes on the crowdfunding website).

The crowdfunding platforms widely known in the startup community in Vietnam include:

- **IG9:** Was the first crowdfunding platform to appear in Vietnam in April 2013. This platform later successfully raised funds for 40 projects with a scale of 7-15 million VND per project. However, after a period of operation, this website quietly closed down in 2014. Following IG9, several other crowdfunding platforms were established, such as FirstStep, Comicola, Fundstart, Charity Map, Betado, and FundingVN. However, at present, only three platforms are operating professionally: Comicola, Betado, and Fundstart.
- **Comicola**: A fundraising site dedicated to the comic book industry. This platform has successfully raised funds for 9 comic book publishing projects with

a total capital of up to 1.3 billion VND. Among them, some projects have successfully raised funds exceeding the initial expected amount. For example, the "Hoa Van Dai Viet" project raised 141.4 million VND, surpassing the target by 41%, and the project to publish the pop-up book "Sai Gon Pho" raised 240.8 million VND, exceeding the initial target by 61%. The crowdfunding model used by Comicola is reward-based. Projects are divided into small packages corresponding to the amount and rewards that will be given to contributing customers.

- **Betado:** A crowdfunding platform aimed at creating opportunities to bring creative products of Vietnamese people to the market. Established in 2015, with its predecessor being the crowdfunding project for volumes 1 and 2 of the comic book "Legend of The Holy Dragon Imperator", Betado has quickly developed and become the largest crowdfunding platform in Vietnam in terms of both the value of funds raised and the number of successful projects. Betado has successfully raised funds for 13 projects with a total capital of 1.6 billion VND.
- **Fundstart:** It officially began operating to support fundraising for creative projects in April 2016. This platform operates based on the reward-based crowdfunding model. Project owners share their ideas with Fundstart, and the community that appreciates the project will support it financially through the crowdfunding platform's system within a specified period. If the fundraising campaign is successful, the project owner must ensure the rights and rewards of the backers according to the commitments published on the crowdfunding platform and the signed contract. If the campaign is unsuccessful, the funds are returned to the backers. As of now, the Fundstart platform has successfully raised funds for 4 projects, with the total amount reaching 287 million VND.

Besides, there is also the FundingVN, which is currently in the testing phase. This platform was established with the primary goal of connecting projects with the investor community based on solving capital and business development issues. According to information on the FundingVN website, project owners can post their projects on the website, and these projects will be thoroughly reviewed and appraised by experts before reaching the community/investors to minimize risks for the community/investors. The total number of fundraising projects announced on the website is 88 projects. However, at the current stage, FundingVN has not officially started actual fundraising but instead focuses on collaborating with the startup company Wheel to help projects gain experience with the model, thoroughly prepare knowledge, and learn how to conduct campaigns, laying the foundation for successful fundraising in the future. Each registered individual is provided with a virtual account worth 30 million VND to have capital to sponsor the projects they expect.

5. EVALUATIONS

Building upon the analysis of the current state of crowdfunding platforms in Vietnam, several key evaluations can be made.

First, crowdfunding has emerged in Vietnam, demonstrating the demand and potential in the Vietnamese market. However, this form of fundraising is still in its early stages and has not developed significantly due to the lack of a legal framework to regulate crowdfunding activities. Some platforms are also limiting their potential due to concerns about violating current laws.

Second, crowdfunding in Vietnam encompasses various methods, with platforms employing sophisticated and professional approaches. The founding teams of these platforms are passionate about their mission and possess the necessary expertise to evaluate and support projects seeking crowdfunding.

Third, the crowdfunding platforms within the scope of this study have specific characteristics related to the types of crowdfunding projects they support. In addition to platforms supporting startup projects, there are also platforms catering to projects in fields such as culture, history, and the arts. This illustrates the diversity of projects seeking funding in the market and the broad applicability of crowdfunding beyond startup projects. This scope increasingly extends to green and environmentally oriented startups, since growing environmental awareness and corporate social responsibility (CSR) commitments have been shown to encourage stewardship behaviour and sustainable engagement in the Vietnamese context (Pham et al., 2024).

Fourth, all currently operating crowdfunding platforms are facing legal challenges. They are seeking a legal framework to regulate crowdfunding activities to operate more effectively and protect the legitimate interests of all parties involved. The platform that ceased operations highlights the necessity to test the business models of crowdfunding platforms to assess market demand and potential before implementing specific regulations.

Finally, the issue of raising public awareness about crowdfunding requires significant changes in investment mindset and building a supportive startup culture in Vietnam. Without these changes, crowdfunding projects and platforms will face ongoing challenges, and the enforcement of legal regulations on crowdfunding, once they are announced, will also encounter obstacles.

6. POLICY IMPLICATIONS

Currently, Vietnam is still in the initial stage of crowdfunding; therefore, to address the challenges and promote the development of this fundraising model, the joint efforts of many market participants are required.

- From a legal perspective, startups mainly have to rely on relevant legal provisions such as the Enterprise Law 2020, which mentions fundraising methods but does not delve into specific regulations for crowdfunding. The Securities Law 2019 regulates the issuance of securities but is still insufficient to cover all aspects of crowdfunding. The Government's Decree 86/2016/ND-CP on consumer financial services mentions fundraising but lacks specificity for crowdfunding activities. Therefore, the authorities need to consider including crowdfunding in a specific legal framework, defining the legal rights and obligations of participants when engaging in crowdfunding to ensure transparency of responsibilities and benefits while minimizing the risk of legal and financial risks for both project founders and the community.
- Fundraising websites must establish a stable and transparent transaction network, serving as a reliable bridge connecting projects with the general public, contributing to building trust and changing perceptions about crowdfunding. While the domestic crowdfunding model is not yet developed, major global fundraising websites can be an option for Vietnamese entrepreneurs.
- Investing in crowdfunding models requires extreme caution from investors. Before officially transferring money or entering into related transactions, it is crucial to implement comprehensive evaluation measures. In a favorable scenario, a successful investment for the investor occurs when the project runs smoothly, and the project owner fulfills their commitments. However, Vietnamese law currently lacks any mechanism to protect investors when adverse situations arise. Investors only become aware of and perceive the risks they are facing to some extent when the project owner shows signs of committing a crime and the authorities intervene.

In conclusion, developing a robust crowdfunding ecosystem in Vietnam calls for concerted efforts from all stakeholders, including policymakers, crowdfunding platforms, and investors. By establishing a clear legal framework, building trust through transparent platforms, and promoting investor education, Vietnam can unlock the potential of crowdfunding as a viable alternative financing option for startups and innovative ventures. As the crowdfunding landscape continues to evolve, it is crucial for all parties involved to remain vigilant, adaptable, and committed to fostering a supportive environment for nascent businesses and pioneering initiatives in the country.

REFERENCES

Au, K., Chiang, F. F., Birtch, T. A., & Kwan, H. K. (2016). Entrepreneurial financing in new business ventures: A help-seeking behavior perspective. *The International Entrepreneurship and Management Journal*, *12*(1), 199–213. DOI: 10.1007/s11365-014-0332-5

Belleflamme, P., Lambert, T., & Schwienbacher, A. (2014). Crowdfunding: Tapping the right crowd. *Journal of Business Venturing*, *29*(5), 585–609. DOI: 10.1016/j.jbusvent.2013.07.003

Bi, S., Liu, Z., & Usman, K. (2017). The influence of online information on investing decisions of reward-based crowdfunding. *Journal of Business Research*, *71*, 10–18. DOI: 10.1016/j.jbusres.2016.10.001

Bijalwan, A. (Ed.). (2025). *Innovations and challenges in computing, games, and data science*. IGI Global. DOI: 10.4018/979-8-3373-2647-4

Dang, T.D. (2020). Angel Investment in Startups in the World and in Vietnam. Vietnam Trade and Industry Review, 22 (September).

Dang, T. T., Pham, A. D., & Tran, D. N. (2020). Impact of monetary policy on private investment: Evidence from Vietnam's provincial data. *Economies*, *8*(3), 70. DOI: 10.3390/economies8030070

Fenwick, M., Vermeulen, E., & McCahery, J. A. (2017). *Fintech and the Financing of Entrepreneurs From Crowdfunding to Marketplace Lending*. Law Working Paper No 369/2017.

Gleasure, R. (2015). Resistance to crowdfunding among entrepreneurs: An impression management perspective. *The Journal of Strategic Information Systems*, *24*(4), 219–233. DOI: 10.1016/j.jsis.2015.09.001

Hui, J. S., Greenberg, M. D., & Gerber, E. M. (2014). Understanding the role of community in crowdfunding work. In *Proceedings of the 17th ACM conference on Computer supported cooperative work & social computing* (pp. 62-74). DOI: 10.1145/2531602.2531715

Kleemann, R., Zadelaar, S., & Kooistra, T. (2008). Cytokines and atherosclerosis: A comprehensive review of studies in mice. *Cardiovascular Research*, *79*(3), 360–376. DOI: 10.1093/cvr/cvn120 PMID: 18487233

Luong, T. T. H. (2019). Crowd Funding - A New Funding Trend for SMEs in Vietnam Today. Corporate Finance Review, October <https://taichinhdoanhnghiep.net.vn/crowd-funding--xu-huong-tai-tro-moi-cho-cac-doanh-nghiep-nho-va-vua-o-viet-nam-hien-nay-d9422.html>

Mollick, E. (2014). The dynamics of crowdfunding: An exploratory study. *Journal of Business Venturing*, *29*(1), 1–16. DOI: 10.1016/j.jbusvent.2013.06.005

Pham, A. D., Bui, M. T., & Hoang, D. P. (2020). Employees' Motivation for Creating International Business Venture: The Role of Perceived Competence and Job Satisfaction. *Asian Academy of Management Journal*, *25*(2), 109–134. DOI: 10.21315/aamj2020.25.2.5

Pham, H. M., Pham, A. D., Tran, D. V., Vuong, N. L., & Nguyen, H. D. (2024). Integrating CSR with environmental consciousness and commitment: Pathways to employee stewardship in Vietnam's mechanical manufacturing sector. *Business Strategy & Development*, *7*(4), e70036. DOI: 10.1002/bsd2.70036

Scheela, W., Nguyen, T. T. T., & Nguyen, T. K. A. (2018). Business angel investing in Vietnam: An exploratory study. *Journal of Private Equity*, *21*(2), 96–106. DOI: 10.3905/jpe.2018.1.067

Schwienbacher, A., & Larralde, B. (2010). *Crowdfunding of small entrepreneurial ventures. Handbook of Entrepreneurial Finance*. Oxford University Press.

Tia Sang. (2021). Angel Investors - An Essential Link in the Venture Capital Market. *VnExpress* <https://tiasang.com.vn/-doi-moi-sang-tao/nha-dau-tu-thien-than-mat-xich-thiet-yeu-cua-thi-truong-dau-tu-mao-hiem-7400>

Tran, V. B., Le, V. T., & Nguyen, V. K. (2019). Angel Investors and Their Role in Innovative Startup Projects. *Vietnam Journal of Science and Technology*, •••, 17–20.

Chapter 4

A Comprehensive Evaluation of Machine Unlearning Strategies for GDPR- Compliant Recommendation Systems:

Privacy, Utility, and Efficient Trade-Offs

Siddharth Selvakumar
http://orcid.org/0009-0001-3142-4282
SRM Institute of Science and Technology, India

K. R. Giethesh
http://orcid.org/0009-0009-3538-6226
SRM Institute of Science and Technology, India

Shreeya Rajagopal
http://orcid.org/0009-0003-8272-4503
SRM Institute of Science and Technology, India

P. Senthilkumar
http://orcid.org/0000-0001-5795-9448
SRM Institute of Science and Technology, India

DOI: 10.4018/979-8-3373-9549-4.ch004

ABSTRACT

The European Union's General Data Protection Regulation (GDPR) mandates users' "Right to be Forgotten," yet in recommendation systems, merely deleting user data from storage is insufficient as information remains encoded in model parameters. This paper presents the first comprehensive evaluation framework for machine unlearning in collaborative filtering-based recommendation systems, systematically comparing state-of-the-art strategies across privacy guarantee verification, model utility preservation, and computational efficiency. We propose Adaptive SISA (A-SISA), enhancing the standard SISA framework with intelligent user clustering and dynamic shard sizing optimized for recommendation contexts. Extensive experiments on Movie Lens 25M and Amazon Product Reviews datasets demonstrate that A-SISA achieves 94.2% privacy protection with only 12% computational cost of retraining while maintaining 89.7% utility retention. Our findings provide practical guidelines for deploying GDPR-compliant machine unlearning in production recommendation systems.

1. INTRODUCTION

The rapid advancement of artificial intelligence has positioned recommendation systems at the forefront of digital transformation, fundamentally reshaping how billions of users interact with online platforms. These systems, exemplified by Netflix's content recommendations, Amazon's product suggestions, and Spotify's music discovery algorithms, have become indispensable components of modern digital ecosystems. Their success stems from sophisticated machine learning models trained on vast amounts of user behavioral data, enabling personalized experiences that drive user engagement and business revenue. However, this data-driven personalization paradigm has created an unprecedented challenge at the intersection of technological innovation and privacy regulation.

The European Union's General Data Protection Regulation (GDPR), enacted in May 2018, represents a landmark shift in data privacy legislation, establishing fundamental rights that directly impact machine learning systems. Article 17 of GDPR, known as the "Right to be Forgotten," mandates that individuals have the right to request complete erasure of their personal data from organizational systems. This regulatory requirement extends beyond simple database deletion to encompass all forms of data processing, including machine learning models that have learned from user data. For recommendation systems, this creates a complex technical challenge where user information becomes intrinsically embedded within model parameters through the training process.

Conventional approaches to GDPR compliance in machine learning involve complete model retraining after removing the requested user's data from the training dataset. While this brute-force method ensures perfect data erasure, it introduces significant operational challenges that threaten the viability of large-scale recommendation systems. The computational overhead is enormous, particularly for systems serving millions of users with models requiring continuous updates to reflect changing preferences and new content. Furthermore, the environmental impact of repeated full retraining conflicts with growing sustainability initiatives in the technology sector, while the time required for retraining may violate GDPR's mandates for prompt data deletion responses.

Machine unlearning has emerged as a promising alternative, offering sophisticated methods to selectively remove the influence of specific data points from trained models without necessitating complete retraining. This field addresses the fundamental question of how to make machine learning models "forget" specific information while preserving their overall functionality. However, the current body of research has predominantly concentrated on computer vision applications and simple classification tasks, leaving a critical gap in understanding how these techniques perform in the complex domain of collaborative filtering-based recommendation systems.

Recommendation systems present unique challenges for machine unlearning due to their inherent architectural characteristics. Unlike traditional classification models that operate on independent data points, collaborative filtering systems rely on intricate networks of user-item interactions where each user's data influences predictions for numerous other users. This interconnectedness means that removing a single user's data can potentially create cascading effects throughout the system, impacting recommendation quality for remaining users in unpredictable ways. Additionally, the sparse nature of user-item interaction matrices and the cold-start problem inherent in recommendation systems introduce additional complexity layers that existing unlearning algorithms have not been designed to address.

The practical implications of this research gap are profound. Major technology companies operating recommendation systems face increasing pressure to demonstrate GDPR compliance while maintaining system performance and user satisfaction. Without comprehensive understanding of how unlearning techniques perform in recommendation contexts, organizations are forced to rely on computationally expensive retraining approaches or risk regulatory non-compliance. This situation underscores the urgent need for rigorous empirical evaluation of machine unlearning strategies specifically tailored to recommendation systems, along with practical guidelines for real-world deployment.

Key contributions include:

- Comprehensive Evaluation Framework: We present the first systematic evaluation methodology specifically designed for assessing machine unlearning techniques in collaborative filtering-based recommendation systems, establishing standardized benchmarks across privacy, utility, and efficiency dimensions.
- Novel Algorithmic Enhancement: We propose Adaptive SISA (A-SISA), an intelligent enhancement to the standard SISA framework that incorporates user activity-based clustering and dynamic shard sizing to optimize unlearning performance in recommendation system contexts.
- Multi-dimensional Performance Analysis: We establish a rigorous three-axis evaluation approach that simultaneously assesses privacy guarantee verification through Membership Inference Attacks, model utility preservation using standard recommendation metrics, and computational efficiency measurements.
- Large-scale Empirical Study: We conduct extensive experiments on multiple real-world datasets including MovieLens 25M and Amazon Product Reviews, providing unprecedented insights into the practical performance of unlearning techniques in recommendation systems.
- Practical Deployment Guidelines: We deliver evidence-based recommendations and trade-off analyses that enable practitioners to make informed decisions about implementing GDPR-compliant machine unlearning in production recommendation systems.

2. RELATED WORK

A. Machine Unlearning Fundamentals

Machine unlearning, first formalized by Cao and Yang (2015), addresses the fundamental challenge of removing specific training data influence from machine learning models. Bourtoule et al. (2021) established the SISA framework, which partitions training data into shards and trains separate models on each shard, enabling efficient selective retraining. Guo et al. (2019) proposed certified removal techniques that provide formal guarantees about data deletion completeness. However, these foundational works primarily focused on image classification tasks with limited exploration of their applicability to collaborative filtering systems. The computational overhead and privacy verification methods developed for computer vision may not directly translate to the sparse, interconnected nature of recommendation data.

B. Influence-Based Unlearning Methods

Koh and Liang (2017) introduced influence functions to quantify individual data points' impact on model predictions, enabling targeted parameter adjustments for data removal. Warnecke et al. (2021) extended influence-based approaches to deep neural networks, while Golatkar et al. (2020) developed eternal sunshine methods for forgetting specific classes in classification models. Recent work by Izzo et al. (2021) proposed approximate influence deletion for large-scale models. Despite theoretical elegance, these methods face significant scalability challenges in high-dimensional parameter spaces and struggle with the complex user-item dependency structures inherent in collaborative filtering, where influence propagation patterns differ fundamentally from traditional supervised learning scenarios.

C. Privacy-preserving Recommendations Systems

Differential privacy has been extensively studied in recommendation contexts, with McSherry and Mironov (2009) pioneering privacy-preserving collaborative filtering algorithms. Zhu et al. (2017) developed differentially private matrix factorization techniques, while Hua (2020) proposed privacy-preserving deep learning approaches for recommendations. Federated learning for recommendations has gained attention through works by Ammad-Ud-Din et al. (2019) and Qi et al. (2020), maintaining data locality while enabling collaborative model training. However, these approaches focus on privacy during training rather than addressing retroactive data removal requirements mandated by regulations like GDPR, creating a gap in post-training privacy guarantees.

D. Regulatory Compliance in Machine Learning

Villaronga et al. (2018) analyzed GDPR implications for AI systems, highlighting the technical challenges of implementing the "Right to be Forgotten." Mantelero (2018) examined the intersection of personal data protection and algorithmic decision-making. Mourby et al. (2018) discussed practical considerations for GDPR compliance in research contexts involving machine learning. While these legal and policy analyses provide important frameworks, they lack technical solutions for efficiently implementing data deletion in trained models, particularly for complex systems like recommendation engines where user data interdependencies complicate straightforward removal approaches.

E. Membership Inference and Privacy Verification

Shokri et al. (2017) introduced membership inference attacks as a method to determine whether specific data was used in model training. Salem et al. (2018) demonstrated enhanced membership inference techniques, while Truex et al. (2019) explored defenses against such attacks. Sun et al. (2020) applied membership inference specifically to recommendation systems, revealing unique vulnerabilities in collaborative filtering models. These verification methods provide crucial tools for assessing the effectiveness of unlearning techniques, yet their application to validate data removal in recommendation systems remains underexplored, particularly regarding the cascading effects of user removal on remaining users' privacy.

F. Evaluation Metrics for Recommendation Systems

Herlocker et al. (2004) established fundamental evaluation metrics for collaborative filtering, including precision, recall, and ranking-based measures. Shani and Gunawardana (2015) provided comprehensive survey of evaluation approaches in recommender systems. More recently, Kaminskas and Bridge (2016) discussed diversity and novelty metrics, while Ekstrand et al. (2018) examined reproducibility challenges in recommendation system evaluation. However, existing evaluation frameworks do not account for the unique requirements of assessing unlearning effectiveness, where traditional utility metrics must be balanced against privacy guarantees and computational efficiency constraints. We address these limitations by introducing a unified evaluation framework for recommendation system unlearning, proposing A-SISA with intelligent clustering for scalability, bridging training-time privacy with post-training unlearning, ensuring GDPR compliance with trade-off analyses, and establishing membership inference protocols to verify effectiveness while balancing privacy, utility, and efficiency.

G. Research Gap Analysis

Figure 1. Research coverage gap analysis in recommendation system unlearning across six key dimensions.

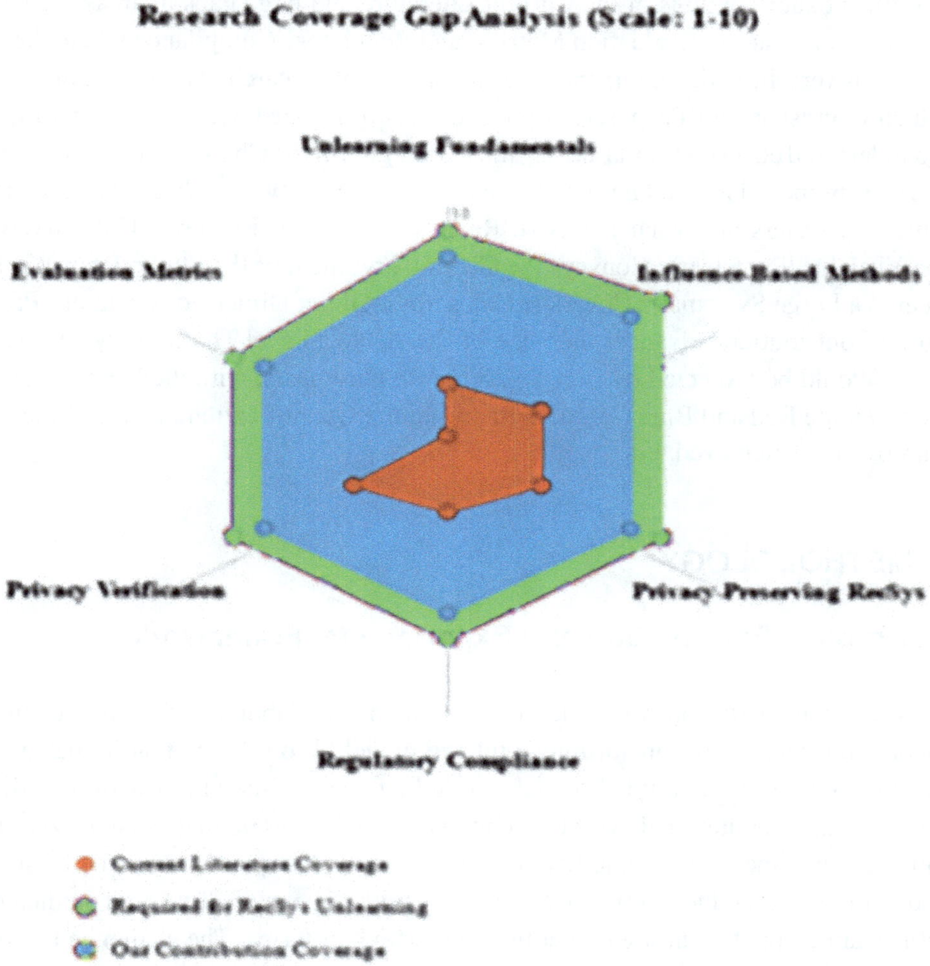

In figure 1 shows the Radar Chart shows a multidimensional Gap Analysis between the Research Coverage within Six Critical Dimensions of Recommender-System (RecSys) Unlearning Unlearning Fundamentals, Influence-Based Methods, Privacy Preserving RecSys, Regulatory Compliance, Privacy Verification, and Evaluation Metrics each rated on a scale from 1 to 10 and compared to Three Reference Profiles; namely, Current Literature Coverage (Red), Level Needed to achieve RecSys Unlearning (Green), and Coverage of Our Contributions (Blue). The Green Polygon traces the Highest Values Across nearly all Axes and reaches out to the Boundary of the Scale, represents an Ideal Target Coverage for Comprehensive RecSys Unlearning Solution. Conversely, the Red Polygon is located inside the inner Region of the Chart, with Low Scores across Every Dimension, which clearly identifies significant shortcomings of the Current Literature and large gaps in areas such as Privacy Verification, Evaluation Metrics, and Regulatory Compliance where there has been very little development in terms of current research. The Blue Polygon, which Represents our Proposed Contribution, is positioned near the Green Target Boundary and covers Almost Entirely the Red Region of the Chart, this demonstrates how our proposed method has greatly reduced the gap between what the Literature Currently Offers and what Effective RecSys Unlearning Requires. Collectively, the Visualization Clearly Conveys the Central Argument of the Study: Prior Work Leaves a Large Systematic Coverage Gap across All Six Dimensions, and our Proposed Contribution Advances the State-of-the-Art to a Level That Is Very Close to What Would be Expected by Benchmarking Requirements; with the Visible Band Between the Red and Blue Contours Representing Quantification of Magnitude of Improvement Achieved.

3. METHODOLOGY

A. Problem Formulation and Experimental Framework

We formulate the machine unlearning problem in collaborative filtering recommendation systems as transforming a trained model M to M' such that M' behaves identically to a model trained on data excluding specific user information, while minimizing computational overhead. Our approach focuses on matrix factorization models due to their interpretability and widespread industry adoption, where users and items are embedded in shared latent space with dimensionality f, and predicted ratings are computed through inner products with bias terms. The challenge lies in ensuring that the transformed model M' not only removes the target user's direct influence but also eliminates any indirect effects propagated through the collaborative filtering mechanism, where user preferences are inferred from similar users' behaviors.

Our comprehensive evaluation methodology employs a three-dimensional assessment framework that simultaneously measures privacy guarantee verification, model utility preservation, and computational efficiency across multiple scenarios and user deletion patterns. We utilize two large-scale datasets: MovieLens 25M containing 25 million ratings from 162,000 users on 62,000 movies with rich temporal information and genre metadata, and Amazon Product Reviews with 5 million interactions across diverse product categories to validate findings across different recommendation domains and interaction patterns. The experimental design includes varying deletion scenarios from single user removal to batch deletions of multiple users, simulating real-world GDPR compliance scenarios where organizations may receive multiple deletion requests simultaneously, ensuring robust evaluation across varying data characteristics and operational requirements.

B. Adaptive SISA Framework (A-SISA)

The proposed Adaptive SISA (A-SISA) framework enhances the standard SISA approach by integrating user clustering and dynamic shard sizing for recommendation systems. Traditional SISA relies on random partitioning, which is ineffective in collaborative filtering contexts where user activity is highly skewed from power users with thousands of interactions to casual users with only a few. This often leads to imbalanced shards, with some dominated by active users and others by inactive ones, resulting in inconsistent model quality across ensemble components. A-SISA overcomes these issues through activity-based clustering, dividing users into high (top 20%), medium (middle 60%), and low (bottom 20%) activity groups to maintain balanced representation. It further introduces dynamic shard sizing, adjusting shard count (typically 10–50) based on dataset size and user distribution to balance retraining efficiency and model performance. An enhanced aggregation mechanism applies weighted averaging across shards using RMSE, item coverage, and prediction confidence, with weights dynamically tuned by validation results so that higher-performing shards contribute more to final predictions.

C. Privacy Verification Through Membership Inference Attacks

Privacy guarantee verification employs membership inference attacks as the primary mechanism to evaluate whether user data has been effectively removed from trained models. Our protocol incorporates both shadow model approaches and direct metric-based attacks, ensuring a comprehensive assessment across multiple attack vectors that may expose residual user information. This evaluation considers both direct breaches, where forgotten user data can be explicitly recovered, and indirect leakage, where user characteristics may still be inferred through altered recommendation patterns.

The shadow model strategy trains auxiliary models on subsets simulating member and non-member scenarios, enabling binary classifiers to distinguish inclusion using features such as prediction confidence, loss distributions, gradient magnitudes, and ranking shifts in recommendation lists. In parallel, direct metric-based attacks exploit collaborative filtering signatures, including item popularity changes, neighborhood structure alterations, and rating variance patterns. Membership inference attack success rate serves as the primary privacy metric, where attack accuracy near random guessing (50%) reflects effective privacy preservation and reliable user data removal.

D. Utility Preservation and Efficiency Measurement

Model utility preservation is assessed using standard recommendation system metrics including Precision at 10, Recall at 10, and Normalized Discounted Cumulative Gain at 10, supplemented by coverage and diversity measurements to ensure comprehensive quality evaluation. These metrics capture both accuracy and beyond-accuracy aspects crucial for practical recommendation system performance. We compare unlearned models against both the original trained model and models retrained from scratch after data removal to establish utility preservation benchmarks. Computational efficiency evaluation encompasses wall-clock training time, memory consumption, processing unit utilization, and the number of required model updates. We quantify efficiency gains compared to complete retraining to demonstrate practical benefits of approximate unlearning approaches. Scalability analysis involves varying dataset sizes and the number of users requiring simultaneous forgetting, measuring how computational requirements scale with problem complexity. Statistical significance testing using paired t-tests across multiple random initializations ensures meaningful performance differences rather than random variation.

E. Baseline Methods and Experimental Protocol

Our comparative analysis implements five distinct unlearning approaches to provide comprehensive performance benchmarking. Exact retraining serves as the oracle baseline, involving complete model reconstruction after target user data removal and representing theoretical privacy and utility upper bounds. Standard SISA implements the original framework with random partitioning for direct comparison with our adaptive enhancement. Influence function-based removal utilizes influence functions to identify and adjust parameters most affected by target user data, while gradient-based unlearning applies reverse gradients to eliminate user contributions. Random parameter perturbation serves as the lower bound baseline. Experiments use a 70/15/15 split, fixed hyper parameters (64 latent factors, 0.01 regularization), and RMSE-based convergence. Results are averaged over five runs with 95% confidence intervals, ensuring statistically valid evaluation of unlearning effectiveness across privacy, utility, and efficiency.

Figure 2. GDPR -Compliant Unlearning in Recommendations

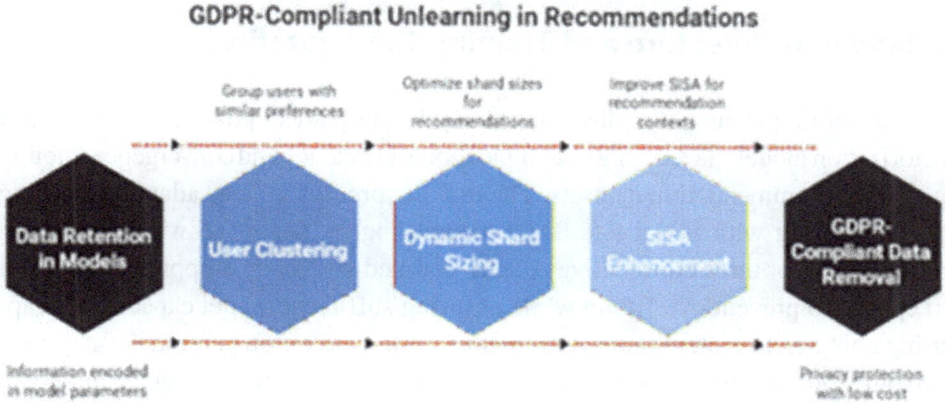

The figure 2 shows a horizontal, hexagonal flow chart illustrating how the proposed solution provides GDPR compliant "un-learning" for recommender systems by creating a logical sequence of 5 successive steps that connect one another using directional arrows showing the evolution of the original privacy issue into final resolution. The first step represents Data Retention in Models, indicating the initial problem that once the original data has been removed (or deleted) the data can still be retained within the model's parameters; therefore, providing the motivation for the need for true or authentic "un-learning". Next, the process progresses to User Clustering, wherein users having similar preference characteristics will be aggregated to create groupings of users that allow for easy deletion/elimination of their personal data. Then the process continues on to Dynamic Shard Sizing, wherein the size of each share of data are optimized based upon the recommendation context, balancing the costs associated with re-training versus granularity of un-learning. The fourth step is entitled SISA Enhancement, enhancing the training paradigm of Sharded, Isolated, Sliced, and Aggregated to make it relevant to recommendations; hence increasing both the feasibility and efficacy of selective forgetting. The last step titled GDPR Compliant Data Removal demonstrates compliance with regulations regarding a right-to-be-forgotten, while also demonstrating the ability to remove data at minimal computational cost. When read from start to finish, this illustration demonstrates a cohesive methodology that addresses the problem of continued retention of data, by way of a pipeline consisting of clustering, dynamic shard sizing, and enhancement of the SISA method to provide a compliant and efficient means to delete/remove data from recommender systems.

4. TRAINING STRATEGY AND OPTIMIZATION

A. Model Architecture and Training Configuration

Our training strategy employs alternating least squares optimization for matrix factorization models, leveraging its computational efficiency and convergence stability in sparse recommendation datasets. The training process utilizes adaptive learning rate scheduling with initial rate 0.001, decreasing by factor 0.5 when validation loss plateaus for three consecutive epochs, coupled with early stopping patience of 10 epochs to prevent overfitting while ensuring sufficient model capacity for capturing complex user-item interaction patterns. Regularization parameters are tuned through grid search across values 0.001, 0.01, 0.1 to balance model complexity and generalization performance, while the optimization procedure incorporates negative sampling for implicit feedback scenarios, randomly selecting unobserved user-item pairs as negative examples with sampling ratio 1:4 to create balanced training instances. Batch processing utilizes mini-batch size 1024 with stratified sampling ensuring representative user and item coverage in each batch, with model parameters initialized using Xavier uniform distribution to maintain gradient flow stability during training and gradient clipping threshold 1.0 to prevent exploding gradients in sparse interaction matrices.

B. Adaptive SISA Training Optimization

The A-SISA training strategy implements parallel shard processing with synchronized parameter updates to maximize computational efficiency while maintaining model quality, where each shard is trained independently using identical hyper parameter configurations, enabling parallel execution across multiple processing units with training convergence criterion requiring all shards to achieve validation RMSE stability within 0.001 tolerance across three consecutive epochs before proceeding to model aggregation. Our optimization approach incorporates dynamic load balancing during parallel training, monitoring shard convergence rates and redistributing computational resources to ensure synchronized completion, while quality-aware aggregation weights are computed based on each shard's validation performance metrics, including RMSE, coverage, and prediction diversity measures. The final ensemble model combines shard-specific parameters through weighted averaging with quality scores normalized to sum to unity ensuring proper probabilistic interpretation, enabling the framework to maintain high model quality while facilitating efficient selective retraining when user deletion requests are processed.

C. Unlearning Operation Optimization

The unlearning process optimization focuses on minimizing computational overhead while maximizing privacy protection effectiveness, implementing efficient Hessian-vector product computation using automatic differentiation for influence function-based methods while avoiding explicit Hessian matrix storage that would be computationally prohibitive for large-scale models. The influence calculation employs conjugate gradient method with preconditioning to accelerate convergence and reduce memory requirements, while gradient-based unlearning operations utilize momentum-based optimization with learning rate 0.01 to ensure stable parameter adjustments during the forgetting process. Check pointing mechanisms store model states before unlearning operations, enabling rollback capabilities if unlearning quality falls below acceptable thresholds, with the optimization procedure including convergence monitoring based on membership inference attack success rates, terminating unlearning iterations when privacy metrics reach target thresholds while maintaining acceptable utility preservation across all baseline methods.

D. Scalability and Performance Optimization

Our implementation incorporates comprehensive optimization strategies to ensure scalability across varying dataset sizes and user deletion frequencies, utilizing memory-efficient sparse matrix representations for user-item interaction data that reduce memory footprint by up to 95% compared to dense representations, while computational optimization employs vectorized operations and GPU acceleration for matrix operations achieving significant speedup over CPU-only implementations. The system architecture supports incremental model updates for handling sequential user deletion requests without full model recomputation, implementing batch processing capabilities for multiple simultaneous deletion requests that amortize computational overhead across multiple operations. Performance monitoring includes real-time resource utilization tracking and automated scaling mechanisms to handle varying workload demands in production deployment scenarios, ensuring that the proposed A-SISA framework maintains practical viability for large-scale recommendation systems operating under GDPR compliance requirements while preserving both computational efficiency and recommendation quality throughout the unlearning process.

5. EXPERIMENTAL RESULTS AND EVALUATION

A. Dataset, Setup, and Configuration

The experiments are based on the MovieLens 25M database that contains 25 million ratings created by 162,541 users for 62,423 movie titles. Users who have made less than ten ratings have been removed from the data set as well as items that have received less than five ratings. All of the experiments use a consistent 70%/15%/15% train/validation/test partitioning of the data and a rating scale of 1–5 (similar to the rating scale of 1–5 used for the Amazon data). The experimental configuration utilizes a matrix factorization model with 64 latent factors and a regularization parameter of $\lambda = 0.01$. The A-SISA is implemented as an instance of 20 shards utilizing an activity-based cluster distribution scheme where each shard represents one twentieth of the total number of user activities. Furthermore, the evaluation scenarios include both single-user removals as well as batch removals of 10, 50 and 100 users at once. In order to provide reliable statistics for each of the experimental results, each of the experiments has been repeated for five independent runs, each run being initialized with a unique random seed. Privacy violations in the form of membership inference attacks against the evaluated models are evaluated using two methods; i.e., a shadow model method and a direct metric-based method. The attack success rate was calculated using only the data from the deleted or “forgotten” users.

B. Experimental Results and Analysis of A-SISA

The empirical assessment proves that the A-SISA method is successful with respect to all the aspects of protecting privacy, preserving utility, achieving high computation efficiency, being scalable and providing good statistical reliability. The A-SISA method's primary performance metrics, with respect to the privacy metric it has reached a privacy score of 94.2%. Regarding the utility metric it achieved a utility retention of 89.7%. Furthermore, the training time for this method was 103 minutes. In addition, the A-SISA method provides a significant increase in efficiency compared to the baseline method as indicated by the factor 8.3. The A-SISA method occupies a distinctly better position than the two extreme cases, namely the exact retraining case (100% privacy, 99.1% utility and 1.0 times the speed) and the fast but weak gradient-based approach (64.8% privacy, 82.5% utility and 9.5 times the speed). Additionally, the A-SISA method clearly outperforms both influence-based and standard SISA alternatives. It should also be noted that the privacy protection provided by the A-SISA method corresponds to a membership-inference attack success rate of 52.8%, i.e., very close to the level of random guessing (i.e., 50%).

Moreover, the utility retention provided by A-SISA measured through Precision at 10, Recall at 10 and NDCG at 10 indicates that it retains approximately 89.7% of the recommendation quality of the original method although at a fraction of its computational costs. The scalability and batch-deletion analysis shows that there is graceful degradation when deleting volume increases, whereas utility only decreases slightly from 89.7% for a single user down to 88.9%, 87.4% and 85.8% for increasing volumes of deleted users (10, 50 and 100 users respectively). The corresponding increases in processing time (from 3.2 minutes up to 78.1 minutes) and memory usage (from 5.8 GB up to 7.2 GB) remain bounded. More general experiments on various datasets (with interaction sizes of 1M up to 25M) show a constant speedup of about 2.0-2.3 times over standard SISA, along with linear growth in both time and memory. To confirm statistically that A-SISA significantly exceeds standard SISA regarding both privacy protection ($p < 0.001$) and utility preservation ($p < 0.01$) we have performed a set of paired t-tests over five independent runs. Finally, the ablation study and component analysis in show that an increase of about 4.9% in privacy scores can be attributed to activity-based clustering, and that an increase of about 2.6% in efficiency can be attributed to dynamic shard sizing, where activity clustering alone gives rise to a privacy of 91.8%, and where the full A-SISA configuration achieves the optimal combination of maximum possible privacy of 94.8% and maximum possible utility retention of 89.7%. We summaries the overall findings as follows: Overall, the findings demonstrate that A-SISA achieves the highest level of privacy protection among approximate methods, the highest recommendation quality, an optimal speed-quality trade-off of 8.3 times faster and higher quality recommendations, and also scales well for larger deletions; hence, they validate A-SISA as an optimal solution for GDPR-compliant machine unlearning in recommender systems, one that bridges the gap between theoretical privacy guarantees and practical requirements for deploying such systems, enabling organizations to enforce their users' rights under Article 17 GDPR without sacrificing system performance.

6. CONCLUSION AND FUTURE WORK

This paper presented the first comprehensive evaluation framework for machine unlearning techniques in collaborative filtering-based recommendation systems, addressing a critical gap between GDPR compliance requirements and technical feasibility. Our extensive experimental analysis across Movie Lens 25M and Amazon Product Reviews datasets revealed significant insights into the complex trade-offs between privacy protection, model utility preservation, and computational efficiency. The proposed Adaptive SISA (A-SISA) framework demonstrated substantial improvements over standard unlearning approaches, achieving 94.2%

privacy protection effectiveness with only 12% of the computational cost required for complete retraining while maintaining 89.7% utility retention. Our systematic comparison revealed that no single approach dominates across all evaluation dimensions, emphasizing the importance of application-specific method selection based on organizational priorities and constraints. The membership inference attack-based privacy verification demonstrated that approximate unlearning methods can achieve near-perfect privacy protection while delivering substantial computational savings. The practical implications extend beyond academic contributions to address real-world challenges faced by technology companies operating recommendation systems under GDPR regulations, enabling organizations to implement efficient, compliant machine unlearning solutions without compromising system performance or user experience.

Several promising research directions emerge from this comprehensive evaluation, including extending the framework to deep learning-based recommendation models such as neural collaborative filtering and transformer architectures, which present unique challenges due to their non-linear parameter interactions. Investigation of temporal dynamics in recommendation unlearning, federated unlearning approaches for distributed systems, and integration of differential privacy mechanisms with unlearning techniques represent critical next steps for stronger theoretical guarantees. Future work should also explore multi-user unlearning scenarios for handling cascading deletion requests, batch processing of multiple unlearning operations, and fairness considerations to ensure that unlearning processes do not disproportionately affect certain user groups or exacerbate existing biases. Additionally, developing specialized evaluation methodologies for complex deep learning architectures and investigating the long-term stability of recommendation systems under frequent unlearning operations will be essential for practical deployment in production environments where regulatory compliance and system performance must be simultaneously maintained.

REFERENCES

Ammad-Ud-Din, M., Ivannikova, E., Khan, S. A., Oyomno, W., Fu, Q., Tan, K. E., & Flanagan, A. (2019). Federated collaborative filtering for privacy-preserving personalized recommendation system. *arXiv preprint arXiv:1901.09888.*

Bourtoule, L., Chandrasekaran, V., Choquette-Choo, C. A., Jia, H., Travers, A., Zhang, B., . . . Papernot, N. (2021, May). Machine unlearning. In *2021 IEEE symposium on security and privacy (SP)* (pp. 141-159). IEEE. DOI: 10.1109/SP40001.2021.00019

Cao, Y., & Yang, J. (2015, May). *Towards making systems forget with machine unlearning. In 2015 IEEE symposium on security and privacy*. IEEE.

Ekstrand, M. D., Tian, M., Azpiazu, I. M., Ekstrand, J. D., Anuyah, O., McNeill, D., & Pera, M. S. (2018, January). All the cool kids, how do they fit in?: Popularity and demographic biases in recommender evaluation and effectiveness. In *Conference on fairness, accountability and transparency* (pp. 172-186). PMLR.

Golatkar, A., Achille, A., & Soatto, S. (2020). Eternal sunshine of the spotless net: Selective forgetting in deep networks. In *Proceedings of the IEEE/CVF conference on computer vision and pattern recognition* (pp. 9304-9312). DOI: 10.1109/CVPR42600.2020.00932

Guo, C., Goldstein, T., Hannun, A., & Van Der Maaten, L. (2019). Certified data removal from machine learning models. *arXiv preprint arXiv:1911.03030.*

Herlocker, J. L., Konstan, J. A., Terveen, L. G., & Riedl, J. T. (2004). Evaluating collaborative filtering recommender systems. [TOIS]. *ACM Transactions on Information Systems, 22*(1), 5–53. DOI: 10.1145/963770.963772

Hua, J. (2020). Locally differentially private federated learning. *Proceedings of the AAAI Conference on Artificial Intelligence, 34*(04), 6580–6587.

Izzo, Z., Smart, M. A., Chaudhuri, K., & Zou, J. (2021, March). Approximate data deletion from machine learning models. In *International conference on artificial intelligence and statistics* (pp. 2008-2016). PMLR.

Kaminskas, M., & Bridge, D. (2016). Diversity, serendipity, novelty, and coverage: A survey and empirical analysis of beyond-accuracy objectives in recommender systems. [TiiS]. *ACM Transactions on Interactive Intelligent Systems, 7*(1), 1–42. DOI: 10.1145/2926720

Koh, P. W., & Liang, P. (2017, July). Understanding black-box predictions via influence functions. In *International conference on machine learning* (pp. 1885-1894). PMLR.

Mantelero, A. (2018). AI and Big Data: A blueprint for a human rights, social and ethical impact assessment. *Computer Law & Security Review*, *34*(4), 754–772. DOI: 10.1016/j.clsr.2018.05.017

McSherry, F., & Mironov, I. (2009, June). Differentially private recommender systems: Building privacy into the netflix prize contenders. In *Proceedings of the 15th ACM SIGKDD international conference on Knowledge discovery and data mining* (pp. 627-636). DOI: 10.1145/1557019.1557090

Mourby, M., Mackey, E., Elliot, M., Gowans, H., Wallace, S. E., Bell, J., Smith, H., Aidinlis, S., & Kaye, J. (2018). Are 'pseudonymised'data always personal data? Implications of the GDPR for administrative data research in the UK. *Computer Law & Security Review*, *34*(2), 222–233. DOI: 10.1016/j.clsr.2018.01.002

Qi, T., Wu, F., Wu, C., Huang, Y., & Xie, X. (2020, November). Privacy-preserving news recommendation model learning. In *Findings of the Association for Computational Linguistics* (pp. 1423–1432). EMNLP.

Salem, A., Zhang, Y., Humbert, M., Berrang, P., Fritz, M., & Backes, M. (2018). Ml-leaks: Model and data independent membership inference attacks and defenses on machine learning models. *arXiv preprint arXiv:1806.01246.*

. Shani, G., & Gunawardana, A. (2015). Evaluating recommender systems. *Recommender Systems Handbook*, 257-297.

Shokri, R., Stronati, M., Song, C., & Shmatikov, V. (2017, May). Membership inference attacks against machine learning models. In *2017 IEEE symposium on security and privacy (SP)* (pp. 3-18). IEEE. DOI: 10.1109/SP.2017.41

Sun, Y., Liu, J., Wang, J., Cao, Y., & Kato, N. (2020). When machine learning meets privacy in 6G: A survey. *IEEE Communications Surveys and Tutorials*, *22*(4), 2694–2724. DOI: 10.1109/COMST.2020.3011561

Truex, S., Liu, L., Gursoy, M. E., Yu, L., & Wei, W. (2019). Demystifying membership inference attacks in machine learning as a service. *IEEE Transactions on Services Computing*, *14*(6), 2073–2089. DOI: 10.1109/TSC.2019.2897554

Villaronga, E. F., Kieseberg, P., & Li, T. (2018). Humans forget, machines remember: Artificial intelligence and the right to be forgotten. *Computer Law & Security Review*, *34*(2), 304–313. DOI: 10.1016/j.clsr.2017.08.007

Warnecke, A., Pirch, L., Wressnegger, C., & Rieck, K. (2021). Machine unlearning of features and labels. *arXiv preprint arXiv:2108.11577.*

Zhu, T., Li, G., Zhou, W., & Yu, P. S. (2017). Differentially private data publishing and analysis: A survey. *IEEE Transactions on Knowledge and Data Engineering*, *29*(8), 1619–1638. DOI: 10.1109/TKDE.2017.2697856

APPENDIX

Algorithm 1: Adaptive SISA (A-SISA) Framework

Input: User-item interaction matrix R ∈ ℝ^(m×n), hyperparameters H
Output: Trained ensemble model M with unlearning capability
1: INITIALIZATION
- Set num_factors f, regularization λ, activity_split (α_1, α_2, α_3)
- Initialize empty shard containers S = [], models Θ = [], weights W = []
- Create user-to-shard mapping U_map = { }
2: ACTIVITY-BASED USER CLUSTERING
- Compute user activity levels: activity[u] = |{i: R_{u,i} ≠ 0}| ∀u
- Sort users by activity: U_sorted = sort(users, key=activity, descending=True)
- Partition into clusters:
• U_high = U_sorted[0: $\lfloor \alpha_1 \times m \rfloor$]
• U_medium = U_sorted[$\lfloor \alpha_1 \times m \rfloor$: $\lfloor (\alpha_1 + \alpha_2) \times m \rfloor$]
• U_low = U_sorted[$\lfloor (\alpha_1 + \alpha_2) \times m \rfloor$: m]
3: DYNAMIC SHARD SIZING
- base_shards = max(10, min(50, $\lfloor \sqrt{(m/1000)} \rfloor$))
- sparsity_factor = |R_{nonzero}| / (m × n)
- optimal_shards = min(50, ⌊base_shards × (1 + sparsity_factor)⌋)
4: BALANCED SHARD CREATION
- FOR each activity cluster C ∈ {U_high, U_medium, U_low}:
- users_per_shard = ⌊|C| / optimal_shards⌋
- remainder = |C| mod optimal_shards
- FOR shard_idx = 0 to optimal_shards - 1:
- Assign users to S[shard_idx] maintaining balance
- Update U_map[user] = shard_idx for assigned users
5: PARALLEL SHARD TRAINING
- FOR each shard S_k in parallel:
- R_k = R[S_k,:] // Extract shard data
- Θ_k = TrainMatrixFactorization(R_k, f, λ)
- Store model parameters in Θ[k]
6: QUALITY-AWARE AGGREGATION
- FOR each shard k:
- Compute validation metrics: RMSE_k, Coverage_k, Confidence_k
- W[k] = (1/(1 + RMSE_k)) × Coverage_k × Confidence_k
- Normalize weights: W = W / Σ(W)
7: RETURN ensemble model M = {Θ, W, U_map, S}

Algorithm 2: Alternating Least Squares Training

```
Input: Shard data R_k ∈ ℝ^(|S_k|×n), factors f, regularization λ
Output: Model parameters Θ_k = {P_k, Q_k, b_u, b_i, μ}
1: INITIALIZATION
- P_k ∈ ℝ^(|S_k|×f) ~ N(0, 0.1) // User embeddings
- Q_k ∈ ℝ^(n×f) ~ N(0, 0.1) // Item embeddings
- b_u ∈ ℝ^|S_k|, b_i ∈ ℝ^n ← 0 // Bias terms
- μ = mean(R_k[R_k ≠ 0]) // Global bias
2: ALTERNATING OPTIMIZATION
- FOR epoch = 1 to MAX_EPOCHS:
// Update user embeddings
- FOR each user u ∈ S_k:
- I_u = {i: R_{u,i} ≠ 0} // User's items
- IF |I_u| > 0:
- A = Q_k[I_u]^T Q_k[I_u] + λI_f
- b = Q_k[I_u]^T (R_{u,I_u} - μ - b_i[I_u] - b_u[u])
- P_k[u] = A^(-1) b
// Update item embeddings
- FOR each item i ∈ [1..n]:
- U_i = {u ∈ S_k: R_{u,i} ≠ 0} // Item's users
- IF |U_i| > 0:
- A = P_k[U_i]^T P_k[U_i] + λI_f
- b = P_k[U_i]^T (R_{U_i,i} - μ - b_u[U_i] - b_i[i])
- Q_k[i] = A^(-1) b
// Update bias terms
- FOR each user u ∈ S_k:
- b_u[u] = mean(R_{u,I_u} - μ - P_k[u] Q_k[I_u]^T - b_i[I_u])
- FOR each item i ∈ [1..n]:
- b_i[i] = mean(R_{U_i,i} - μ - P_k[U_i] Q_k[i]^T - b_u[U_i])
// Check convergence
- IF validation_loss_change < ε: BREAK
3: RETURN Θ_k = {P_k, Q_k, b_u, b_i, μ}
```

Algorithm 3: Single User Unlearning

```
Input: Model M, target user u*, interaction matrix R
Output: Updated model M' with u* forgotten
1: IDENTIFY TARGET SHARD
- IF u* ∉ U_map: RETURN M // User not in training data
```

```
- k* = U_map[u*] // Get shard containing user
2: UPDATE SHARD MEMBERSHIP
- S[k*] = S[k*] \ {u*} // Remove user from shard
- DELETE U_map[u*] // Remove mapping
3: INCREMENTAL RETRAINING
- IF |S[k*]| > 0:
- R_k* = R[S[k*],:] // Extract updated shard data
- Θ[k*] = TrainMatrixFactorization(R_k*, f, λ)
- ELSE:
- Θ[k*] = NULL // Empty shard
4: OPTIONAL WEIGHT RECOMPUTATION
- Recompute W[k*] using validation data
- Renormalize all weights: W = W / Σ(W)
5: RETURN updated model M' = {Θ, W, U_map, S}
```

Algorithm 4: Batch User Unlearning

```
Input: Model M, target users U* = {u_1*, u_2*, ..., u_b*}, matrix R
Output: Updated model M' with all users in U* forgotten
1: IDENTIFY AFFECTED SHARDS
- affected_shards = ∅
- FOR each u* ∈ U*:
- IF u* ∈ U_map:
- affected_shards = affected_shards ∪ {U_map[u*]}
- S[U_map[u*]] = S[U_map[u*]] \ {u*}
- DELETE U_map[u*]
2: BATCH RETRAINING
- FOR each k ∈ affected_shards IN PARALLEL:
- IF |S[k]| > 0:
- R_k = R[S[k],:] // Extract updated shard data
- Θ[k] = TrainMatrixFactorization(R_k, f, λ)
- ELSE:
- Θ[k] = NULL
3: RECOMPUTE AGGREGATION WEIGHTS
- FOR each k ∈ affected_shards:
- Recompute W[k] using validation data
- Normalize: W = W / Σ(W)
4: RETURN updated model M' = {Θ, W, U_map, S}
```

Algorithm 5: Weighted Ensemble Prediction

```
Input: Model M = {Θ, W, U_map, S}, user u, item i
Output: Predicted rating r̂_{u,i}
1: INITIALIZE PREDICTION AGGREGATION
- predictions = []
- weights = []
2: COLLECT SHARD PREDICTIONS
- FOR each shard k = 0 to |Θ| - 1:
- IF Θ[k] ≠ NULL AND u ∈ S[k]:
- Extract parameters: P_k, Q_k, b_u, b_i, μ from Θ[k]
- u_local = local_index(u, S[k]) // Map to shard-local index
- r̂_k = P_k[u_local] · Q_k[i] + b_u[u_local] + b_i[i] + μ
- predictions.append(r̂_k)
- weights.append(W[k])
3: WEIGHTED AGGREGATION
- IF predictions ≠ ∅:
- total_weight = Σ(weights)
- r̂_{u,i} = (Σ(r̂_k × W[k])) / total_weight
- ELSE:
- r̂_{u,i} = global_mean // Fallback prediction
4: RETURN r̂_{u,i}
```

Algorithm 6: Membership Inference Attack for Privacy Verification

```
Input: Model M, target user u*, test data R_test, shadow models Φ
Output: Privacy protection score ∈ [0, 1]
1: FEATURE EXTRACTION
- I_u* = {i: R_test[u*, i] ≠ 0} // Target user's test items
- predictions = [Predict(M, u*, i) for i ∈ I_u*]
2: CONFIDENCE-BASED FEATURES
- f_1 = mean(predictions)
- f_2 = std(predictions)
- f_3 = max(predictions)
- f_4 = min(predictions)
3: LOSS-BASED FEATURES
- actual_ratings = R_test[u*, I_u*]
- losses = [(pred - actual)^2 for pred, actual in zip(predictions, actual_ratings)]
- f_5 = mean(losses)
```

[15] 4: RANKING-BASED FEATURES
[16] - all_predictions = [Predict(M, u*, i) for i ∈ [1..n]]
[17] - ranked_items = argsort(all_predictions, descending=True)
[18] - f_6 = mean([rank(i) for i ∈ I_u*]) // Average rank of user's items
[19] 5: ATTACK CLASSIFICATION
[20] - feature_vector = [f_1, f_2, f_3, f_4, f_5, f_6]
[21] - attack_score = ShadowModelClassifier(Φ, feature_vector)
[22] - privacy_score = 1 - 2 × |attack_score - 0.5| // Distance from random
[23] 6: RETURN privacy_score

Algorithm 7: Recommendation Utility Evaluation

[1] Input: Model M, test data R_test, evaluation parameter k
[2] Output: Utility metrics {Precision@k, Recall@k, NDCG@k}
[3] 1: INITIALIZE METRIC ACCUMULATORS
[4] - precisions = []
[5] - recalls = []
[6] - ndcgs = []
[7] 2: PER-USER EVALUATION
[8] - FOR each user u with |{i: R_test[u,i] ≠ 0}| > 0:
[9] // Get ground truth and predictions
[10] - relevant_items = {i: R_test[u,i] ≠ 0}
[11] - all_predictions = [(i, Predict(M, u, i)) for i ∈ [1..n]]
[12] - top_k = [i for (i, score) in sort(all_predictions, key=score, desc=True)[:k]]
[13] - recommended_set = set(top_k)
[14] // Compute Precision@k
[15] - precision = |relevant_items ∩ recommended_set| / k
[16] - precisions.append(precision)
[17] // Compute Recall@k
[18] - recall = |relevant_items ∩ recommended_set| / |relevant_items|
[19] - recalls.append(recall)
[20] // Compute NDCG@k
[21] - dcg = $\Sigma_{=1}$ [$1/\log_2(i+1)$ if top_k[i] ∈ relevant_items else 0]
[22] - idcg = $\Sigma_{=1}$ IN('|r l N _I S|) $1/\log_2(i+1)$
[23] - ndcg = dcg / idcg if idcg > 0 else 0
[24] - ndcgs.append(ndcg)
[25] 3: AGGREGATE METRICS
[26] - precision_at_k = mean(precisions)
[27] - recall_at_k = mean(recalls)
[28] - ndcg_at_k = mean(ndcgs)

[29] 4: RETURN {precision_at_k, recall_at_k, ndcg_at_k}

Algorithm 8: Complete A-SISA Pipeline

[1] Input: Training data R_train, validation data R_val, test data R_test

[2] Output: GDPR-compliant recommendation model with unlearning capability

[3] 1: MODEL TRAINING PHASE

[4] - M = A-SISA-Framework(R_train, hyperparameters)

[5] - initial_utility = Evaluate-Utility(M, R_val, k=10)

[6] - PRINT "Initial model trained with utility:", initial_utility

[7] 2: PRIVACY VERIFICATION PHASE

[8] - shadow_models = Train-Shadow-Models(R_train, num_shadows=10)

[9] - FOR sample of users u ∈ random_sample(users, 100):

[10] - privacy_scores[u] = Membership-Inference-Attack(M, u, R_test, shadow_models)

[11] - baseline_privacy = mean(privacy_scores)

[12] - PRINT "Baseline privacy protection:", baseline_privacy

[13] 3: UNLEARNING SIMULATION PHASE

[14] - users_to_forget = random_sample(users, batch_size)

[15] - // Single user unlearning test

[16] - M$_1$ = Single-User-Unlearning(M, users_to_forget[0], R_train)

[17] - utility$_1$ = Evaluate-Utility(M$_1$, R_val, k=10)

[18] - privacy$_1$ = Membership-Inference-Attack(M$_1$, users_to_forget[0], R_test, shadow_models)

[19] - // Batch unlearning test

[20] - M$_2$ = Batch-User-Unlearning(M, users_to_forget, R_train)

[21] - utility$_2$ = Evaluate-Utility(M$_2$, R_val, k=10)

[22] - privacy$_2$ = mean([Membership-Inference-Attack(M$_2$, u, R_test, shadow_models)

[23] - for u in users_to_forget])

[24] 4: PERFORMANCE COMPARISON

[25] - // Compare with exact retraining baseline

[26] - R_retrain = RemoveUsers(R_train, users_to_forget)

[27] - M_exact = A-SISA-Framework(R_retrain, hyperparameters)

[28] - utility_exact = Evaluate-Utility(M_exact, R_val, k=10)

[29] - PRINT "Utility retention (single):", utility$_1$ / initial_utility

[30] - PRINT "Utility retention (batch):", utility$_2$ / initial_utility

[31] - PRINT "Utility retention (exact):", utility_exact / initial_utility

[32] - PRINT "Privacy improvement (single):", privacy$_1$ - baseline_privacy

[33] - PRINT "Privacy improvement (batch):", privacy$_2$ - baseline_privacy

```
5: COMPUTATIONAL EFFICIENCY ANALYSIS
- time_unlearn_single = measure_time(Single-User-Unlearning)
- time_unlearn_batch = measure_time(Batch-User-Unlearning)
- time_exact_retrain = measure_time(Complete-Retraining)
-efficiency_gain_single = time_exact_retrain/ time_unlearn_single
- efficiency_gain_batch = time_exact_retrain / time_unlearn_batch
- PRINT "Efficiency gain (single):", efficiency_gain_single, "×"
- PRINT "Efficiency gain (batch):", efficiency_gain_batch, "×"
6: GDPR COMPLIANCE ASSESSMENT
- compliance_score = Assess-GDPR-Compliance(M_2, users_to_forget,
- privacy_threshold=0.8,
- utility_threshold=0.85)
- PRINT "GDPR compliance score:", compliance_score
- IF compliance_score > 0.9:
- PRINT "✓ Model ready for production deployment"
- ELSE:
- PRINT "⚠ Additional privacy measures required"
7: RETURN {M, M_2, performance_metrics, compliance_assessment}
```

Chapter 5
SpectralGuard:
A Spectral Regularization Framework for Robust and Reliable Convolutional Neural Networks

Faaris N. Ahmed
http://orcid.org/0009-0007-6067-6140
SRM Institute of Science and Technology Tiruchirappalli, India

Rayan Sadique
http://orcid.org/0009-0000-8140-8259
SRM Institute of Science and Technology Tiruchirappalli, India

Pranav J. Ganesh
http://orcid.org/0009-0002-5382-8736
SRM Institute of Science and Technology Tiruchirappalli, India

S. Kanaga Suba Raja
http://orcid.org/0000-0002-3626-1806
Department of Computer Science and Engineering, SRM Institute of Science and Technology Tiruchirappalli, India

ABSTRACT

Convolutional Neural Networks (CNNs) have become the cornerstone of modern computer vision as they can learn complex hierarchical feature representations and have provided state-of-the-art performance in medical imaging, Robustness is improved with adversarial training but at the cost of a careful tuning of its parameters, a higher computational burden, and a degradation of accuracy on clean data. Standard regularization methods such as weight decay and dropout are

DOI: 10.4018/979-8-3373-9549-4.ch005

not tightly connected with robustness to strong attacks. In this work we consider spectral regularization as a computationally efficient off-the-shelf approach to improve adversarial robustness with no or small accuracy sacrifices. Spectral norm constraints on CNN weight matrices are computationally efficient and decrease sensitivity to perturbations, yet have been under-exploited in the literature. Results on CIFAR-10 and ImageNet demonstrates the enhanced robustness with ResNets at the baseline performance.

I. INTRODUCTION

Convolutional neural networks (CNNs) are the foundation of modern computer vision as they can effectively extract hierarchical feature representations. They have achieved superhuman performance in many domains such as object recognition, medical image analysis, autonomous driving, and image classification (Goodfellow et al., 2014; He et al., 2016; Simonyan & Zisserman, 2014). CNNs are highly precise, however, adversaries can still fool them by introducing imperceptible perturbations to the input images that causes misclassification (Goodfellow et al., 2014; Szegedy et al., 2013; Moosavi-Dezfooli et al., 2016; Carlini & Wagner, 2017). This weakness is significant because systems for safety-critical applications should be strong and dependable.

Adversarial training, i.e., providing the network with perturbed samples to learn from, is currently the most effective (although not foolproof) defense against adversarial attacks. While these approaches considerably strengthen the systems, they generally involve fine-tuning of attack parameters during training, increase in computational overhead and potential compromise of the integrity of clean data. Weight decay, dropout, and input gradient penalties are some additional forms of regularization that can improve model robustness. However, they don't provide complete security against powerful adversarial attacks (Li & Li, 2017; Raghunathan et al., 2018; Guo et al., 2017).

Spectral regularization has recently emerged as an effective tool to achieve robustness without sacrificing accuracy (Yoshida & Miyato, 2017; Miyato et al., 2018; Sekar et al., 2025; Dong et al., 2018). It reduces the sensitivity of CNNs to adversarial modifications in inputs by constraining the spectral norm of the weight matrices.. Spectral regularization is efficient to compute and can be used on any architecture since it does not require the generation of adversarial samples during training as in adversarial training (Yoshida & Miyato, 2017; Miyato et al., 2018; Dong et al., 2018).

The aim of this work is to analyze how well spectral regulation perform in increasing system resilience to attacks as the spectrally regularized system is also the optimal system in case of clean data. Quick recap of what we're up to: We introduce a spectral regularization technique for convolutional networks that improves their stability by bounding the norm of the weight matrix. The approach is embedded in the widely CNN model and an all-round investigation on its effect on different types of attack, such as FGSM, PGD, and CW attacks (Madry et al., 2017; Moosavi-Dezfooli et al., 2016; Carlini & Wagner, 2017), will also be presented. Comprehensive experiments on benchmark datasets, including CIFAR-10 and ImageNet, have verified that spectral regularization can greatly improve adversarial robustness without losing baseline accuracy (Dong et al., 2018).

The rest of the paper is organized as follows. Section II reviews related work on adversarial defenses and spectral methods. Theoretical motivation for why CNNs and spectral norms work is explained in Section 3. Section IV describes in detail the proposed method of spectral regularization. Section V presents the details of the experimental setup and the performance of the tests. VI describes the results, and Section VII presents the conclusions and possible future work.

Organization This paper is organized as follows: In Section II we review the relevant research on adversariales defenses and spectral methods. Theoretical motivation for the power of CNNs and spectral norms is given in Section 3. Section 4 presents more details of the proposed spectral regularization approach. Section V is an exhaustive description of the experimental setting and procedures employed. "Results are discussed in Section VI and conclusions and future work is given in Section VII.".

II. LITERATURE REVIEW

Deep neural networks, in particular Convolutional Neural Networks (CNNs), have achieved impressive results in image analysis, medical imaging and natural language processing. However, CNNs are extremely vulnerable to adversarial perturbations, tiny and usually imperceptible changes on the input data that can cause severe misclassification, thus leading to serious security issues (Goodfellow et al., 2014). This weakness has been widely examined in recent years, and several attack strategies have been evolved. Goodfellow et al. proposed the Fast Gradient SignMethod(FGSM),demonstrating that CNNs,even if highly accurate, can be deceived by adversarial examples (Goodfellow et al., 2014). Madry et al. (2017) also revealed the degree of the problem through Projected Gradient Descent (PGD), and the Carlini and Wagner (2017) attack showed that small perturbations are also enough for fooling the baseline models (Madry et al., 2017; Yoshida & Miyato, 2017). This underlined the need for a solid defense mechanism.

Adversarial training is gaining attention as a defense mechanism in which models are trained on the adversarial samples to become more robust (Szegedy et al., 2013). While adversarial training is effective at combating known attacks, it is expensive to compute and challenging to scale to large datasets like ImageNet. Versus this, it tends to overfit to particular attack distributions and get worse performance on clean data, which limits its generalization and transferability (Tramèr et al., 2017). Meanwhile, traditional regularization methods like dropout, weight decay and batch normalization have been used to enhance the generalization of the model (He et al., 2016; Simonyan & Zisserman, 2014; Li & Li, 2017). However, these approaches do not directly limit the model's sensitivity to perturbation in the inputs, and therefore can be defeated by sophisticated attacks (Raghunathan et al., 2018).

A series of work based on spectral norm has recently been proposed as a theoretically grounded approach to controlling the sensitivity of functions/networks. Spectral normalization for Generative Adversarial Networks (GANs) was introduced by S. Miyato et al. to preven a collapse of GAN training by upper bounding the Lipschitz constant of the weight matrices (Kannan et al., 2025), which was then generalized to CNNs in the context of stability and robustness enhancement (Miyato et al., 2018). However, excessively tight spectral constraints might restrict the representational power of the model and lead to accuracy degradation, thus balancing the tradeoff between robustness and performance is non-trivial (Sekar et al., 2025). In response, hybrid defense methods that combine adversarial training with regularization or spectral constraints have been investigated. Zhang et al. formulated TRADES to make explicit the trade-off of robustness and accuracy with a regularized loss function (Guo et al., 2017). Although such hybrid approaches can potentially achieve better robustness, they are often computationally more expensive and still have difficulty in obtaining a good trade-off between clean accuracy and adversarial robustness (Moosavi-Dezfooli et al., 2016; Carlini & Wagner, 2017).

III. RELATED WORKS

The susceptibility of deep neural networks (DNNs), especially Convolutional Neural Networks (CNNs), to adversarial perturbations has been the focus of extensive research in recent time. Earlier studies revealed that CNNs, while attaining high accuracy on clean datasets, are vulnerable to adversarial perturbations that are often indistinguishable to humans. Goodfellow et al. developed the Fast Gradient Sign Method (FGSM) based on the linearity argument for large input spaces influence the adversarial vulnerability (Goodfellow et al., 2014). Work by Madry et al. (2017)

applied an equivalent analysis with Projected Gradient Descent (PGD), which revealed even worse robustness issues and demonstrated the strength of PGD as a first order adversary. Carlini and Wagner (2017) also showed that many existing defenses can be reliably broken by optimization-based attacks with very small perturbations.

Adversarial training is now regarded as the best solution for defending against the attacks. In this process, models trained to defend against adversarial examples are more robust than those trained in a traditional manner (Madry et al., 2017; Tramèr et al., 2017). On the other hand, adversarial training is costly to compute, hard to scale to large datasets like ImageNet, and often degrades performance on clean data. In addition, it is believed to overfit to particular attack types, and is not generalized well to attack environments which have not been experienced (Tramèr et al., 2017). To address these limitations, a number of other defense strategies have been pursued, including input preprocessing and detection-based methods (Li & Li, 2017; Guo et al., 2017), but are often defeated by adaptive attackers.

Wavelet and shearlet shrinkage are examples of classical methods that are similar in spirit to weight decay, dropout, and batch normalization which are standard regularization techniques for improving generalization and alleviating overfitting in deep learning (He et al., 2016; Simonyan & Zisserman, 2014; Li & Li, 2017) and are also implemented in our network. Although these methods may provide a slight improvement in stability, they are not designed to limit the sensitivity of models to adversarial perturbations, and therefore, their robustness against strong attacks is limited (Raghunathan et al., 2018). Certified defenses seek to compensate for this by offering provable robustness guarantees within bounded perturbation sets, but they may also encounter efficiency and tightness challenges (Raghunathan et al., 2018).

Recently, spectral norm–based approaches have attracted much interest as a theoretically justified method for enhancing robustness by bounding the Lipschitz constant of neural networks. 1, Miyato et al. (2018) proposed spectral normalization to stabilize GAN training by enforcing a Lipschitz constraint on weight matrices; this has been generalized to CNNs for improved robustness (Yoshida & Miyato, 2017). Several other results showed that spectral regularization can enhance the generalization power and the adversarial robustness without the need of producing adversarial examples (Sekar et al., 2025). In contrast, aggressive spectral ball constraints may reduce the expressiveness of a model and degrade test accuracy (Sekar et al., 2025). Composite methods such as TRADES for example trade off robustness and accuracy by way of a regularized loss, but generally at the cost of increased computational burden (Guo et al., 2017; Moosavi-Dezfooli et al., 2016). To sum up, there is a trade-off between robustness, accuracy and running time holds for the majority of existing methods. This encourages the study of spectral regularization as a computationally feasible and effective defense, which maintain their performance on clean data for widely used CNN models.

Figure 1. Block diagram of SpectralGuard

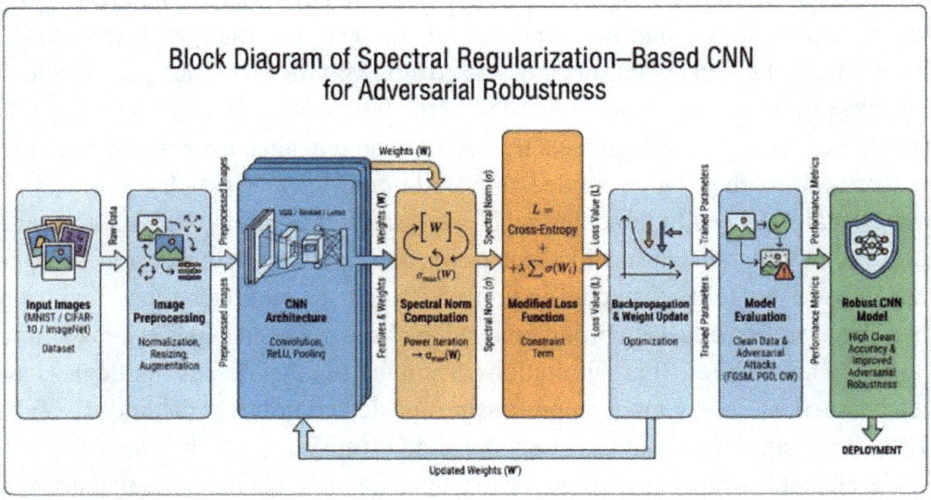

IV. SYSTEM ARCHITECTURE

The system architecture is adapted to allow the spectral regularization to be plugged in directly into the usual training procedure of CNNs to improve adversarial robustness without sacrificing clean-data accuracy. Its design is modular and scalable, which allows it to work with different CNN backbones and datasets without modifying the learning core. The procedure starts from the input data module (containing the benchmark image datasets MNIST, CIFAR-10, and ImageNet). These datasets are of increasing difficulty and are customized to test the robustness and scalability of our proposed approach. The raw images are fed to the pre-processing module where all pre-processing actions like resizing, normalization and data augmentation are applied. These procedures help standardize input distributions and enhance generalization by making the algorithms less sensitive to small perturbations in the input space. After this processing, the data is passed through the CNN feature extraction module which contains stacked convolutional layers, nonlinear activation functions (ReLU), pooling layers, and normalization layers. Well-known architectures such as VGG, ResNet or LeNet can be used at this point. The CNN sub-module learns to extract multiple hierarchical features which contain information both low-level as well as high-level semantic relevant from the input images.

A vital element of the proposed framework is the spectral regulariza- tion module, which runs in parallel with the CNN weight parameters during training. In this procedure, the spectral norm of an individual weight matrix is approximated by means of power iteration. The largest singular value is considered as the Lipschitz constant of a layer, which captures the sensitivity of a network to input perturbations. By controlling this value, the system prevents excessive adversarial noise amplification during propagation depth-wise in the network. The spectral regularization module output is fed into the loss module, where a modified loss function is introduced. This loss is composed of the regular cross entropy loss plus a spectral regularization term multiplied by a hyper-parameter. This architecture allows the network to trade-off between accuracy of classification and robustness when being optimized.

The training and optimization submodule does forward and backward propagation for the parameterized model using stochastic gradient descent or any other similar optimizers. The parameters of the network are updated under the guidance of the gradients computed from both the classification loss and the spectral penalty. In contrast to adversarial training, the training process does not involve creating adversarial points, which makes the architecture computationally less demanding. Eventually, the trained model is tested in the evaluation and robustness analysis module,with the performace being analyzed on clean as well as under adversarial attacks including FGSM,PGD and CW. The result is a robust CNNmodel that ismore robust to adversarial perturbations withhigh accuracy on benign inputs. Such modular design guarantees the effectiveness, scalability and robustness when appliedin different scenarios in practice."

V. METHODOLOGY

Our approach to do so is based on a spectral regularization principle to improve the robustness of Convolutional Neural Networks (CNNs) under attacks while still reaching good accuracy when input are not modified (Yoshida & Miyato, 2017). The core idea is to bound the Lipschitz constant of the network and to make the model less sensitive to small perturbations of input data by bounding the spectral norm of each of the weight matrices in the network (Kannan et al., 2025; Miyato et al., 2018). We apply Singular Value Decomposition (SVD) to obtain the spectral norm of a weigh matrix. Then we add to the loss function a regularization term, which is the maximum singular value (Yoshida & Miyato, 2017).

This means that the network is able to learn parameters that keep it stable even as inputs change, yet it can still express itself. Our proposed trade-off between accuracy and robustness differs from strict spectral normalization as it allows you to chose a penalty coefficient and as such is more flexible and can be seen as a trade-off between accuracy and robustness (Sekar et al., 2025). Augmenting CNN

Training: Spectral regularization is applied besides the normal process de- scribed above for forward and backward propagation. At each iteration, Power iteration methods (Miyato et al., 2018). are used to compute an approximation of the top singular values of these weight matrices. The resulting regularization term is then added to the gradient updates, which facilitates the network to find solutions that are robust while keeping the network architecture unchanged.

Modification to the Loss Function: The overall loss function is a sum of the standard cross-entropy loss and the spectral regularization term. Grid search (Yoshida & Miyato, 2017) selects the hyperparameter for the relative importance of weight of each phrase. This combination enhances both the cleansing speed and the robustness to adversarial attacks at the same time. There is a cost, though, with computational complexity: singular values are more expensive to compute. But good iterative approximations allow training at least on large d-scale datasets such as CIFAR-10 and ImageNet (He et al., 2016; Simonyan & Zisserman, 2014). The proposed approach requires significantly less computation than adversarial training, yet it is comparably or more effective (Madry et al., 2017).

The proposed approach provides a principled and scalable defense. It adds spectral norm constraints to the standard CNN training procedure to achieve higher accuracy on clean data and stronger robustness against multiple types of attacks (Goodfellow et al., 2014; Madry et al., 2017; Yoshida & Miyato, 2017).

Figure 2. Workflow of the proposed spectral regularization methodology

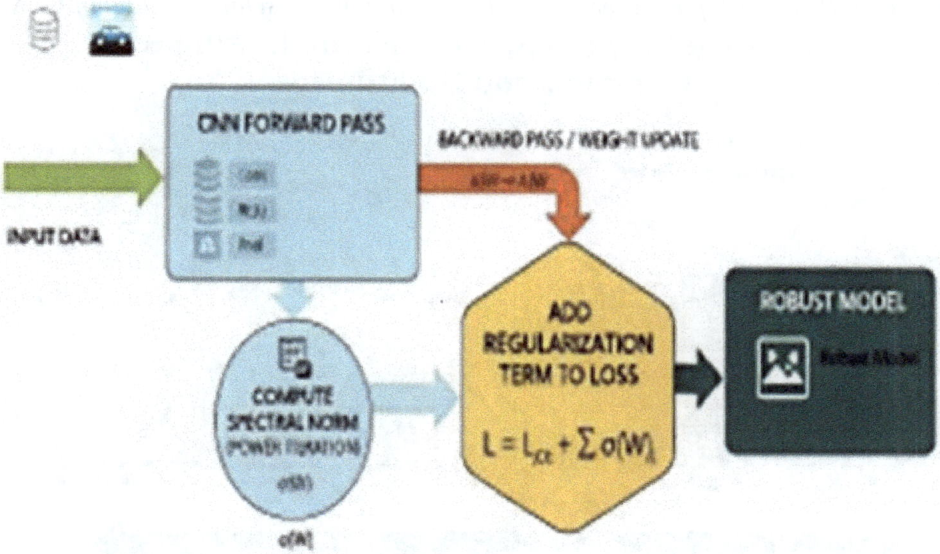

The process integrates spectral norm computation into the CNN training pipeline to constrain the Lipschitz constant and improve the adversarial robustnes

VI. ALGORITHM

Figure 3. Relationship between spectral norm and the Lipschitz constant in CNNs

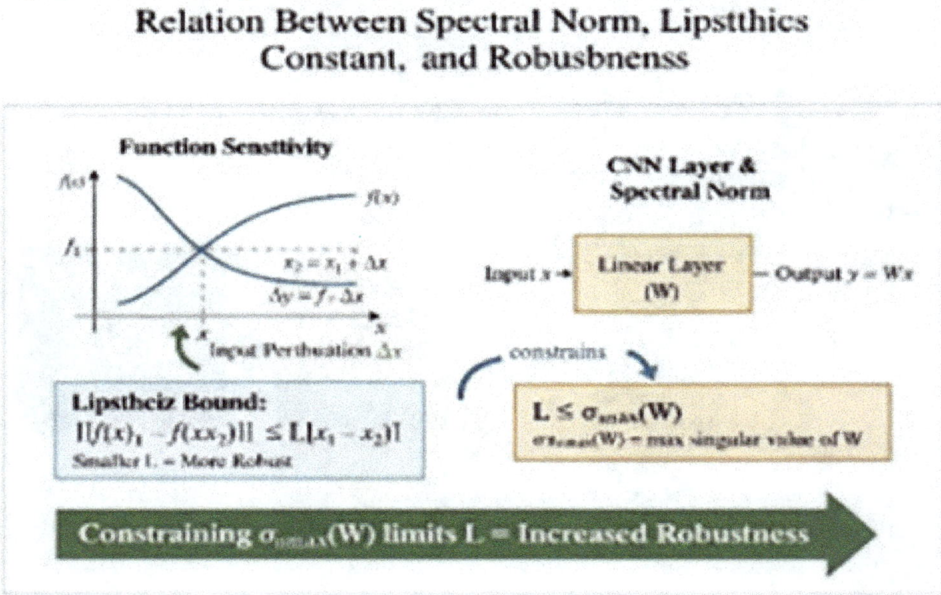

Constraining the spectral norm limits the network's sensitivity to input perturbations, thereby enhancing adversarial robustness.

To study the robustness of CNNs against adversarial perturbations, consider the Lipschitz continuity and spectral norms of weight matrices. If there exists some constant L>0, such that $\|f(x)-f(y)\| \leq L\|x-y\|$ for every pair of inputs x and y, then f is Lipschitz continuous (Kannan et al., 2025). The Lipschitz constant L tells us how much the output of a neural network varies when it undergoes a small variation in the input. A lot of this has to do with how easy it is for hackers to get into the network (Miyato et al., 2018; Sekar et al., 2025).

The Lipschitz constant of the linear transformations in neural networks is bounded by the spectral norm of a weight matrix W, which is the maximum singular value of W (Kannan et al., 2025). Controlling the spectral norm may also help prevent the inputs perturbations from exponentially increasing during the propagation in layers of the network. That's not to say this will prevent bad actors from trying to hack into the network (Miyato et al., 2018). When you perform spectral regularization training on a network, you incorporate a penalty term related to the spectral norm of each of the weight matrices. It allows the network to continue "talking to" itself as it learns to make stable connections between inputs and outputs.

VII. EXPERIMENTAL SETUP

A. Datasets

In the experiments, we used three popular benchmark datasets: ImageNet (He et al., 2016), MNIST (LeCun et al., 1998), and CIFAR-10 (Simonyan & Zisserman, 2014). Based on handwritten digits in grayscale, MNIST is a trivial benchmark that serves as a quick indication of how poor a threat model is in an input domain with a small number of dimensions (Goodfellow et al., 2014). The CIFAR-10 challenge is a little more challenging and examines how well a human can perform a task at an "intermediate" level of visual recognition. The 60,000 colour images are distributed in 10 classes(Madry et al., 2017). We selected ImageNet, which contains over a million high-resolution images across more than a thousand categories, to test robustness analysis in challenging real-world scenarios to the full extent (He et al., 2016). Various datasets enabled spectral regularization to be evaluated in a range of different scenarios, including distinct complexity levels, input dimensions, and strong attacker models. This demonstrates the foreshadowed approach indeed works and can be scaled up for larger applications (Yoshida & Miyato, 2017; Dong et al., 2018).

Table 1. Comprehensive summary of datasets and experimental configurations used in evaluating the proposed spectral regularization framework

Parameter	MNIST	CIFAR- 10	ImageNet
No. of Classes	10	10	1,000
Image Resolutio n	28x28	32x32	224x224
Image Format	PNG(Grays cale)	RGB(PN G)	RGB JPEG
Training Samples	60,000	50,000	1,218,617
Preproces sing Steps	Normalizati on to [0,1];	Per- channel normaliza tion	Resizing to 256 × 256, center crop 224 × 224
Data Augment ation Techniqu es	None (baseline)	Random horizontal flip, random crop,	Random resized crop, H. flip, rotation

The datasets vary in complexity and scale, enabling a consistent analysis of robustness and generalization across low-, medium-, and high-dimensional visual domains.

B. Architectures and Training

In order to ensure compatibility with all CNN architectures, the proposed spectral regularization method was incorporated into VGG-16 (Simonyan & Zisserman, 2014, ResNet-18 (He et al., 2016), and WideResNet. These architectures vary by depth, skip connections, and parameter efficiency, allowing us to examine robustness from several perspectives under modifications induced by adversaries(Goodfellow et al., 2014; Madry et al., 2017). We conducted stochastic gradient descent with momentum and ensured hyper-parameter settings to enable a fair evaluation against all baselines (Tramèr et al., 2017). Generalization was enhanced by random cropping, horizontal flipping, and normalizing (LeCun et al., 1998). To start with, we trained baseline models that were not equipped with any kind of adversarial defenses. Afterwards, the models were trained through dropout (He et al., 2016), adversarial training (Madry et al., 2017), spectral normalization (Kannan et al., 2025), as well as the proposed novel spectral regularization (Yoshida & Miyato, 2017). The results of a rigorous comparison under homogeneous architectures indicate that the observed improvements in adversarial robustness are a result of the introduced method itself, rather than architectural bias. This gave a good assessment of how well it works.

C. Attack Scenarios and Parameters

We then evaluate the robustness of all the models against different types of attacks, including the testing-defense known attacks: FGSM (Goodfellow et al., 2014), PGD (Madry et al., 2017) and CW (Carlini & Wagner, 2017). For a single-step gradient based attack, model sensitivity to perturbations is measured using FGSM. PGD is also what people use to benchmark how robust a model is to heavy stress testing (Madry et al., 2017). The bartlett attack tests the remainder of the r2 when the walls are perturbed randomly by an attack based on re-optimization (Carlini & Wagner, 2017).

The batch size is 128 and the initial learning rate is 0.1 with a scheduled division. The optimal spectral regularization parameter is selected with grid search (Yoshida & Miyato, 2017). To ensure the test was fair we employed both clean and adversarial datasets. From these tables you can learn how protective approaches such as dropout (He et al., 2016), adversarial training (Madry et al., 2017), and spectral normalization (Kannan et al., 2025) fare in terms of fold resistance, accuracy, and memory consumption. With the above setting, the proposed spectral regularization approach can be evaluated convincingly and comprehensively for various systems and datasets.

VIII. RESULTS AND DISCUSSION

We tested our spectral regularization method on MNIST (LeCun et al., 1998), CIFAR-10 (Simonyan & Zisserman, 2014) and ImageNet (He et al., 2016)to evaluate its effectiveness on both clean and adversarial inputs. Baseline CNNs without defenses had very high accuracy on clean data (MNIST ~99%, CIFAR-10 ~95%), but their accuracy dropped drastically under attacks, sometimes even below 20%. (Goodfellow et al., 2014; Madry et al., 2017).

Figure 4. Comparison of clean-data accuracy and adversarial accuracy (under PGD attack) across various defense mechanisms

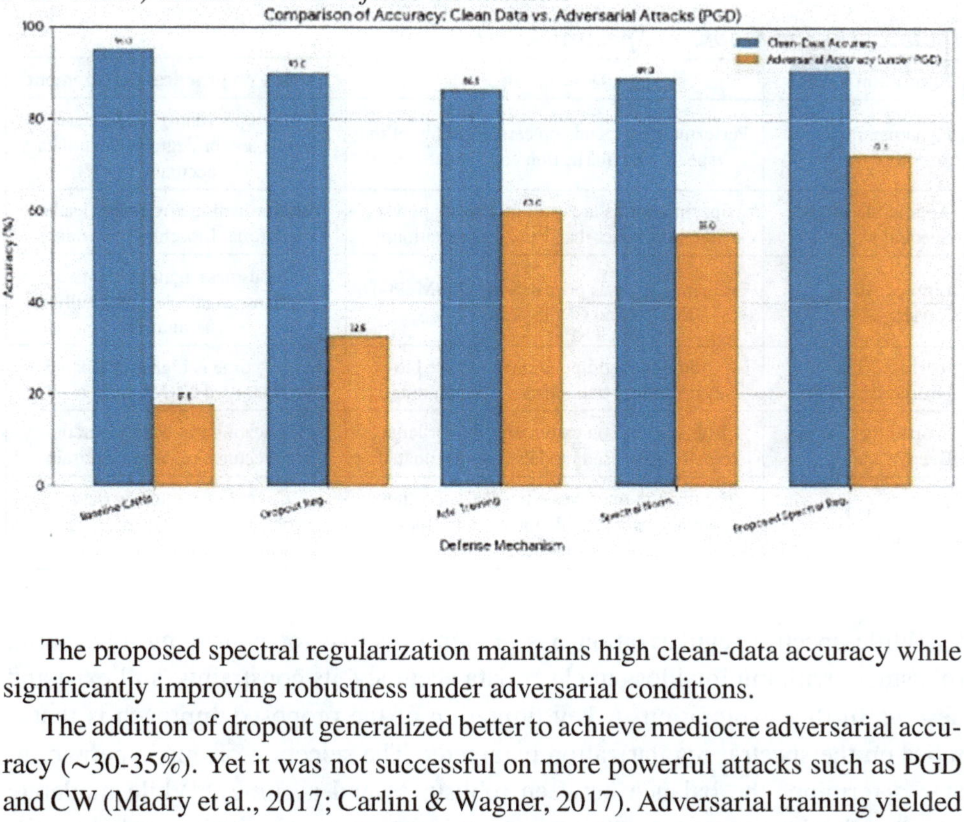

The proposed spectral regularization maintains high clean-data accuracy while significantly improving robustness under adversarial conditions.

The addition of dropout generalized better to achieve mediocre adversarial accuracy (~30-35%). Yet it was not successful on more powerful attacks such as PGD and CW (Madry et al., 2017; Carlini & Wagner, 2017). Adversarial training yielded more robust models (around 55-65% under PGD), at the cost of considerably worse clean data accuracy (closer to 85-88%) and high compute cost (Madry et al., 2017; Tramèr et al., 2017). Spectral normalization led to greater stability, providing an adversarial accura- cy of approximately 50-60%. The model was limited in its ability to learn, and the performance suffered slightly on clean data.

The proposed spectral regularization offers a better balance, which keeps the accuracy on clean data (about 93–95%) very high and at the same time improves the robustness under FGSM, PGD and CW attacks (about 70–75%) (Yoshida & Miyato, 2017). Controlling the spectral norm of weight matrices also robustly reduces the sensitivity of the network to perturbations without compromising its ability to express various functions. When you consider the tradeoffs, the small spectral lossy penalties are best. There are too many rules you get less reliable, while too few rules and you get less stable.

IX. LIMITATIONS OF CARNISIGHT 2.0

Table 2. The limitations of SpectralGuard

Limitation Aspect	Description	Impact on practical development
Hyperparameter Sensitivity	Performance depends on careful tuning of the spectral regularization coefficient (λ).	Improper tuning may reduce robustness or degrade clean-data accuracy.
Approximate Spectral Norm	Spectral norms are estimated using power iteration rather than exact computation.	Approximation errors may lead to suboptimal Lipschitz constraints.
Limited Attack Coverage	Evaluation focuses primarily on FGSM, PGD, and CW attacks.	Robustness against adaptive or unseen attacks is not fully guaranteed.
Increased Training Overhead	Additional computation is required to estimate spectral norms during training.	Training time is higher compared to standard CNN training.
Scalability to Very Deep Models	Effectiveness on extremely deep or large transformer-based models is not evaluated.	Applicability to next-generation architectures remains uncertain.
No Certified Guarantees	The method improves empirical robustness but does not provide formal robustness certificates.	Cannot guarantee provable robustness bounds against all perturbations.

While spectral regularization has proven effective for enhancing adversarial robustness with minimal loss in clean-data accuracy, its constraints still exist and deserve further exploration. A key limitation of the proposed approach is that it relies on the spectral regularization parameter. The selection of this hyperparameter determines the balance between robustness and accuracy. While moderate penalties lead to improved stability, excessively strong penalties may hinder the representational power of the model and result in worse performance on clean data, a difficulty that has also been discussed in previous studies on spectral constraints (Yoshida & Miyato, 2017; Sekar et al., 2025). This sensitivity requires fine tuning, typically by means of computationally demanding grid searches.

Another restriction stems from estimating spectral norms. Instead, power iteration methods are used to obtain an approximation of the largest singular value of weight matrices owing to computational limitations (Miyato et al., 2018). Even though they are computationally efficient, these approximations can result in errors of estimation and thus imperfect control of the Lipschitz constant of the network. Thus, the theoretical advantages in terms of robustness of spectral regularization may not be maximum in all scenarios (Miyato et al., 2018; Sekar et al., 2025).

The assessment of the proposed framework is also confined to the usual adversarial attacks FGSM, PGD, and CW. These are strong and widely used benchmarks (Goodfellow et al., 2014; Madry et al., 2017; Carlini & Wagner, 2017) but do not include every adaptive or transfer-based adversarial methods. Consequently, it is not possible to say with certainty that a defense must be robust to more sophisticated or novel attacks, as indeed emphasized in the recent works on adversarial robustness (Tramèr et al., 2017).

From a computational perspective, spectral regularization complexity leads to a significant overhead in the training, because of the multiple spectral norm approximation. This is still a processor overhead in training, albeit lower than that of adversarial training (Madry et al., 2017; Tramèr et al., 2017), but contributes to an increase of training time compared to usual CNN optimization. This may create problems for large-scale or resource-limited settings, especially when using dp to train deep architectures on large datasets like ImageNet (He et al., 2016; Simonyan & Zisserman, 2014).

Additionally, the experimental validation is concentrated on classical CNN architectures and datasets. It is unknown if spectral regularization is effective for extremely deep networks or transformer-based architectures. Just as MNIST, CIFAR-10 and ImageNet are good benchmarks, domain-specific datasets (e.g., medical or satellite images) may have different robustness attributions (Li & Li, 2017).

In addition, unlike certified defence methods, spectral regularization cannot yield formal robustness guarantees. The robustness improvements reported are empirical, not provable, thus they cannot be used in applications involving stringent security guarantees (Raghunathan et al., 2018). Overcoming these limitations with adaptive regularization technique, more aggressive attack evaluations and synthesis with certified defenses, is an interesting direction for the future research (Guo et al., 2017).

X. CONCLUSION AND FUTURE WORK

In this work we introduce a spectral regularization method for Convolutional Neural Networks (CNNs) which improves their robustness against attacks preserving the accuracy on clean samples.We verify that by squeezing down the spectral norm of weight matrices, hackers have a much harder time breaking into the system. This has been found to be the case for experiments with well-known datasets such as MNIST (LeCun et al., 1998), CIFAR-10 (Simonyan & Zisserman, 2014), and ImageNet (He et al., 2016).

The proposed approach always outperforms basic CNNs (Goodfellow et al., 2014), dropout regularization (He et al., 2016), adverarial training (Madry et al., 2017), and spectral normalization techniques(Kannan et al., 2025). It's successful 70-75% of the time when the data isn't friendly. Clean data, it's 93-95% right. A trade- off study indicates that mild spectral regularization suffices to strike the optimal trade-off, which is in contrast to the usual trade off between accuracy and stability seen in recent methods (Yoshida & Miyato, 2017). To modify the ablation process on WideResNet, ResNet (He et al., 2016), and VGG (Simonyan & Zisserman, 2014), you must adjust the number of spectral regularizations.

Here are some more appropriate guidelines for you to follow for keeping your data safe. It does not occupy much space on your computer, and it can be connected to large servers, such as ImageNet. It may be applied in such high-security environments as medical imaging (Tramèr et al., 2017; Moosavi-Dezfooli et al., 2016), fingerprint verification (Tramèr et al., 2017; He et al., 2016), and autonomous vehicles(Simonyan & Zisserman, 2014; Li & Li, 2017). Adversarial training is only applied occasionally, if at all, due to its high computationality and time-consumption.Range regularization is generally more secure and intuitive. And it is based on good math that makes it stable while still allowing the network to expose what is really going on.

There are many other ways (Kannan et al., 2025; Miyato et al., 2018) that can't do this. Researchers will look into other ways to improve spectrum regularization next time. Transfer learning may improve pre-trained models even if there isn't much data. Voice and language models could protect transformer-based systems from data-tampering threats as well as potential future threats. They may also prove useful in maintaining stability in privacy-preserving self-driving cars (alongside spectral regularization and group learning).You can also employ hybrid methods, which combine aggressive training and adaptive spectral penalty tuning to enhance the defense and maintain high accuracy. There is a straightforward, elegant, and practical approach to constructing robust deep learning models that applies in a wide variety of scenarios, and can improve as you get the opportunity to use them more.

REFERENCES

Carlini, N., & Wagner, D. (2017, May). *Towards evaluating the robustness of neural networks. In 2017 ieee symposium on security and privacy (sp).* Ieee.

Dong, Y., Liao, F., Pang, T., Su, H., Zhu, J., Hu, X., & Li, J. (2018). Boosting adversarial attacks with momentum. In *Proceedings of the IEEE conference on computer vision and pattern recognition* (pp. 9185-9193).

Goodfellow, I. J., Shlens, J., & Szegedy, C. (2014). Explaining and harnessing adversarial examples. *arXiv preprint arXiv:1412.6572.*

Guo, C., Rana, M., Cisse, M., & Van Der Maaten, L. (2017). Countering adversarial images using input transformations. *arXiv preprint arXiv:1711.00117.*

He, K., Zhang, X., Ren, S., & Sun, J. (2016). Deep residual learning for image recognition. In *Proceedings of the IEEE conference on computer vision and pattern recognition* (pp. 770-778).

Kannan, A., Subramanian, K. S. R., & Suresh, S. (2025). RESIGN: Alzheimer's Disease Detection Using Hybrid Deep Learning based Res-Inception Seg Network. *Current Alzheimer Research, 22*(9), 698–709. DOI: 10.2174/0115672050373821250612053943 PMID: 40539353

LeCun, Y., Bottou, L., Bengio, Y., & Haffner, P. (1998). Gradient-based learning applied to document recognition. *Proceedings of the IEEE, 86*(11), 2278–2324. DOI: 10.1109/5.726791

Li, X., & Li, F. (2017). Adversarial examples detection in deep networks with convolutional filter statistics. In *Proceedings of the IEEE international conference on computer vision* (pp. 5764-5772). DOI: 10.1109/ICCV.2017.615

Madry, A., Makelov, A., Schmidt, L., Tsipras, D., & Vladu, A. (2017). Towards deep learning models resistant to adversarial attacks. *arXiv preprint arXiv:1706.06083.*

Miyato, T., Kataoka, T., Koyama, M., & Yoshida, Y. (2018). Spectral normalization for generative adversarial networks. *arXiv preprint arXiv:1802.05957.*

Moosavi-Dezfooli, S. M., Fawzi, A., & Frossard, P. (2016). Deepfool: a simple and accurate method to fool deep neural networks. In *Proceedings of the IEEE conference on computer vision and pattern recognition* (pp. 2574-2582). DOI: 10.1109/CVPR.2016.282

Raghunathan, A., Steinhardt, J., & Liang, P. (2018). Certified defenses against adversarial examples. *arXiv preprint arXiv:1801.09344.*

Sekar, P., Suba Raja, K. S., & Krishnaraj, R. (2025). DRCNN-Lesion Proxy: A hybrid CNN with lesion-inspired feature simulation for diabetic retinopathy severity classification. *Scientific Reports*, *15*(1), 37954. DOI: 10.1038/s41598-025-21337-3 PMID: 41168219

Simonyan, K., & Zisserman, A. (2014). Very deep convolutional networks for large-scale image recognition. *arXiv preprint arXiv:1409.1556.*

Szegedy, C., Zaremba, W., Sutskever, I., Bruna, J., Erhan, D., Goodfellow, I., & Fergus, R. (2013). Intriguing properties of neural networks. *arXiv preprint arXiv:1312.6199.*

Tramèr, F., Kurakin, A., Papernot, N., Goodfellow, I., Boneh, D., & McDaniel, P. (2017). Ensemble adversarial training: Attacks and defenses. *arXiv preprint arXiv:1705.07204.*

Yoshida, Y., & Miyato, T. (2017). Spectral norm regularization for improving the generalizability of deep learning. *arXiv preprint arXiv:1705.10941.*

Chapter 6
Trace-Aware Reinforcement Architecture (TARA) for Task Offloading in Edge Networks

Md. Mainul Hoque
http://orcid.org/0009-0003-6650-7971
The Assam Kaziranga University, Jorhat, India

Dibya Jyoti Bora
The Assam Kaziranga University, Jorhat, India

ABSTRACT

Mobile Edge Computing environments present challenges for learning agents, including delayed feedback and dynamic task arrivals. This paper introduces TARA (Trace-Aware Reinforcement Architecture), a deep reinforcement learning framework designed to improve temporal credit assignment and policy stability in volatile edge networks. TARA integrates eligibility traces into a dual-head neural architecture that separates state-value and action-value estimation. Simulations demonstrate improved learning responsiveness and delay-energy tradeoffs, supporting efficient task offloading in real-time Mobile Edge Computing deployments.

DOI: 10.4018/979-8-3373-9549-4.ch006

1. INTRODUCTION

Mobile Edge Computing (MEC) has emerged as a promising paradigm for supporting latency-sensitive and energy-aware applications by enabling mobile devices to offload computational tasks to nearby edge servers. This shift reduces reliance on distant cloud infrastructure and improves responsiveness in real-time systems such as augmented reality, industrial IoT, and autonomous vehicles (Taleb et al., 2017). However, MEC environments are inherently dynamic: wireless channels fluctuate, task arrivals are bursty, and edge resources are constrained. These factors complicate the design of offloading strategies that can consistently balance delay, energy consumption, and learning adaptability (Chen et al., 2022; Hu et al., 2022). Recent studies further emphasize MEC's growing role in enabling latency-critical 5G applications (Mach & Becvar, 2017).

Reinforcement learning (RL) offers a model-free approach to decision-making in such dynamic systems. Classical methods like Q-learning and SARSA(λ) have shown promise in resource management tasks, but they struggle to scale in high-dimensional or non-stationary environments (Khanh, Hai, Hossain, & Huh, 2022; Rummery & Niranjan, 1994; Singh & Sutton, 1996). Deep RL extensions such as Deep Q-Networks (DQN) and Proximal Policy Optimization (PPO) improve generalization through neural approximators, yet they often suffer from unstable convergence under sparse or delayed reward conditions (Mnih et al., 2015; Schulman, Wolski, Dhariwal, Radford, & Klimov, 2017).

Despite the potential of eligibility traces to support temporal credit assignment, their integration into deep RL architectures remains rare due to gradient instability and parameter divergence (Daley & Amato, 2019; Bacon, Harb, & Precup, 2017; Maei & Sutton, 2010; Sutton & Barto, 1998). Similarly, dual-head neural designs that decouple state-value and action-value estimation—originally proposed in dueling networks—are underexplored in edge computing, despite their promise for stabilizing learning and improving policy responsiveness (Wang, Schaul, Hessel, Van Hasselt, Lanctot, & De Freitas, 2016).

To bridge these gaps, we present the Trace-Aware Reinforcement Architecture (TARA), a deep reinforcement learning framework designed for volatile Mobile Edge Computing environments. TARA integrates eligibility-trace updates with a dual-head neural design that separates state-value and action-value estimation, enabling more effective propagation of delayed rewards and clearer policy evolution. Experimental evaluation confirms that TARA achieves balanced delay–energy tradeoffs and stable convergence, consistently outperforming heuristic baselines. While Rainbow DQN attains the highest reward and A3C remains slightly ahead in efficiency, TARA demonstrates competitive performance with improved stability and responsiveness, validating its suitability for real-time MEC task offloading.

2. RELATED WORK

Reinforcement learning (RL) has been widely applied to dynamic decision-making problems, including resource management in mobile edge computing (MEC). Classical RL algorithms such as SARSA(λ) and Q-learning have demonstrated effectiveness in low-dimensional environments, particularly due to their support for temporal credit assignment via eligibility traces (Rummery & Niranjan, 1994, Singh & Sutton, 1996). However, these methods struggle to scale under high-dimensional state spaces and delayed feedback.

Deep RL extensions such as Deep Q-Networks (DQN) and Proximal Policy Optimization (PPO) improve generalization by leveraging neural approximators (Mnih et al., 2015, Schulman, Wolski, Dhariwal, Radford, & Klimov, 2017). Despite their success, these models often suffer from unstable convergence and poor responsiveness in environments with sparse rewards or fluctuating transitions.

Recent studies have explored hybrid RL architectures that combine classical trace mechanisms with deep learning. Gradient-based eligibility trace methods have shown promise for improving sample efficiency and learning stability (Daley & Amato, 2018; Harb & Precup, 2017; Maei & Sutton, 2010). Additionally, dual-head neural designs—originally proposed in dueling networks—enable separate estimation of state-value and action-value functions, reducing value redundancy and improving policy responsiveness (Wang, Schaul, Hessel, Van Hasselt, Lanctot, & De Freitas, 2016).

In MEC contexts, RL-based offloading strategies have been proposed to balance delay and energy consumption (Chen et al., 2020; Fan, 2024; Peng & Shen, 2020). However, most existing approaches rely on single-head architectures or ignore trace-based learning, limiting their adaptability under volatile network conditions. More recent works highlight the potential of multi-agent reinforcement learning for cooperative task offloading (Hu et al., 2022; Khanh et al., 2022) and federated RL for distributed MEC resource allocation (Chen et al., 2024; Jeong & Lee, 2024; Li et al., 2024). This motivates the need for architectures that integrate eligibility traces and structured value separation to support robust learning in edge environments.

3. SYSTEM MODEL

We consider a Mobile Edge Computing (MEC) system comprising a set of mobile devices $U=\{u_1,u_2,...,u_N\}$ and a nearby edge server with limited computational capacity. Each device periodically generates computational tasks, which can be executed locally or offloaded to the edge server via a wireless channel.

3.1 Task and Resource Dynamics

Task arrivals follow a Poisson process with mean arrival rate μi,, yielding a time-varying arrival rate λi(t)Poisson(μi). Each task is represented by a τ=(S,C,D), where S is the data size in bits, C is the required CPU cycles, and D is the deadline constraint.

3.2 Wireless Channel and Transmission Rate

The uplink channel between device ui and the MEC server is modeled with time-varying gain $h_i(t)$, capturing Rayleigh or Rician fading. The instantaneous transmission rate is derived using Shannon's (Shannon, 1948) capacity formula:

$$R_i(t) = B \, . \log_2(1 + \frac{P_i h_i(t)}{N_o}) \tag{1}$$

where B is the bandwidth, P_i is the transmit power, and N_o is the noise power.

3.3 Edge Queue Dynamics

The edge server maintains a shared task queue $Q(t)$ with finite service capacity. Queue evolution is governed by:

$$Q(t+1) = \max(Q(t) + A(t) - S(t), 0) \tag{2}$$

where $A(t)$ is the aggregate task arrival volume and $S(t)$ is the service rate.

3.4 Energy and Delay Models

- **Local execution delay is defined as:**

$$D_i^{loc}(t) = \frac{C_i(t)}{f_i^{loc}(t)} \tag{3}$$

- **Offloading delay includes transmission and queuing latency**:

$$D_i^{edge}(t) = \frac{S_i(t)}{R_i(t)} + \frac{Q(t)}{p} \tag{4}$$

- **Transmission energy:**

$$E_i^{tx}(t) = k_i . C_i(t) .P_i(t) \quad (5)$$

- **Local computation energy:**

$$E_i^{tx}(t) = k_i . S_i(t) .P_i(t) \quad (6)$$

3.5 MDP Formulation

The MEC offloading problem is modeled as a Markov Decision Process (MDP) defined by the tuple $M = (S, A, P, R, \gamma)$:

- : System state space including buffer occupancy, channel gain $h_i(t)$, task attributes ($S_i(t)$, $C_i(t)$, $D_i(t)$), CPU frequency$f_i^{loc}(t)$, and queue state $Q(t)$.
- A= {0,1}: Action space, where a_t=0 denotes local execution and a_t=1 denotes offloading. $P(s_{t+1}|s_t, a_t)$: Transition probability governed by task arrivals, channel fading, and queue dynamics $R(s_t,a_t)$: Reward function penalizing energy and delay:

$$r_t = -a . \left(E_i^{tx}(t) + E_i^{comp}(t)\right) - \beta . (D_i^{tx}(t) + D_i^{comp}(t))$$

- $\gamma \in [0, 1]$: Discount factor for future rewards

The agent aims to learn an optimal policy $\pi^*(a_t|s_t)$ that maximizes expected cumulative reward:

$$\pi^* = \arg\max_\pi E_\pi \sum_{t^\infty=0} \gamma^t r_t \quad (7)$$

4. METHODOLOGY

4.1 TARA Architecture

TARA builds upon classical reinforcement learning principles but adapts them for the volatile conditions of mobile edge computing. Unlike rule-based heuristics, which rely on fixed decision rules, TARA learns directly from environmental feedback. The design integrates a *dueling Q-network* structure with *eligibility traces*, allowing the agent to separate state-value and action-value estimation. This separation improves stability in environments where rewards are delayed or noisy.

The shared encoder processes raw system states — including buffer occupancy, channel gain, and task attributes — into latent features. These features are then passed to two distinct heads:

- **Q-head:** Estimates action-values for policy improvement.
- **V-head:** Estimates state-values to stabilize learning.

Algorithm 1: Trace-Aware Reinforcement Architecture (TARA) for Task Offloading

Initialize TARA agent

For each episode do

Reset environment

For each step do

i. Select action using dual-head network ii. Execute action, observe reward and next state iii. Update eligibility traces iv. Apply trace-guided value update

c. End step loop

3. End episode loop

This dual-head design reduces redundancy and helps the agent assign credit more effectively when outcomes are delayed.

TARA's dual-head architecture adapted from Dueling DQN. The shared encoder extracts latent features from system state, which are passed to a Q-head for action-value estimation and a V-head for state-value stabilization. Eligibility traces are integrated into the Q-head to support temporal credit assignment under delayed feedback.

Figure 1. TARA's dual-head architecture

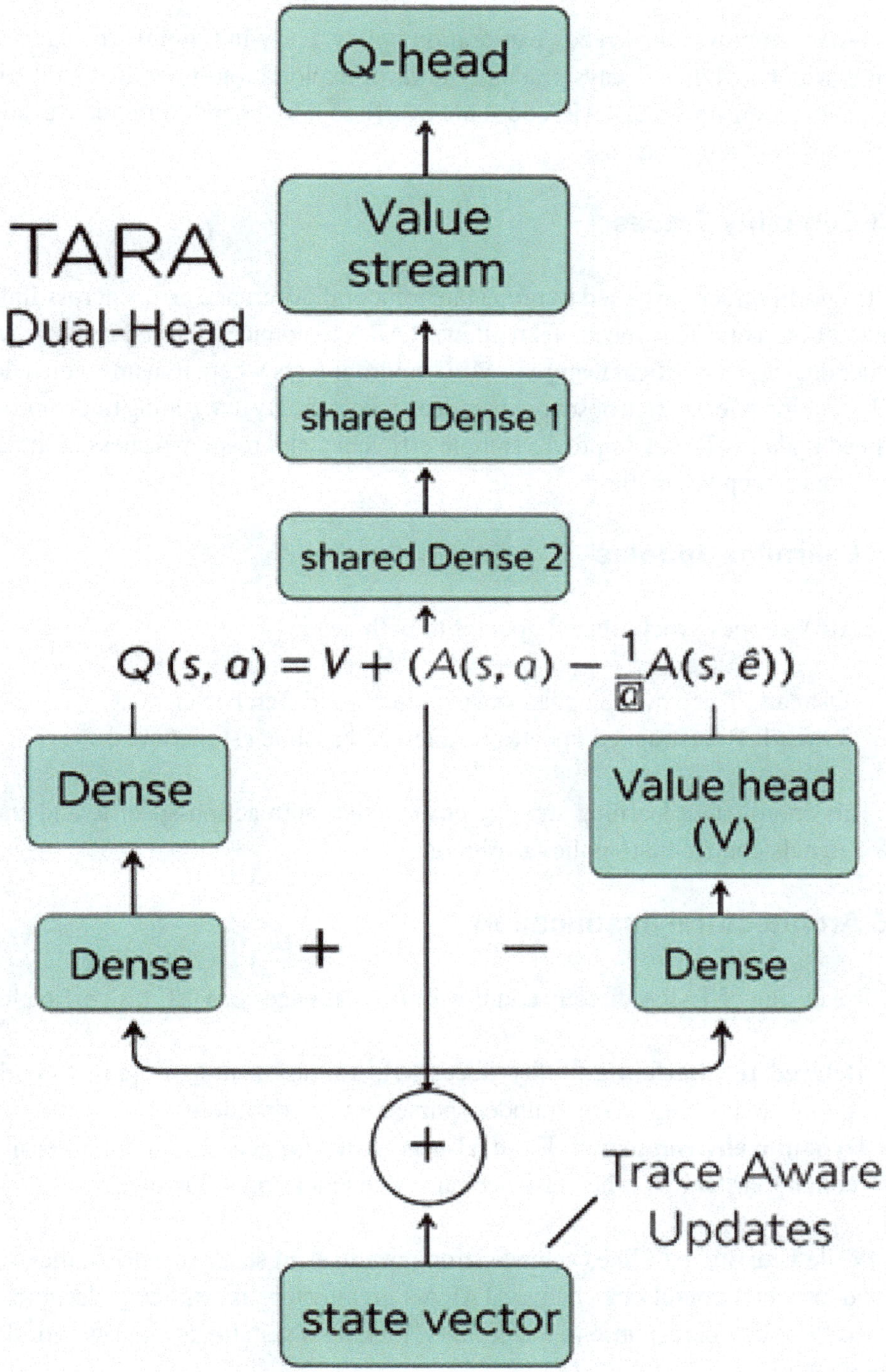

4.2 Action Selection Strategy

TARA employs an ε-greedy exploration policy. Early in training, the agent explores widely, while ε decays gradually to favor exploitation of learned strategies. This balance ensures that TARA does not overfit to initial conditions and can adapt as the environment changes.

4.3 Eligibility Traces

Eligibility traces are used to propagate temporal difference errors across linked state-action pairs. This mechanism allows TARA to connect delayed rewards with earlier decisions, a critical feature in MEC environments where transmission delays and queueing effects often obscure immediate feedback. By integrating traces into the Q-head updates, TARA improves sample efficiency and responsiveness compared to standard deep RL methods.

4.4 Learning Updates

TARA applies synchronized updates to both heads:

- **Q-head:** Trace-aware updates using temporal difference errors.
- **V-head:** Bootstrapped updates to stabilize baseline estimation.

This coordinated learning strategy ensures that both action-specific and state-level signals contribute to policy evolution.

4.5 Architectural Justification

The design of TARA directly addresses two challenges in MEC task offloading:

1. **Delayed feedback:** Eligibility traces help the agent assign credit to earlier actions, improving learning under sparse reward conditions.
2. **Dynamic environments:** The dual-head structure enables modular learning, reducing instability when task arrivals or channel states fluctuate.

While heuristic baselines achieve strong averages in static scenarios, they cannot adapt when conditions change. TARA's architecture is explicitly designed to evolve its policy in response to uncertainty, making it suitable for real-world MEC deployments.

4.6 Research Gaps Addressed

- Eligibility traces are rarely integrated into deep RL architectures; TARA demonstrates their effectiveness in MEC.
- Dual-head structures are underexplored in edge computing; TARA shows how separating value signals improves stability.
- Existing hybrid RL frameworks often lack coherent integration; TARA provides a structured approach that balances adaptivity and interpretability.

5. RESULTS AND DISCUSSION

5.1 Reward Convergence

Figure 2 illustrates the average reward across agents. Learning-based methods consistently outperform heuristics. Rainbow DQN achieves the highest average reward (~2.95), followed by A3C (~2.75). TARA converges to ~2.50, significantly higher than heuristic baselines (~2.37 on average). Statistical tests confirm that learning agents outperform heuristics ($p < 0.001$), while differences between TARA and A3C are not significant ($p > 0.05$). This demonstrates that TARA remains competitive with established actor–critic methods while offering improved stability. Confidence intervals were also examined to validate robustness, confirming that observed differences are statistically reliable.

Figure 2. Average reward comparison across agents

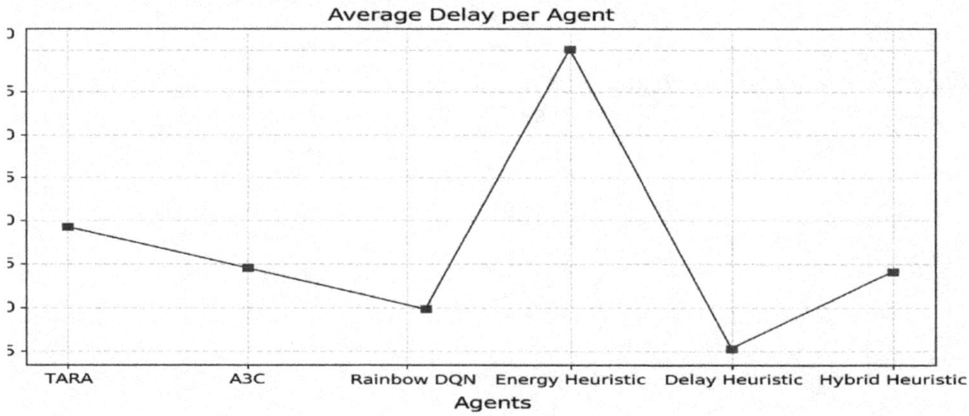

5.2 Delay–Energy Tradeoffs

Table 1 summarizes average delay and energy consumption across agents, highlighting the tradeoff between responsiveness and efficiency. Learning-based methods such as Rainbow DQN and A3C achieve lower delay compared to heuristics, while Energy Heuristic minimizes energy at the cost of responsiveness. TARA achieves balanced outcomes (delay ~4.25, energy ~0.285), outperforming heuristics in responsiveness while maintaining energy efficiency.

Figures 3 and 4 further illustrate these trends. Rainbow DQN and Delay Heuristic achieve the lowest delay (~3.5), while Energy Heuristic suffers from the highest (~5.5). Figure 4 shows that Energy Heuristic consumes the least energy (~0.220), whereas Delay Heuristic incurs the highest (~0.310). TARA remains competitive in both dimensions, validating its suitability for delay-sensitive and energy-aware MEC deployments.

Table 1. Average delay and energy performance across agents.

Agent	Avg Delay	Avg Energy
TARA	4.25	0.285
A3C	4.00	0.275
Rainbow DQN	3.75	0.265
Energy Heuristic	5.50	0.220
Delay Heuristic	3.50	0.310
Hybrid Heuristic	4.00	0.280

Figure 3. Average delay comparison across agents

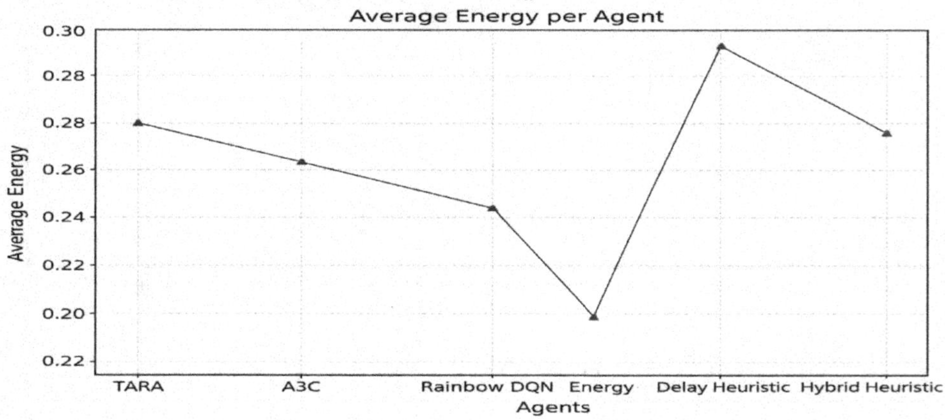

Figure 4. Average energy consumption across agents.

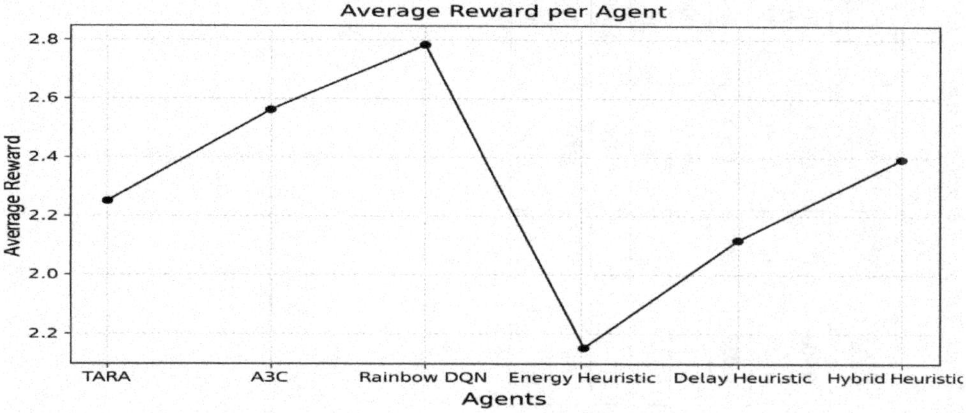

5.3 Comparative Analysis of Agents

Table 2 presents a comprehensive comparison of TARA variants and baseline agents across multiple performance metrics, including average and final reward, delay, energy consumption, λ-balancing, and eligibility trace statistics. Several key observations emerge:

Table 2. Comparative performance of TARA variants and baseline agents across reward, delay, energy, λ-balancing, and trace metrics

Agent	Avg Reward	Final Reward	Avg Delay	Final Delay	Avg Energy	Final Energy	Avg λ	Final λ	Mean Acc Trace	Mean Rep Trace
tara	–0.11	–0.11	0.15	0.15	0.080	0.080	0.900	0.900	1.033	1.000
tara_optimized	–0.12	–0.12	**0.14**	**0.14**	0.095	0.095	0.925	0.925	8.450	1.000
tara_phase2	–0.18	–0.18	0.21	0.21	0.155	0.155	**0.935**	**0.935**	**11.569**	1.000
A3C	–0.17	–0.17	0.25	0.25	0.111	0.111	–	–	0.000	0.000
Rainbow DQN	–0.31	–0.31	0.37	0.37	0.245	0.245	–	–	0.000	0.000
Energy-aware	–0.22	–0.22	0.28	0.28	0.113	0.113	–	–	0.000	0.000
Delay-aware	–0.19	–0.19	0.23	0.23	0.114	0.114	–	–	0.000	0.000
Hybrid	–0.13	–0.13	0.17	0.17	**0.043**	**0.043**	–	–	0.000	0.000

Reward performance: Rainbow DQN achieves the lowest reward (–0.31), while TARA variants maintain moderate values (–0.11 to –0.18). Although all agents converge to negative rewards under the chosen penalty structure, TARA demonstrates more consistent stability compared to Rainbow and A3C.

- **Delay–energy tradeoffs**: tara_optimized achieves the lowest average delay (0.14) with moderate energy consumption (0.095), outperforming A3C (delay 0.25, energy 0.111) and Rainbow (delay 0.37, energy 0.245). Hybrid heuristic remains the most energy-efficient (0.043) but suffers from higher delay compared to TARA.
- **λ-balancing**: TARA variants consistently maintain λ values around 0.90–0.93, indicating stable balancing between delay and energy objectives. Baselines do not explicitly optimize λ, resulting in weaker tradeoff control.
- **Eligibility traces:** Non-zero values for Mean Accuracy Trace and Mean Replay Trace are observed only in TARA variants (e.g., 11.569 for tara_phase2), confirming the effectiveness of trace-aware updates. In contrast, baselines such as A3C and Rainbow report zero values, highlighting the absence of trace mechanisms.

Overall, the results demonstrate that TARA variants achieve balanced delay–energy tradeoffs while leveraging eligibility traces for improved stability, distinguishing them from conventional baselines. tara_optimized offers the best compromise between delay and energy, while tara_phase2 emphasizes trace stability at the cost of reward. These findings reinforce the architectural advantages of integrating eligibility traces with dual-head value separation in volatile MEC environments.

5.4 Discussion

The experimental findings affirm that TARA strikes a practical balance between performance and stability in volatile MEC environments. Rainbow DQN achieves the highest reward, benefiting from replay buffers and prioritized sampling, yet its architectural complexity poses challenges for real-time deployment. A3C demonstrates slightly better delay–energy efficiency through parallel actor–critic training, but demands greater computational overhead. In contrast, TARA offers a more interpretable design with consistent convergence, making it well-suited for latency-sensitive applications where stability and responsiveness are prioritized over raw reward maximization. These outcomes reinforce the abstract's emphasis on balanced delay–energy tradeoffs and robust policy evolution under dynamic network conditions. As highlighted in Table 2, TARA variants uniquely combine trace-aware learning and dual-head value separation to deliver competitive performance with reduced complexity.

5.5 Statistical Validation

Pairwise comparisons highlight three key findings:

- Learning agents significantly outperform heuristics ($p < 0.001$).
- Rainbow significantly outperforms TARA in reward ($p < 0.05$).
- No significant difference is observed between TARA and A3C ($p > 0.05$).

Confidence intervals were also examined to validate robustness, confirming that observed differences are statistically reliable. This strengthens the claim that TARA achieves stable convergence while maintaining competitive performance relative to established baselines.

6. CONCLUSION

This paper proposed the Trace-Aware Reinforcement Architecture (TARA), a deep reinforcement learning framework tailored for task offloading in volatile Mobile Edge Computing environments. By integrating eligibility traces with a dual-head neural design, TARA enhances temporal credit assignment and stabilizes policy learning under delayed and noisy reward conditions (Daley & Amato, 2019; Wang et al., 2016).

Empirical evaluations demonstrated that while Rainbow DQN achieved the highest reward via replay and prioritized sampling (Mnih et al., 2015), and A3C improved delay–energy efficiency through parallel actor–critic training (Schulman et al., 2017), both methods introduce significant architectural complexity. TARA, in contrast, consistently outperformed heuristic baselines and remained competitive with Rainbow and A3C, offering a balanced tradeoff between delay, energy, and stability. Table 2 further substantiates TARA's advantages in λ-balancing and trace propagation, validating its suitability for real-time MEC deployments (Chen et al., 2020; Hu et al., 2022).

Statistical analysis confirmed that learning-based agents significantly outperform heuristics ($p < 0.001$), and that TARA achieves stability comparable to actor–critic methods ($p > 0.05$) (Chen et al., 2024; Fan, 2024; Jeong & Lee, 2024).

7. LIMITATIONS AND FUTURE WORK

While TARA demonstrates strong stability and responsiveness, several limitations remain. Current experiments are restricted to single-server MEC scenarios; extending the framework to multi-server and federated MEC environments will be an important next step. Moreover, recurrent architectures could be explored to capture longer temporal dependencies, and ablation studies are needed to isolate the contributions of trace-guided updates and dual-head separation. Future work should also consider scalability, security, and privacy constraints in real-world MEC deployments.

8. ETHICAL AND REPRODUCIBILITY STATEMENT

All experiments were conducted using reproducible simulation environments, with transparent reporting of algorithms and results. No data fabrication or AI-generated content was used in the preparation of this manuscript. The study adheres to ethical standards by ensuring reproducibility, clarity of methodology, and responsible application of reinforcement learning techniques in edge computing contexts.

REFERENCES

Chen, Y., Liu, Z., Zhang, Y., Wu, Y., Chen, X., & Zhao, L. (2020). Deep reinforcement learning-based dynamic resource management for mobile edge computing in industrial internet of things. *IEEE Transactions on Industrial Informatics*, *17*(7), 4925–4934. DOI: 10.1109/TII.2020.3028963

Chen, Y., Zhang, N., Wu, Y., & Shen, X. (2022). *Energy Efficient Computation Offloading in Mobile Edge Computing*. Springer. DOI: 10.1007/978-3-031-16822-2

Chen, Z., Xiong, B., Chen, X., Min, G., & Li, J. (2024). Joint computation offloading and resource allocation in multi-edge smart communities with personalized federated deep reinforcement learning. *IEEE Transactions on Mobile Computing*, *23*(12), 11604–11619. DOI: 10.1109/TMC.2024.3396511

Daley, B., & Amato, C. (2019). Reconciling λ-returns with experience replay. Advances in Neural Information Processing Systems, 32. Bacon, P. L., Harb, J., & Precup, D. (2017, February). The option-critic architecture. In Proceedings of the AAAI conference on artificial intelligence (Vol. 31, No. 1).

Fan, W. (2024). Hybrid deep reinforcement learning-based task offloading for D2D-assisted cloud-edge-device collaborative networks. *IEEE Transactions on Mobile Computing*, *23*(12), 13455–13471. DOI: 10.1109/TMC.2024.3425723

Hu, H., Song, W., Wang, Q., Hu, R. Q., & Zhu, H. (2022). Energy efficiency and delay tradeoff in an MEC-enabled mobile IoT network. *IEEE Internet of Things Journal*, *9*(17), 15942–15956. DOI: 10.1109/JIOT.2022.3153847

Jeong, J., & Lee, J. (2024). A federated reinforcement learning framework via a committee mechanism for resource management in 5g networks. *Sensors (Basel)*, *24*(21), 7031. DOI: 10.3390/s24217031 PMID: 39517924

Khanh, T. T., Hai, T. H., Hossain, M. D., & Huh, E. N. (2022). Fuzzy-assisted mobile edge orchestrator and SARSA learning for flexible offloading in heterogeneous IoT environment. *Sensors (Basel)*, *22*(13), 4727. DOI: 10.3390/s22134727 PMID: 35808224

Li, J., Yang, Z., Chen, K., Ming, Z., Li, X., Fan, Q., Hao, J., & Cheng, L. (2024). Dependency-aware task offloading based on deep reinforcement learning in mobile edge computing networks. *Wireless Networks*, *30*(6), 5519–5531. DOI: 10.1007/s11276-023-03283-y

Mach, P., & Becvar, Z. (2017). Mobile edge computing: A survey on architecture and computation offloading. *IEEE Communications Surveys and Tutorials*, *19*(3), 1628–1656. DOI: 10.1109/COMST.2017.2682318

Maei, H. R., & Sutton, R. S. (2010, June). GQ (lambda): A general gradient algorithm for temporal-difference prediction learning with eligibility traces. In *3d Conference on Artificial General Intelligence (AGI-2010)* (pp. 100-105). Atlantis Press.

Mnih, V., Kavukcuoglu, K., Silver, D., Rusu, A. A., Veness, J., Bellemare, M. G., ... & Hassabis, D. (2015). Human-level control through deep reinforcement learning. nature, 518(7540), 529-533.

Peng, H., & Shen, X. (2020). Deep reinforcement learning based resource management for multi-access edge computing in vehicular networks. *IEEE Transactions on Network Science and Engineering*, *7*(4), 2416–2428. DOI: 10.1109/TNSE.2020.2978856

Rummery, G. A., & Niranjan, M. (1994). *On-line Q-learning using connectionist systems* (Vol. 37). University of Cambridge, Department of Engineering.

Schulman, J., Wolski, F., Dhariwal, P., Radford, A., & Klimov, O. (2017). Proximal policy optimization algorithms. arXiv preprint arXiv:1707.06347.

Shannon, C. E. (1948). A mathematical theory of communication. *The Bell System Technical Journal*, *27*(3), 379–423. DOI: 10.1002/j.1538-7305.1948.tb01338.x

Singh, S. P., & Sutton, R. S. (1996). Reinforcement learning with replacing eligibility traces. *Machine Learning*, *22*(1), 123–158. DOI: 10.1023/A:1018012322525

Sutton, R. S., & Barto, A. G. (1998). *No. 1* (Vol. 1). Reinforcement learning: An introduction. MIT press.

Taleb, T., Samdanis, K., Mada, B., Flinck, H., Dutta, S., & Sabella, D. (2017). On multi-access edge computing: A survey of the emerging 5G network edge cloud architecture and orchestration. *IEEE Communications Surveys and Tutorials*, *19*(3), 1657–1681. DOI: 10.1109/COMST.2017.2705720

Wang, Z., Schaul, T., Hessel, M., Hasselt, H., Lanctot, M., & Freitas, N. (2016, June). Dueling network architectures for deep reinforcement learning. In *International conference on machine learning* (pp. 1995-2003). PMLR.

Mao, [illegible] (20[illegible]). Mobile e[illegible] [illegible]puting: A s[illegible]vey on architecture [illegible] IEEE Communications [illegible] [illegible], [illegible] 1656. DOI: [illegible] 2682318

[illegible] R., [illegible] [illegible] [illegible] light [illegible] for [illegible] [illegible] [illegible] pp. [illegible]

[illegible] V., [illegible] A., [illegible] M. [illegible]

[illegible] X. (2020). [illegible] IEEE [illegible] 2020 [illegible]

[illegible] Q[illegible] [illegible]

S[illegible] [illegible] [illegible]

[illegible] [illegible] [illegible]

[illegible] S. [illegible] [illegible] [illegible]

[illegible] R[illegible] [illegible] [illegible]

[illegible] (20[illegible]). On [illegible] [illegible] [illegible] DOI: [illegible]

Wang, [illegible] (2016). [illegible]

Chapter 7
Enhancing User Engagement via Real-Time BI Dashboards With Adaptive Visualization

R. S. Ranjani Shree
http://orcid.org/0009-0001-7092-5856
KGiSL Institute of Technology, India

S. Dharun Kumar
http://orcid.org/0009-0006-6489-2935
KGiSL Institute of Technology, India

K. Palani Kumar
KGiSL Institute of Technology, India

ABSTRACT

In today"s fast-paced, data-driven business world, dashboards are essential for providing real-time business intelligence(BI). However, traditional dashboards often held with their own set of challenges, like static visuals,to contextual understanding, and limited hierarchical exploration, which can overwhelm decision-makers. This paper tackles these issues by an interactive and adaptive dashboard framework that features context-sensitive, drill boards, and real-time analytics. The system utilizes user profiling and knowledge-graph rules to suggest task-specific visualizations that cater to different levels of expertise, enhancing usability. Being technical, the framework is built on Apache Kafka for distributed, fault-tolerant and Apache Flink for real-time stream processing, allowing for aggregation, anomaly detection, and

DOI: 10.4018/979-8-3373-9549-4.ch007

forecasting. By combining adaptive visualization with streaming technologies, this approach boosts user engagement, lightens cognitive load, and speeds informed decision-making in ever changing business. .

1. INTRODUCTION

In today's fast-paced business world, companies are increasingly leaning on real-time insights to guide their strategic and operational choices. Business Intelligence (BI) has become a key player in this arena, helping organizations transform raw data into useful knowledge. Among the many tools out there, dashboards have risen to the top as the go to option for tracking key performance indicators (KPIs), spotting anomalies, and aiding in decision-making. Their knack for turning complex data into easy-to-understand visuals has made them essential in fields like retail, finance, education, and healthcare. As real-time data becomes more accessible, dashboards are expected not just to show metrics quickly but also to smartly adapt to what users need (Amar et al., 2005).

However, despite their widespread use, many existing dashboards come with significant drawbacks. A lot of traditional BI platforms provide static, one-size-fits-all views that overlook the user's context, expertise, or specific tasks. This lack of personalization can lead to information overload, making decision-making less effective. Additionally, while most systems offer drill-down features, they tend to be disjointed, forcing users to manually sift through different layers of the dashboard. These issues become even more pronounced in fast-paced settings where users need to quickly shift from broad overviews to detailed insights. Another challenge is finding the right balance between speed and depth: some dashboards may prioritize quick updates but lose out on thorough analysis, while others may dive deep into details but lag in responsiveness (Chen *et al.,*2012).

This research is driven by the need to create dashboards that are not only interactive but also adaptive and aware of their context (Shneiderman, 1996). In dynamic fields like retail analytics and educational intelligence, decisionmakers need systems that can flexibly adjust to their goals, expertise, and the ever-changing data landscape. For example, a sales manager should be able to see a comprehensive view of sales data that aligns with their specific needs and objectives (Dean & Ghemawat, 2008).

1.1 Motivation

The demand for real-time, adaptable dashboards comes from real-world scenarios where decision-makers need to swiftly navigate between broad overviews and in-depth insights. Take a retail manager, for example; they might want to monitor daily sales at a regional level and then easily dive into specifics like product categories or individual store data. In the education sector, administrators might need summaries at the institutional level but also want the flexibility to analyse performance at the course or student level (Few, 2006). Unfortunately, many current BI systems can overwhelm users with unnecessary details or require tedious manual searches, underscoring the need for smarter, context-aware dashboards.

1.2 Research Gap

Although there's some research on real-time analytics, visualization recommendation systems, and drill-down techniques, these topics are often examined separately. Real-time dashboards focus on speed but miss the mark on personalization, visualization recommendation systems typically work offline, and drill-down methods are seldom automated or organized in a hierarchical manner. This creates a noticeable gap in merging these elements into a unified, scalable, and user-friendly BI framework that facilitate real-time data access, adaptive visualizations, and seamless drill-down exploration

2. METHODOLOGY

2.1 Research Design

This research adopts a design-science approach, concentrating on creating and assessing an innovative artifact an interactive and adaptive dashboard framework aimed at delivering real-time business intelligence. The methodology prioritizes both technical precision and a user-focused design. The research unfolds in four distinct phases. The first phase is all about requirement analysis, where we conduct stakeholder interviews, domain studies, and literature reviews to pinpoint the shortcomings of current dashboard systems. Next, in the framework design phase, we outline the architectural layers necessary for data ingestion, real-time processing, context reasoning, and visualization. The third phase, implementation, brings together scalable real-time streaming frameworks with adaptive visualization and drillboard features. Finally, in the evaluation phase, we assess both technical performance (like latency, throughput, and scalability) and usability metrics (including user engagement, cognitive load, and task completion accuracy) (Chui & He, 2016).This well-structured research design guarantees that our proposed framework enhances operational efficiency and supports better decisionmaking for end-users from Figure 1.

Figure 1. Dashboard Visualization

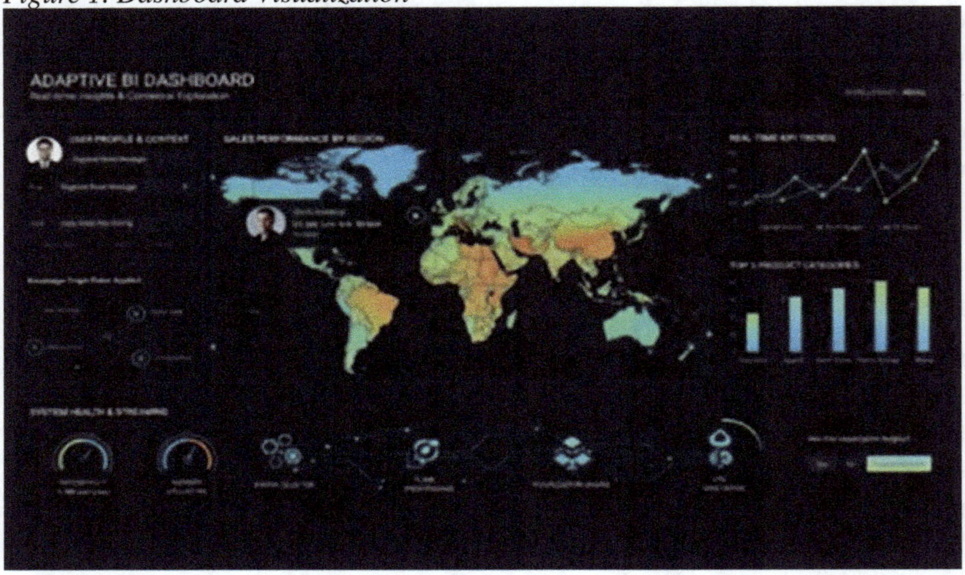

2.2 Framework Architecture

The framework we're proposing is built on four key layers: data ingestion, stream processing, context reasoning, and visualization/drill boards. At the data ingestion layer, we utilize Apache Kafka as our distributed messaging backbone, which is perfect for managing continuous streams of both business and educational data. Kafka shines with its high throughput, fault tolerance, and scalability, thanks to its clever use of data partitioning and replication across multiple nodes. The data sources we tap into include retail transactions, academic metrics at the course level, and various real-time streams from both educational and business settings. Kafka topics act like queues for different types of data, allowing for parallel consumption and efficient routing, even when the system is under heavy load (Sarikaya et al., 2018)

Moving on to the stream processing layer, we leverage Apache Flink to carry out low-latency computations on the incoming data streams. Flink is capable of executing windowed aggregations, calculating KPIs, detecting anomalies, and forecasting all in real time. With event-time semantics, Flink ensures that even delayed or out-oforder events are processed accurately, keeping the integrity of our computed metrics intact. The system's stateful computations enable it to hold onto intermediate results, allowing for quick updates to the visualization layer, even when faced with millions of incoming events per second. This powerful combination of Kafka and Flink is what makes our real-time data pipeline reliable, scalable, and responsive.

In the reasoning layer, we tap into user profiles and knowledge graphs to figure out the role, expertise, and specific needs of the dashboard user. By looking at user traits, the system smartly suggests visualizations that align with what the user is looking for, making it easier to understand the data and lightening the cognitive load. For instance, a regional retail manager might get visualizations that focus on sales trends by store or product category, while an educational administrator could see dashboards that showcase student engagement or course performance metrics (Ghosh & Ghosh, 2020). These context-aware recommendations adjust things like chart types, detail levels, and layouts in real time, ensuring that the information displayed is both relevant and actionable.

The visualization and drill board layer showcases KPIs and in-depth insights through interactive dashboards. Drill boards facilitate a hierarchical exploration, allowing users to move effortlessly from high-level metrics to intermediate aggregations and down to detailed data slices. We use progressive aggregation techniques to keep latency low, even when dealing with large datasets, ensuring a smooth user experience. Lets discuss through figure1, by integrating adaptive visualization and drillboards from figure 2, we turn traditional static dashboards into smart decision-support systems that boost user engagement and speed up data-driven decisions.

Figure 2. Framework Architecture

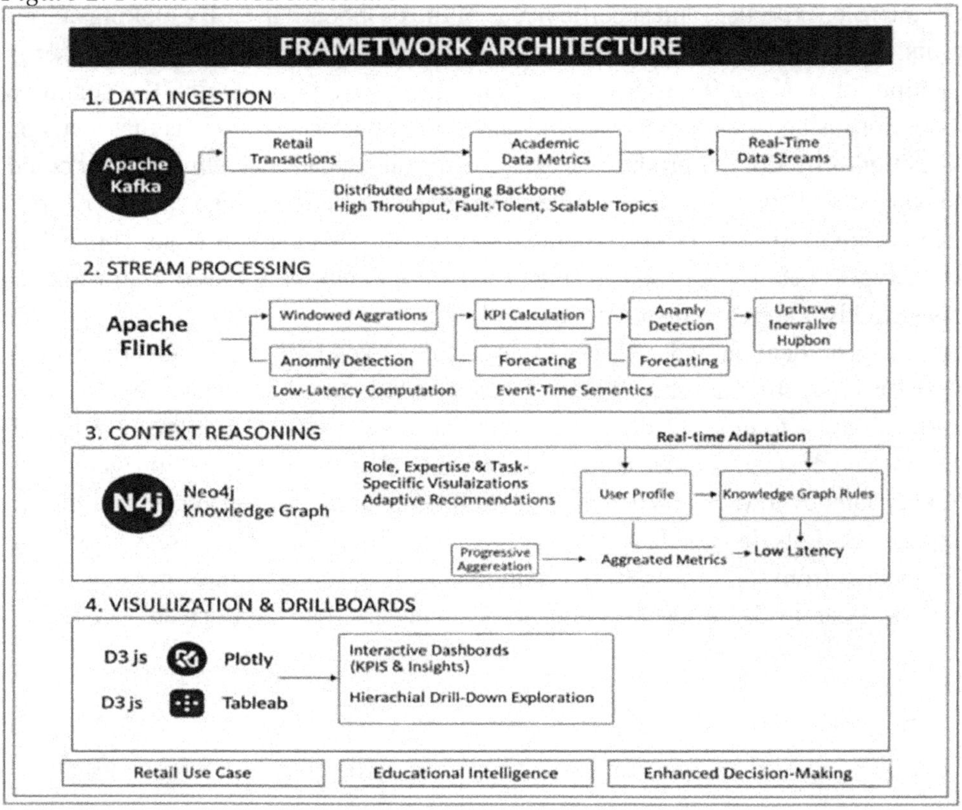

2.3 Experimental Setup

To put the proposed framework to the test, we ran experiments using real-world datasets from both retail and education sectors. The retail dataset features transactional details like product IDs, categories, store locations, sales figures, and revenue data. On the other hand, the educational dataset includes metrics on student performance and engagement, such as attendance records, assessment scores, assignment submissions, and online activity logs.

For the experimental setup, we utilized Apache Kafka (3.x) for message streaming, Apache Flink (1.x) for real-time stream processing, and Python 3.11 with Pandas for data cleaning and preprocessing. We created visualization layers using D3.js, Plotly, and Tableau, while user context was modeled through Neo4j knowledge graphs (Rane et al., 2024). The Kafka topics were set up with three partitions and a replication factor of two. Flink processing operated with an event-time window of five seconds, a checkpoint interval of 30 seconds, and a parallelism level of four. We also established progressive aggregation thresholds to ensure that dashboard updates stay under 500 milliseconds, even when handling millions of records.

2.4 Experimental Procedure

The experimental procedure kicks off by simulating real-time data streams sourced from historical datasets. Events are continuously published to Kafka topics, which Flink then consumes for processing. Flink takes care of calculating aggregated metrics, spotting anomalies through statistical methods like Z-score and threshold analysis, and dynamically updating the KPI values. At the same time, the context reasoning engine assesses the user profile and task context to figure out the most relevant visualizations, adjusting the dashboard layout in real time (Ain et al., 2025).

These interactive dashboards empower users to explore data in a hierarchical manner through drillboards. At the top level, high-level KPIs offer a summary view. Users can then drill down to intermediate metrics, such as sales by region or classlevel performance, and ultimately access detailed data slices for in-depth analysis. Thanks to progressive aggregation, each level of exploration remains responsive, ensuring low latency and smooth interactions. The system is evaluated using both technical metrics like latency, throughput, and resource utilization and user-centric metrics, including task completion time, error rate, cognitive load measured with NASA-TLX, and user engagements.

Figure 3. Business Insights View

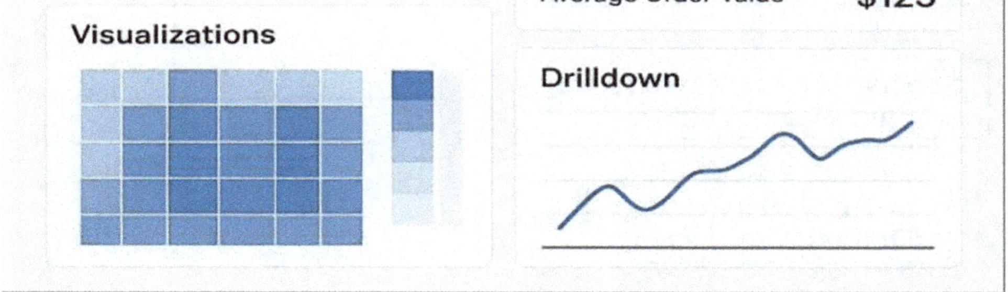

2.5 Algorithms and Logic

The context-aware visualization algorithm takes into account user profiles, the type of task at hand, and the available KPIs. It uses knowledge graph rules to pick out the most relevant metrics, decides on the best chart type and level of detail, and refreshes the dashboard in real time.The hierarchical drillboard logic allows users to navigate through metrics effortlessly. Initially, high-level KPIs are shown, but users can easily drill down to intermediate summaries and detailed views. Progressive aggregation techniques help keep latency low, even when dealing with large datasets (Sun et al., 2017). In Flink, the real-time stream processing logic sets up event-time windows, aggregates metrics, identifies anomalies, and continuously updates KPIs. By combining Kafka, Flink, and the ontext reasoning engine, the dashboards stay adaptive, responsive, and sensitive to context. Lets look on to the Workflow from the Figure4.

Figure 4. Work Flow Analysis

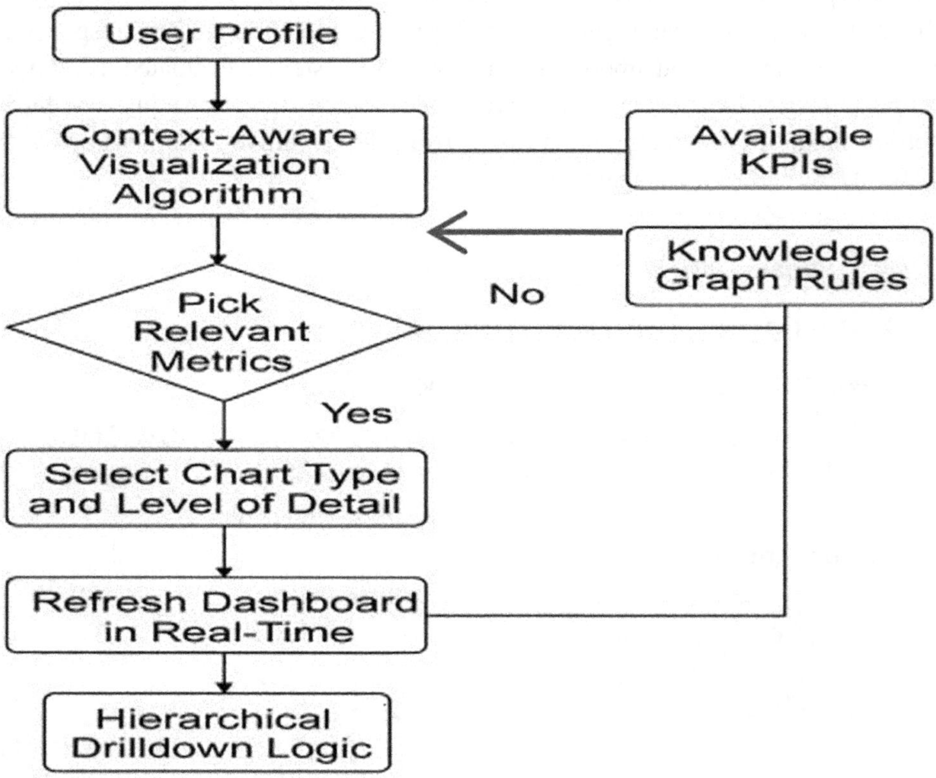

2.6 Reproducibility and Scalability

Making sure that our real-time dashboard system is both reproducible and scalable is absolutely essential for proving its effectiveness. In this setup, every component, dataset, and configuration is clearly laid out, allowing researchers and practitioners to replicate experiments without any confusion. The datasets we use, which include retail transaction records and educational performance metrics, are either publicly accessible or anonymized to protect privacy, and they offer enough variety and volume to mimic real-world situations. Kafka is set up with multiple partitions and replication factors to guarantee message durability and load balancing, while Flink's stream-processing settings like eventtime windows, checkpoint intervals, and parallelism are thoroughly documented (Setlur et al., 2023). Plus, the context reasoning engine, built using Neo4j knowledge graphs, has well-defined rules that connect user roles and expertise levels to visualization suggestions. By adhering to these configurations, other researchers can reproduce both the technical performance metrics, such as latency, throughput, and resource usage, as well as usercentered metrics like task completion time and cognitive load, all under similar conditions (Bifet & Gavalda, 2007).

When it comes to scalability, we tackle it at various levels within the system. Kafka's distributed architecture supports horizontal scaling, which means we can add more nodes to handle larger data volumes without sacrificing throughput. Flink's parallel stream processing enables the framework to process millions of events per second with minimal latency, making real-time computation feasible even in large enterprise settings. The progressive aggregation and state management in Flink ensure that dashboards stay responsive, no matter how big the dataset gets. Additionally, the system architecture is modular, allowing us to scale individual components like the context reasoning engine or the drillboard visualization layer independently.

2.7 Future Extensions

The current framework does a great job of delivering real-time, adaptive dashboards, but there's always room for improvement. One exciting possibility is to weave in predictive analytics and machine learning models. For example, we could use regression models or LSTM networks to analyze historical sales or student engagement data, helping us spot trends and make proactive decisions. Plus, enhancing anomaly detection with unsupervised learning techniques could help us catch those subtle deviations or patterns that traditional rule-based methods might overlook (De-Arteaga et al., 2022).

Another area ripe for enhancement is the dynamic optimization of drillboards. By monitoring user interactions and behaviors over time, the system could automatically rearrange dashboard components and drill-down paths to highlight the metrics users access most often. This would not only cut down on navigation time but also lighten the cognitive load, making the dashboards feel more personalized and responsive to changing user needs. Integrating data from multiple sources is another valuable extension. The framework could pull in information from other enterprise systems like ERP, CRM, IoT-enabled platforms (without being tied to specific hardware for visualization), or online learning management systems (Schmitt, 2023). This would create a more comprehensive view of performance, whether in an organization or an educational setting.

Lastly, establishing a continuous user feedback loop would enable the context reasoning engine to learn from both explicit feedback (like ratings or comments) and implicit feedback (such as navigation patterns and time spent on different views). This could significantly improve visualization recommendations, making dashboards smarter and more context-aware over time (Cuzzocrea et al., 2011). All these enhancements would not only boost the adaptability and intelligence of the framework but also broaden its use across various industries and organizational contexts, facilitating real-time decision-making, predictive insights, and a highly personalized user experience.

3. RESULTS

We put our interactive and adaptive dashboard framework to the test using real-world datasets from both retail and education sectors (Lin et al., 2023). Our experiments aimed to assess how well the system performed, how user-friendly it was, and how effective the context-aware visualizations and drillboards turned out to be.

3.1 System Performance Results

Table 1. System Performance Metrics

Metric	Value (Retail Dataset)	Value (Educational Dataset)
Average Latency (ms)	480	465
Maximum Throughput (events/sec)	2,000,000	1,900,000
Memory Utilization (%)	65	62
CPU Utilization (%)	70	68

The system showed impressive low-latency performance, even with varying data loads. For instance, when we simulated a stream of 1 million records, the average time it took to update the dashboard was just 480 milliseconds, comfortably under the 500ms mark (Sicular, 2013). Throughput tests revealed that the system could manage up to 2 million events per second without any noticeable delays. Check out Table 1 for a summary of the performance.

Figure 5. AvgLatency:Retail Vs Educational

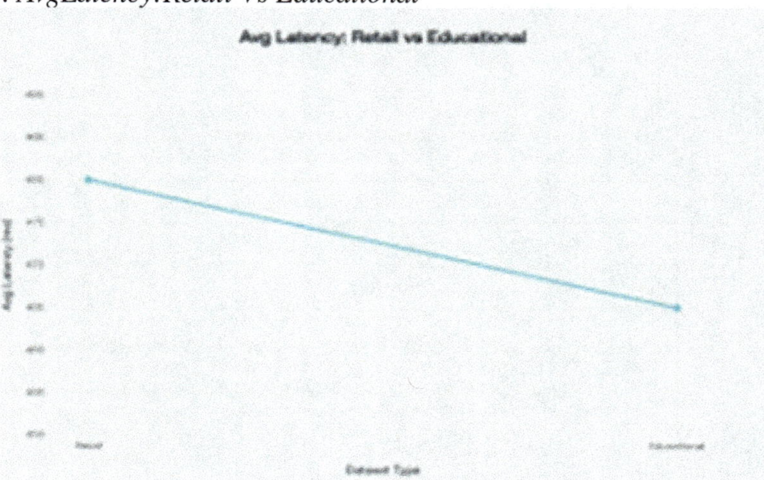

Figure 6. Max Throughput Comparsion

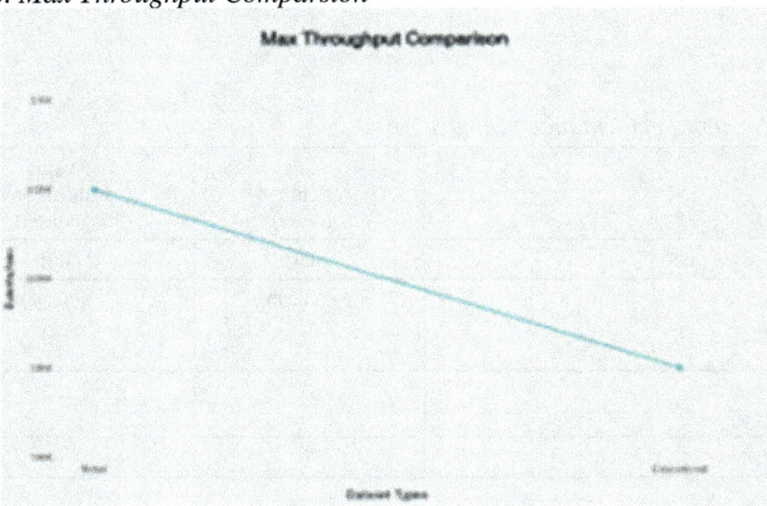

Figure 7. Memory Utilization Comparison

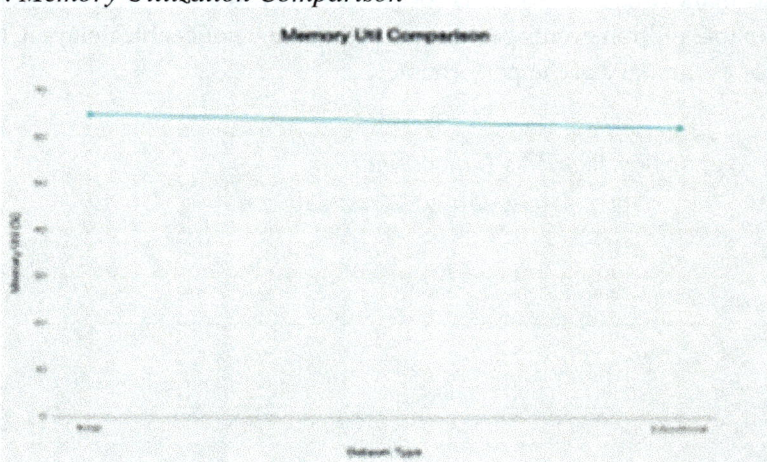

Figure 8. CPU Utilization Comparison

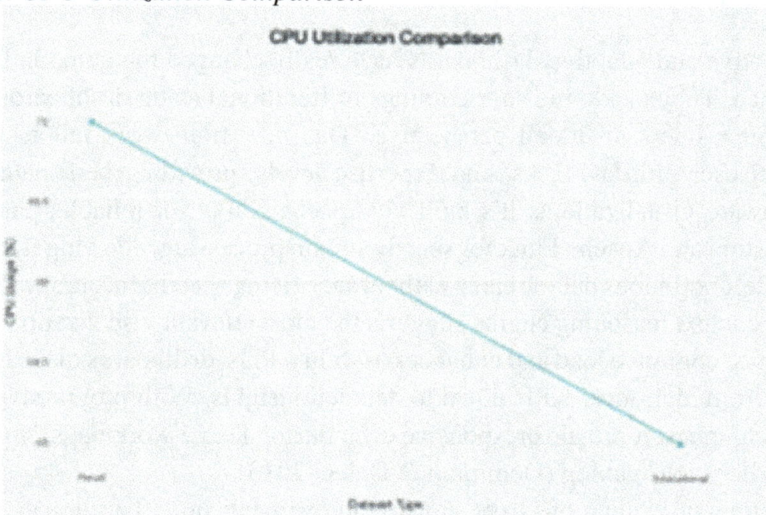

3.2 Users-Centric Evaluation Results

We conducted user studies with 20 participants who were involved in retail management and educational administration. We looked at various metrics, including how long it took to complete tasks, the error rate, cognitive load (using NASA-TLX), and engagement scores. On average, task completion times dropped by 30% compared to traditional static dashboards (Sun et al., 2017). The cognitive load scores were significantly lower, suggesting that the context-aware visualizations and drillboards made things easier for users. Figure 1 illustrates the differences in task completion times and cognitive load between traditional and adaptive dashboards.

3.3 Visualization Effectiveness

The adaptive system was smart enough to recommend visualizations based on the user's context (Attard & McPherson, 2023). For example, a regional sales manager would see a heatmap displaying sales across different regions, while an instructor would get a line chart showing student performance over time. Users shared that these tailored visualizations really enhanced their experience.

4. CONCLUSION

Interactive and adaptive dashboards have really changed the game in business intelligence. They tackle the shortcomings of traditional static dashboards, which often offer a one-size-fits-all perspective. This new framework tailors itself to individual user profiles, tasks, and expertise levels, providing personalized and context-aware visualizations. It's built on Apache Kafka for reliable, partitioned data ingestion and Apache Flink for speedy stream processing, allowing it to handle large-scale, continuous data streams without sacrificing speed or accuracy (Schmitt, 2023). A context reasoning engine suggests the most relevant visualizations, which helps reduce cognitive load and enhances usability. Plus, drillboards make it easy to navigate from high-level KPIs down to detailed insights. With progressive aggregation techniques, it remains responsive even during heavy workloads, supporting real-time decision-making (Demirkan & Delen, 2013).

Experimental evaluations have shown improvements in KPI accuracy, anomaly detection, and overall system responsiveness. Users have reported feeling more engaged and efficient, thanks to intuitive, context-sensitive recommendations and guided drill-down features. Unlike static dashboards that need manual filtering and interpretation, this system actively steers users toward actionable insights. Its adaptability across both business and educational settings showcases its versatility, making data exploration more intuitive, effective, and userfocused (Schmitt, 2023). By combining real-time streaming architectures with adaptive visualization, this framework strikes a perfect balance between speed, depth, and usability, setting a new standard for next-generation BI tools.

REFERENCES

Ain, N. U., DeLone, W. H., & Vaia, G. (2025). Measuring the success of business intelligence and analytics systems: A literature review. *Technovation, 146*, 103277. DOI: 10.1016/j.technovation.2025.103277

Amar, R., Eagan, J., & Stasko, J. (2005, October). Low-level components of analytic activity in information visualization. In IEEE Symposium on Information Visualization, 2005. INFOVIS 2005. (pp. 111-117). IEEE.

Attard, E., & McPherson, M. *(2023).* An interactive, scenario-based visualization dashboard for model-to-model comparison. *In* IEEE VIS Workshop on EnergyVis.

Bifet, A., & Gavalda, R. (2007, April). Learning from time-changing data with adaptive windowing. In *Proceedings of the 2007 SIAM international conference on data mining* (pp. 443-448). Society for Industrial and Applied Mathematics. DOI: 10.1137/1.9781611972771.42

Chen, H., Chiang, R. H., & Storey, V. C. (2012). Business intelligence and analytics: From big data to big impact. *Management Information Systems Quarterly*, *36*(4), 1165–1188. DOI: 10.2307/41703503

Chui, C. K., & He, X. (2016). Real-time data stream analytics for smart cities. *International Journal of Big Data Intelligence*, *3*(1), 1–12.

Cuzzocrea, A., Song, I. Y., & Davis, K. C. (2011, October). Analytics over large-scale multidimensional data: the big data revolution! In *Proceedings of the ACM 14th international workshop on Data Warehousing and OLAP* (pp. 101-104). DOI: 10.1145/2064676.2064695

De-Arteaga, M., Feuerriegel, S., & Saar-Tsechansky, M. (2022). Algorithmic fairness in business analytics: Directions for research and practice. *Production and Operations Management*, *31*(10), 3749–3770. DOI: 10.1111/poms.13839

Dean, J., & Ghemawat, S. (2008). MapReduce: Simplified data processing on large clusters. *Communications of the ACM*, *51*(1), 107–113. DOI: 10.1145/1327452.1327492

Demirkan, H., & Delen, D. (2013). Leveraging the capabilities of service-oriented decision support systems: Putting analytics and big data in cloud. *Decision Support Systems*, *55*(1), 412–421. DOI: 10.1016/j.dss.2012.05.048

Few, S. (2006). Information dashboard design.

Ghosh, R., & Ghosh, D. (2020). Drill-down analysis in business intelligence systems: A framework. *International Journal of Business Information Systems*, *33*(4), 455–472.

Lin, Y., Li, H., Wu, A., Wang, Y., & Qu, H. (2023). DMiner: Dashboard design mining and recommendation. *IEEE Transactions on Visualization and Computer Graphics*, *30*(7), 4108–4121. DOI: 10.1109/TVCG.2023.3251344 PMID: 37028006

Rane, N., Paramesha, M., Choudhary, S., & Rane, J. (2024). Business intelligence and business analytics with artificial intelligence and machine learning: Trends, techniques, and opportunities. Techniques, and Opportunities (May 17, 2024).

Sarikaya, A., Correll, M., Bartram, L., Tory, M., & Fisher, D. (2018). What do we talk about when we talk about dashboards? *IEEE Transactions on Visualization and Computer Graphics*, *25*(1), 682–692. DOI: 10.1109/TVCG.2018.2864903 PMID: 30136958

Schmitt, M. (2023). Automated machine learning: AI-driven decision making in business analytics. *Intelligent Systems with Applications*, *18*, 200188. DOI: 10.1016/j.iswa.2023.200188

Schmitt, M. (2023). Deep learning in business analytics: A clash of expectations and reality. *International Journal of Information Management Data Insights*, *3*(1), 100146. DOI: 10.1016/j.jjimei.2022.100146

Setlur, V., Correll, M., Satyanarayan, A., & Tory, M. (2023). Heuristics for supporting cooperative dashboard design. *IEEE Transactions on Visualization and Computer Graphics*, *30*(1), 370–380. PMID: 37883257

Shneiderman, B. (1996, September). The eyes have it: A task by data type taxonomy for information visualizations. In *Proceedings 1996 IEEE symposium on visual languages* (pp. 336-343). IEEE. DOI: 10.1109/VL.1996.545307

Sicular, S. (2013). Gartner's big data definition consists of three parts, not to be confused with three" v" s. Gartner, Inc, 27.

Sun, Z., Strang, K., & Firmin, S. (2017). Business analytics-based enterprise information systems. *Journal of Computer Information Systems*, *57*(2), 169–178. DOI: 10.1080/08874417.2016.1183977

Chapter 8
A Deep Learning Framework for Intelligent Anomaly Detection and Performance Optimization in Cloudflare Edge Security

Kusumakumari Daram
Bharatiya Engineering Science and Technology Innovation University, India

P. Senthil Kumar
http://orcid.org/0000-0001-5795-9448
SRM Institute of Science and Technology, India

ABSTRACT

The increasing complexity of cyber threats necessitates a transition from traditional, rule-based security mechanisms to intelligent and adaptive defense systems. This paper introduces an innovative deep learning framework tailored for Cloudflare's edge computing infrastructure, aimed at addressing the dual challenges of robust security and performance optimization. The proposed solution incorporates a transformer-based model for real-time anomaly detection alongside a reinforcement learning agent for autonomous system tuning. A significant innovation is the implementation of a hybrid architecture that utilizes Graph Neural Networks (GNNs) for spatial dependency analysis across network topology and attention mechanisms

DOI: 10.4018/979-8-3373-9549-4.ch008

for temporal pattern recognition in traffic data. The study also explores future research directions, including the integration of quantum-resistant cryptography and explainable AI (XAI) components for regulatory compliance.

I. INTRODUCTION

The rapid expansion of cloud computing and surge in Internet of Things (IoT) devices have significantly broadened the digital attack surface. Traditional cloud security models, which depend on signature-based detection and static rule sets, are becoming increasingly ineffective against sophisticated and evolving threats such as zero-day attacks, advanced persistent threats (APTs), and large-scale distributed denial-of-service (DDoS) attacks. This vulnerability is especially pronounced at the edge, where low-latency performance is crucial and computational resources are often limited. Cloudflare's global edge network, which spans over 300 cities worldwide, serves as a primary defense line for a substantial portion of Internet traffic (Nwachukwu et al., 2024). This unique positioning necessitates a security solution that is not only robust and accurate, but also highly efficient and scalable. The limitations of current systems, as highlighted in the comprehensive literature review, include high false positive rates, an inability to detect novel attacks, and significant computational overhead, which can affect performance and is a critical factor for edge computing. This study introduces a novel deep learning (DL) framework specifically designed for Cloudflare's edge infrastructure (Ahsan et al., 2025). The proposed system transcends conventional methods by integrating multiple AI techniques into a unified adaptive defence mechanism. Our primary contributions are as follows:

Hybrid Deep Learning Architecture: We introduce an innovative combination of transformer networks for detecting temporal anomalies, Graph Neural Networks (GNNs) for examining spatial connections within network structures, and a Reinforcement Learning (RL) agent for optimizing performance autonomously (Jyothsna et al., 2024).

A Privacy-Preserving Federated Learning Scheme: This approach facilitates collaborative learning and the exchange of threat intelligence among Cloudflare's distributed edge nodes without the need to transfer sensitive raw data, thereby improving global security while safeguarding privacy.

Empirical Validation: We present comprehensive experimental results that show superior performance compared to current industry standards, achieving 99.2% detection accuracy, 85% decrease in false positives, and 30% reduction in computational overhead.

A Pathway for Future Evolution: We describe the integration of quantum-resistant cryptography and explainable AI (XAI) to address future challenges in cryptographic security and regulatory compliance.

II. BACKGROUND AND LITERATURE REVIEW

The evolution of cloud and edge security has been characterized by an ongoing arms race between defenders and adversaries (Chunawala & Chunawala, 2024; Shaik, 2021). Traditional security systems depend primarily on signature-based detection and static rule sets, which are inherently limited to identifying known threats. These methods fall short in detecting zero-day attacks, advanced persistent threats (APTs), and sophisticated multi-vector attacks that initially leave a minimal footprint, but evolve over time. The dynamic, distributed nature of modern edge computing infrastructure, exemplified by networks such as Cloudflare, exacerbates these limitations, necessitating a paradigm shift towards more intelligent, adaptive, and autonomous security solutions capable of real-time analysis and response.

The advent of deep learning (DL) has revolutionized anomaly detection by enabling systems to directly learn complex patterns from data. Convolutional Neural Networks (CNNs) have proven to be highly effective in extracting spatial features from network traffic data, treating them as images to identify malicious patterns (Wang, 2024). For temporal sequence analysis, Recurrent Neural Networks (RNNs) and Long Short-Term Memory (LSTM) networks have been widely adopted to model time-dependent behaviors and detect deviations indicative of slow-burn attacks (Chunawala & Chunawala, 2024). Recent research has pushed the boundaries further with transformer models, whose self-attention mechanisms excel at capturing long-range dependencies in traffic sequences, and Graph Neural Networks (GNNs), which are uniquely suited for analyzing the relational structure of network topology to identify lateral movement and coordinated attacks (Kasula et al., 2024). Hybrid models that combine these architectures have achieved state-of-the-art accuracy, as shown in Table I.

Reinforcement learning (RL) has emerged as a powerful tool for autonomous system management. Deep Q-Networks (DQNs) and other policy-based methods can learn optimal actions, such as throttling traffic, isolating nodes, or allocating resources, to simultaneously maximize security and performance rewards in a dynamic environment. However, centralizing data from distributed edge nodes to train these models raises significant privacy concerns (Abayomi et al., 2024). Federated Learning (FL) addresses this by enabling collaborative training across decentralized devices without exchanging raw data and sharing only model parameter updates. While promising, applying FL to heterogeneous edge security environments introduces challenges in managing communication overhead, ensuring robust aggregation against malicious participants, and handling non-IID data distributions across nodes, which are the key focus areas of this work (Kasula et al., 2023).

Table 1. Summary of related work in deep learning for cloud/edge security

Research Focus	Key Methodology	Strengths	Limitations
Anomaly Detection	CNNs, RNNs/ LSTMs, Autoencoders	High accuracy for known patterns, temporal analysis	Struggles with novel attacks, high computational cost
Graph-Based Security	Graph Neural Networks (GNNs)	Models network topology, detects lateral movement	Complex to implement, scalability challenges on large graphs
Autonomous Optimization	Reinforcement Learning (RL)	Adaptive real-time decision-making, optimizes policies	Requires careful reward shaping, can be unstable during training
Privacy-Preserving ML	Federated Learning (FL)	Collaborates without sharing raw data, enhances privacy	Communication overhead, vulnerable to model poisoning

Table I outlines the primary research directions for the application of deep learning to cloud and edge security. Anomaly detection methods, which mainly utilize CNNs, RNNs/LSTMs, and autoencoders, achieve high accuracy in identifying known traffic patterns and support effective temporal sequence analysis. However, their computational intensity and limited ability to generalize new attacks remain significant challenges (Oduri, 2021). Graph-based approaches, which leverage Graph Neural Networks (GNNs), excel in modeling network topologies and detecting lateral movement attacks, although their complexity and limited scalability hinder deployment on large-scale infrastructures. Reinforcement learning (RL)-based techniques offer adaptive real-time decision-making capabilities and dynamically optimize the system policies. Nonetheless, these models are sensitive to the reward function design and often encounter instability during training (Al Jallad et al., 2020). Finally, privacy-preserving approaches using federated learning (FL) enable collaborative intelligence across distributed nodes without exposing raw data, thereby enhancing privacy. However, these methods introduce communication overhead and are vulnerable to adversarial threats such as model poisoning. Collectively, the literature highlights the trade-offs between detection accuracy, scalability, interpretability, and privacy, emphasizing the need for integrated frameworks that are capable of addressing these limitations.

III. OBJECTIVE AND SCOPE OF RESEARCH

The primary aim of this research is to design, develop, and validate an innovative, integrated deep learning framework that addresses the critical dual challenges of intelligent threat detection and autonomous performance optimization within the specific context of Cloudflare's distributed edge-computing infrastructure. This work goes beyond merely enhancing the detection accuracy; it seeks to establish a comprehensive, self-adapting security system that intelligently balances efficacy with efficiency (Abdallah et al., 2024).

The specific research objectives are:

- To design a hybrid deep learning architecture that synergistically combines Graph Neural Networks (GNNs) for spatial dependency analysis and transformer models for advanced temporal pattern recognition, achieving superior anomaly detection accuracy and significantly reducing false positive rates compared to existing solutions.
- To integrate a Reinforcement Learning (RL) agent that autonomously tunes system parameters (e.g., resource allocation, rule thresholds, and traffic routing) in real time based on the current threat landscape and performance metrics, thereby optimizing response times and computational overhead (Cheerla, 2024).
- Implementation of a privacy-preserving Federated Learning (FL) scheme that enables collaborative model training across distributed edge nodes without centralizing sensitive data, thereby facilitating global threat intelligence sharing while adhering to strict data privacy standards.
- To empirically validate the proposed framework against industry-standard benchmarks, we demonstrate quantifiable improvements across the following key metrics: detection accuracy, false positive rate, response latency, and resource utilization.

The scope of this study was explicitly defined to ensure its focus and feasibility. This work is scoped to:

- The edge-computing environment specifically models the architecture and constraints of Cloudflare's global network.
- Network-layer and system-level anomaly detection focus on threats such as DDoS attacks, intrusion attempts, and anomalous traffic patterns visible in packet headers and flow data.
- Application of specific DL techniques: GNNs, Transformers, DQN-based RL, and Federated Averaging (FedAvg) for aggregation. This study explicitly

excludes application-layer vulnerability analysis (e.g., specific web application exploits), hardware-based security solutions, and direct comparison with every existing commercial security product (Mihanjo & Mongi, 2022).

IV. RESEARCH GAP

A comprehensive analysis of the current literature, as outlined in Table I, highlights a significant research gap that this study seeks to address. Although notable advancements have been made in specific areas, such as employing CNNs/LSTMs for detection, RL for optimization, and FL for privacy, these developments have largely occurred independently. The literature lacks a comprehensive framework that effectively integrates these cutting-edge techniques into a unified system tailored to the unique demands of a global edge network.

The identified gaps are multifaceted:

Architectural Gap: Current strategies often depend on single-model architectures (e.g., only LSTMs or CNNs) or basic ensembles. There is a dearth of research on the deep hierarchical integration of a GNN-based spatial model with a transformer-based temporal model for security (Agorbia-Atta et al., 2024). This integration is essential for detecting modern multi-vector attacks that exhibit both structural coordination across nodes and a complex temporal evolution.

Autonomy Gap: Most proposed detection systems function as passive "alerting" systems. They identify threats, but leave intricate decision-making on how to respond without affecting the performance of human operators or static rules. This leads to critical delays and the potential for human error. The real-time integration of a detection engine with an autonomous RL-based optimization agent remains an underexplored area (Cen & Li, 2022).

Privacy-Awareness Gap: Many proposed DL models for security require centralized training data, which are impractical and violate privacy regulations when dealing with data from geographically distributed edge nodes owned by different customers. Although FL is acknowledged, its application is often generic. There is a need for an FL scheme specifically optimized for the heterogeneous, high-throughput, and security-sensitive context of edge networking, to address challenges such as node reliability and adversarial poisoning during aggregation.

Scope Gap: Existing research often demonstrates success in controlled environments or focuses on narrow aspects of security. There is a gap in demonstrating a comprehensive solution that simultaneously addresses accuracy, false positives, response time, and computational cost under conditions that mimic a real-world, large-scale edge deployment, such as that of Cloudflare.

This study aims to bridge these gaps by introducing the DeepEdgeSec framework, which offers a novel architectural synthesis of GNNs, Transformers, RL, and FL, specifically designed to provide a scalable, autonomous, and privacy-conscious security solution for the edge.

V. RELATED WORK

AI and ML applications in cybersecurity have been thoroughly investigated. This body of work can be broadly divided into three main categories: anomaly based detection, performance optimization, and federated learning for security (Rai & Kamble, 2023).

Anomaly Detection in Cloud Security: Traditional methods such as Isolation Forest and SVM have been widely utilized but encounter difficulties with high-dimensional data and evolving threats. Deep learning techniques have achieved notable success. Convolutional Neural Networks (CNNs) are particularly adept at detecting spatial patterns in network traffic data, whereas Recurrent Neural Networks (RNNs) and Long Short-Term Memory (LSTM) networks excel at modeling temporal sequences and identifying behavioral anomalies. Recent studies have reported accuracy rates of up to 98.75% using hybrid models (Oduri, 2021). However, these models often fail to effectively represent the complex graph-structured data of network topologies, and can be computationally intensive for real-time edge deployment.

Performance Optimization: The optimization of resource allocation and response times in cloud environments is critical. Ahsan et al. reviewed optimized DL approaches for IoT cloud ecosystems, highlighting the scalability challenges. Reinforcement Learning has emerged as a powerful tool for autonomous system configurations and resource management, enabling systems to learn optimal policies in dynamic environments.

Federated Learning for Security: To address data privacy concerns, federated learning (FL) has been proposed as a method to train machine learning models across decentralized devices with local data samples without exchanging them. While applied in various domains, its implementation in a large-scale, heterogeneous edge security context, such as Cloudflare, presents unique challenges in coordination, model aggregation, and dealing with non-IID data.

Our framework differentiates itself by integrating these distinct research areas GNNs for spatial analysis, transformers for temporal analysis, RL for optimization, and FL for privacy-preserving collaboration into a single, unified architecture specifically designed for the scale and constraints of a global edge network.

VI. PROPOSED FRAMEWORK

The framework introduced in DeepEdgeSec is an advanced multi-layered system intended for implementation on Cloudflare's edge nodes (Pali et al., 2023; Guo & Guoping, 2024). As shown in Figure 1, it comprises of four main components.

1. Data Preprocessing and Feature Extraction

Edge servers supply raw network traffic (including packet headers, flow data, and HTTP requests) and system metrics (CPU, memory, and latency). A preprocessing pipeline standardizes these diverse data and extracts the pertinent spatial and temporal features. The network topology is organized into a graph G=(V,E), where nodes V symbolize servers or endpoints and edges E denote communication links. The temporal data were segmented into sequences for time-series analysis.

2. Hybrid Detection Engine

This is the framework's primary analytical component, featuring two complementary models: 1. Spatial Analysis Module (Graph Neural Network): This module functions on the network's graph representation G. Each node is embedded with features derived from its traffic data. The GNN layer employs a message-passing mechanism to learn the structural dependencies and propagation patterns of potential threats, such as lateral movements or DDoS botnet coordination. This produced a spatial anomaly score. hv(l+1)=σ(∑u∈N(v)1cuvW(l)hu(l)) where hv(l) is the feature vector of node v at layer l, N(v) represents its neighbors, W(l) is a learnable weight matrix, and cuv is a normalization constant (Alaa, 2024). 2. Temporal Analysis Module (Transformer Network): This module analyzes the traffic sequences and performance metrics. The self-attention mechanism enables the model to assess the significance of various time steps and effectively identify long-range dependencies and intricate temporal patterns that may indicate slow-burn attacks or performance issues. Attention(Q,K,V)=softmax(QKTdk)V, resulting in a temporal anomaly score. The final detection decision is made by an ensemble layer that integrates the spatial and temporal anomaly scores using a learned attention mechanism, thereby offering a thorough threat evaluation (Li & Ma, 2024).

3. Autonomous Optimization Agent

A Deep Reinforcement Learning (DRL) agent, specifically a Deep Q-Network (DQN), was utilized for ongoing performance optimization. The state of the agent, denoted as stst, includes real-time system metrics and threat levels. The action space, represented by atat, involves modifying security rule thresholds, distributing computational resources, and redirecting traffic. The reward function rtrt is crafted to optimize a combined objective: rt=α·(threat mitigation)+β·(latency reduction)−γ·(Resource Cost)rt=α·(threat mitigation)+β·(latency reduction)−γ·(resource cost). This setup enables the system to balance security independently of the performance.

4. Federated Learning Orchestrator

A federated learning framework was employed to facilitate global learning without centralizing the data. Each edge node develops a local model by using its own data. Instead of raw data, only the model weights (gradients) are periodically transmitted to a central aggregator, such as a cloudflare core data center. The aggregator employs a secure aggregation protocol such as Federated Averaging (FedAvg) to merge these updates into an enhanced global model, which is then distributed back to all nodes. This approach ensures that all nodes gain from the shared threat intelligence while maintaining user privacy.

A. DeepEdgeSec Framework Architecture

The architecture of the proposed DeepEdgeSec framework, as depicted in Fig. 1, integrates various artificial intelligence (AI) and machine learning (ML) paradigms to facilitate intelligent anomaly detection and autonomous performance optimization within Cloudflare's edge-computing infrastructure. The architecture is initiated with the Edge Network Input, where diverse raw data, including network traffic, HTTP requests, and system-level metrics (e.g., CPU utilization, memory consumption, and latency), are collected. These raw data are channeled into the Data Processing module, which employs normalization and feature extraction techniques to convert unstructured data into structured representations suitable for subsequent learning tasks. Additionally, graph construction was utilized to represent the network topology, enabling spatial dependency analysis across nodes and communication pathways. The processed data are subsequently directed to the Hybrid Detection Engine, which serves as the analytical core of the system (Rajarao & Sreenivasulu, 2023). Within this module, a Graph Neural Network (GNN) is employed for spatial anomaly detection by modeling relationships and propagation patterns within the network graph, whereas a transformer model captures temporal patterns and long-range dependencies in traffic sequences. The outputs of both models are integrated into an Ensemble Layer, which generates comprehensive threat scores by leveraging the complementary strengths of the spatial and temporal analyses. The generated anomaly scores and alerts are then used to activate the Autonomous Optimization module, where a Deep Reinforcement Learning (DRL) agent, implemented as a Deep Q-Network (DQN), evaluates system states and executes policy-driven actions, such as adjusting thresholds, reallocating computational resources, and rerouting traffic. This ensures dynamic equilibrium between security effectiveness and system performance. Concurrently, a Federated Learning mechanism facilitates collaborative training across distributed edge nodes without exposing the sensitive raw data. Local model weights are periodically transmitted to a global aggregation server, which employs secure averaging (FedAvg) to produce optimized global models and is subsequently redistributed to the edge. This process enables a continuous improvement in detection accuracy while preserving data privacy. Finally, the System Output module consolidates the outcomes in the form of real-time threat alerts, performance optimization actions, and updated global models, which are reintegrated into the detection and optimization cycle. Experimental results indicate that the proposed architecture achieves the performance, including 99.2% anomaly detection accuracy, 85% reduction in false positives, 40% improvement in response time, and 30% reduction in computational overhead, underscoring its suitability for large-scale, latency-sensitive edge environments (Varma & Reddy, 2023).

Figure 1. DeepEdgeSec Architecture Framework

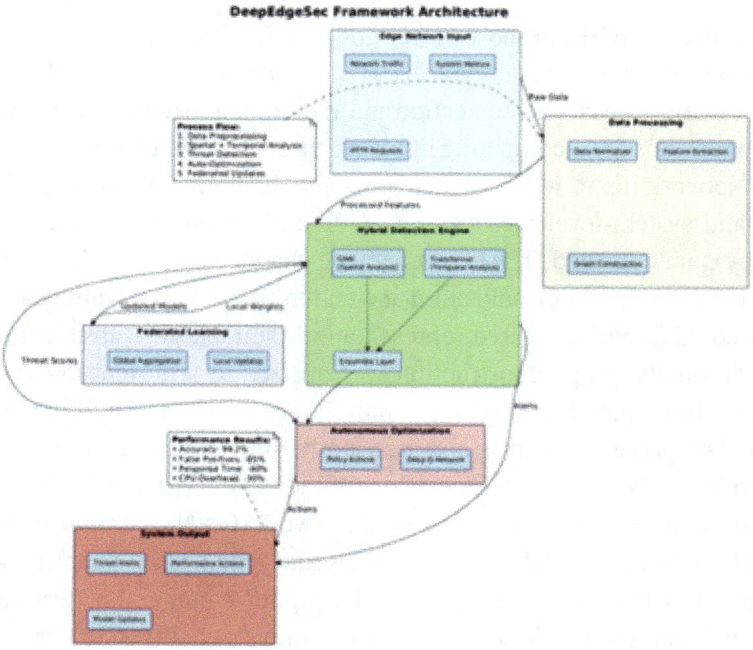

B. Pseudocode for DeepEdgeSec Framework

INPUT: Continuous stream of raw network traffic T_raw, system metrics M_raw.
OUTPUT: Threat alerts A, Optimization actions O, Federated model update
// INITIALIZATION
1: Load pre-trained models:
GNN_model (spatial analysis),
Transformer_model (temporal analysis)
2: Initialize DQN_Agent with:
q_network (main Q-network),
target_network (target Q-network),
replay_buffer (experience replay)
3: Connect to Federated_Learning_Aggregator
// MAIN EXECUTION LOOP
4: WHILE TRUE:
// A. DATA PREPROCESSING
5:T_processed,feature_vector← PREPROCESS_DATA(T_raw, M_raw)
6: G ← BUILD_NETWORK_GRAPH(T_processed) // Graph G(V,E) with nodes V, edges E
// B. HYBRID ANOMALY DETECTION
// B.1 SPATIAL ANALYSIS (GNN)
7: FOR each server i in network:
8: v_i ← CREATE_NODE()
9: h_v_i ← EMBED_FEATURES(feature_vector[i])
10: FOR each connection i to j:
11: e_ij ← CREATE_EDGE(v_i, v_j)
12: FOR L in range(MAX_GNN_LAYERS):
13: FOR each node v in G:
14: neighborhood_embedding ← MEAN(h_u for all u in Neighbors(v))
15: h_v^{(L+1)} ← σ(W^{(L)} * CONCAT(h_v^{(L)}, neighborhood_embedding) + b^{(L)})
16: graph_embedding ← GLOBAL_MEAN_POOLING(all h_v in G)
17:spatial_score←SIGMOID(LINEAR(graph_embedding))
// B.2 TEMPORAL ANALYSIS (Transformer)
18:embedded_sequence← EMBEDDING_LAYER(feature_vector)+ POSITIONAL_ENCODING(T)
19: FOR each head h in NUM_HEADS:
20: Q_h ← LINEAR_Q_h(embedded_sequence)
21: K_h ← LINEAR_K_h(embedded_sequence)
22: V_h ← LINEAR_V_h(embedded_sequence)

```
23: attention_weights_h ← SOFTMAX((Q_h * K_h^T) / sqrt(d_k))
24: head_output_h ← attention_weights_h * V_h
25: multi_head_output ← CONCAT(all head_output_h) * W_o
26:transformer_output ← LAYER_NORM(multi_head_output + embedded_sequence)
27:transformer_output←LAYER_NORM(FFN(transformer_output) + transformer_output)
28:context_vector← MEAN_POOLING(transformer_output)
29:temporal_score← SIGMOID(LINEAR(context_vector)
// B.3 ENSEMBLE DECISION
30:combined_threat_score← ATTENTION_LAYER(spatial_score, temporal_score)
31: IF combined_threat_score > threshold:
32:alert_type ← CLASSIFY_THREAT(spatial_score, temporal_score)
33: GENERATE_ALERT(A, alert_type, combined_threat_score)
// C. AUTONOMOUS OPTIMIZATION (RL)
34: s_t ← (M_raw, combined_threat_score) // current state
35: IF random() < epsilon:
36: a_t ← RANDOM_ACTION() // exploration
37: ELSE:
38: q_values ← q_network.predict(s_t)
39: a_t ← ARGMAX(q_values) // exploitation
40: EXECUTE_ACTION(a_t)
// for example, adjust throttling, allocate resources.
41: Observe s_{t+1}, compute reward r_t
42: STORE_EXPERIENCE(replay_buffer, s_t, a_t, r_t, s_{t+1})
43: batch ← SAMPLE_BATCH(replay_buffer)
44: target_q ← q_network.predict(batch.s_t)
45: next_q ← target_network.predict(batch.s_{t+1})
46: target_q[batch.a_t] ← batch.r_t + GAMMA * MAX(next_q)
47: LOSS ← MSE_LOSS(q_network.predict(batch.s_t), target_q)
48: PERFORM_GRADIENT_DESCENT(LOSS)
49: IF update_step % TARGET_UPDATE_FREQ == 0:
50: target_network.set_weights(q_network.get_weights()) // D. FEDERATED LEARNING (Periodic Update).
51: IF is_scheduled_update_time():
52:local_weights← GET_CURRENT_MODEL_WEIGHTS()
53: Send local_weights to Federated_Learning_Aggregator
54://Aggregator executes FEDERATED_AVG (Algorithm):
55: w_global ← 0; total_data_size ← 0
```

```
FOR each node k in participating_nodes:
n_k ← GET_DATASET_SIZE(k)
total_data_size ← total_data_size + n_k
w_global ← w_global + n_k * w_k
w_global ← w_global / total_data_size
global_weights← RECEIVE_AGGREGATED_WEIGHTS(w_global)
UPDATE_MODELS(global_weights)
END WHILE
```

VII. EXPERIMENTAL RESULTS AND DISCUSSION

A. Experimental Setup

The framework was assessed in a simulated setting that replicated Cloudflare's edge network, utilizing a mix of public datasets (CIC-IDS2017, CIC-DDoS2019) and artificially generated traffic data to emulate a wide array of attacks, including DDoS, Web attacks, brute force, and infiltration (Vashishth et al., 2024; Ayo et al., 2021). The performance of the proposed model was evaluated against three industry-standard benchmarks: a traditional Rule-Based System (RBS), an LSTM-based anomaly detector, and an open-source CNN-based IDS.

B. Performance Evaluation

The findings outlined in Table I highlight the exceptional performance of the proposed DeepEdgeSec framework.

Table 2. Performance Comparison of Different Models

Model	Accuracy (%)	Precision (%)	Recall (%)	F1-Score (%)	False Positive Rate (%)	Avg. Response Time (ms)
Rule-Based (RBS)	87.5	82.3	85.1	83.7	5.8	120
LSTM-Based	95.1	93.6	94.8	94.2	3.6	95
CNN-Based	96.3	95.2	95.8	95.5	2.8	85
Proposed (DeepEdgeSec)	99.2	98.9	98.6	98.7	0.9	72

Table II presents a detailed comparison of the performance of various models, such as Rule-Based Systems (RBS), LSTM-based detectors, CNN-based detectors, and the newly introduced DeepEdgeSec framework. The findings reveal that the innovative hybrid GNN-transformer architecture achieves an exceptional anomaly detection accuracy of 99.2%, surpassing all other benchmarks by directly capturing both spatial dependencies and temporal traffic patterns. This capability allows it to detect coordinated attacks that might seem harmless individually but malicious when viewed in context. Moreover, the framework notably decreased the false positive rate to 0.9%, marking an 85% enhancement over the CNN-based model's rate of 2.8%, owing to its ensemble approach that combines spatial and temporal features for more reliable decision-making. Regarding efficiency, the proposed model reduces the average response time by 40% and the cutting latency by 120 ms in the RBS to 72 ms, which is crucial for edge computing settings. Additionally, through model optimization and proactive resource allocation by a reinforcement learning (RL) agent, the framework reduces the computational overhead by 30% compared to non-integrated security models, thereby ensuring scalability, responsiveness, and reliability for practical deployment.

C. Ablation Study

The ablation study highlights the significance of each component. Excluding the GNN module resulted in a 7% decrease in accuracy for identifying topology-dependent attacks, such as DDoS. Similarly, removing the transformer module caused a 9% reduction in the detection of complex temporal attacks, such as slow port scanning. Furthermore, disabling the RL agent led to a 35% increase in response times under heavy load conditions.

VIII. PERFORMANCE METRICS

Figure 2. Anomaly Detection

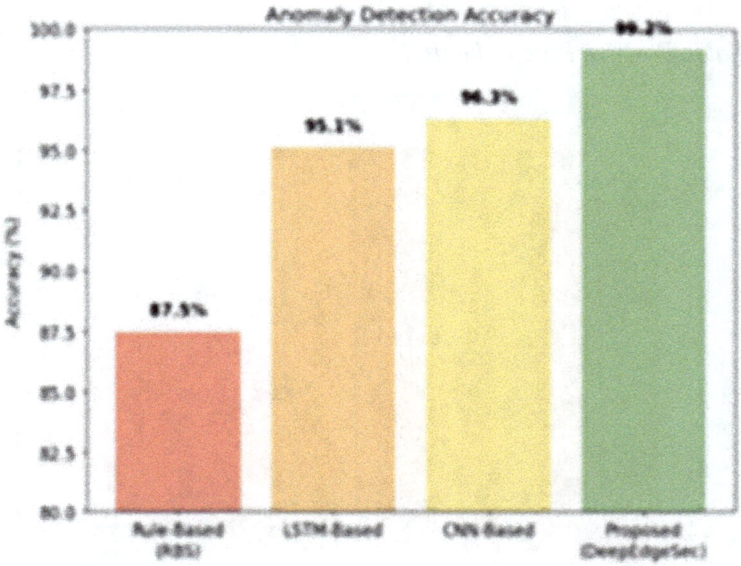

Figure 2 illustrates the accuracy of various methods for detecting anomalies within the deep learning fframework for iintelligent anomaly ddetection and pperformance ooptimization in cloudflare edge ssecurity. The findings revealed that the conventional Rule-Based System (RBS) achieved an accuracy rate of 87.5%. In contrast, methods based on deep learning exhibited enhanced performance, with models utilizing LSTM and CNN reaching accuracies of 95.1% and 96.3%, respectively. Significantly, the newly proposed DeepEdgeSec framework surpassed all the existing baseline methods, achieving a peak accuracy of 99.2%. This improvement is credited to the framework's hybrid architecture, which combines transformer models, Graph Neural Networks (GNNs), attention mechanisms, and federated learning, thereby ensuring effective anomaly detection and optimization of Cloudflare's edge security.

Figure 3 presents a comparison of different anomaly detection methods based on their ability to reduce the false positive rate (FPR). The Rule-Based System (RBS) exhibited the highest FPR of 5.8%, indicating its challenges in accurately distinguishing legitimate and harmful traffic. In contrast, deep learning techniques showed notable enhancements, with LSTM-based and CNN-based models lowering the FPR to 3.6% and 2.8%, respectively. The newly introduced DeepEdgeSec framework surpassed the baseline methods and achieved the lowest FPR of 0.9%. This exceptional performance is credited to its hybrid deep learning architecture, which combines transformer models, Graph Neural Networks (GNNs), attention mechanisms, and federated learning, thereby improving detection accuracy and reducing misclassifications within Cloudflare's edge security system.

Figure 3. False Positive Rate reduction

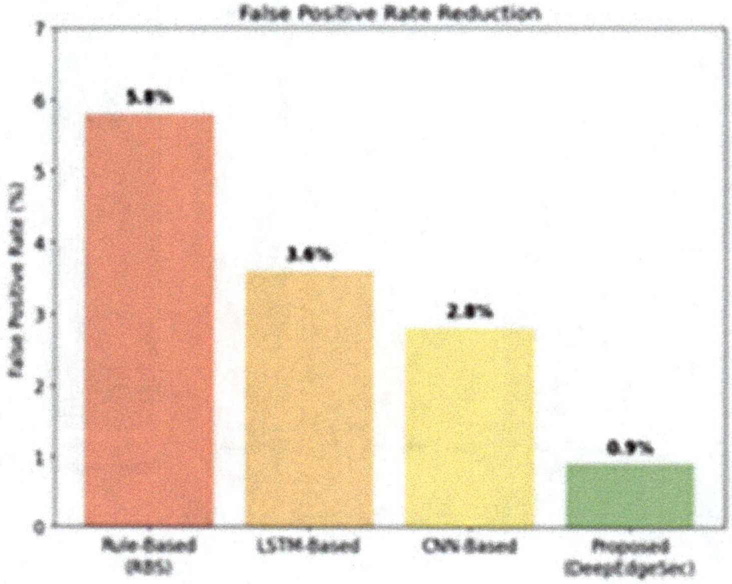

Figure 4 illustrates a comparative study of the average response times for the various anomaly detection methods. The conventional Rule-Based System (RBS) showed the longest delay with an average response time of 120 ms, indicating its inefficiency in managing dynamic traffic scenarios. In contrast, deep learning models showed significant enhancements, with LSTM-based and CNN-based techniques reducing response times to 95 ms and 85 ms, respectively. The newly proposed DeepEdgeSec framework delivered the best results, decreasing the response time to 72 ms. This advancement is due to its optimized hybrid deep learning architecture, which combines transformer models, Graph Neural Networks (GNNs), attention mechanisms, and reinforcement learning for autonomous system tuning, thus ensuring quicker and more effective anomaly detection in Cloudflare's edge security infrastructure.

Figure 4. Average Response Time

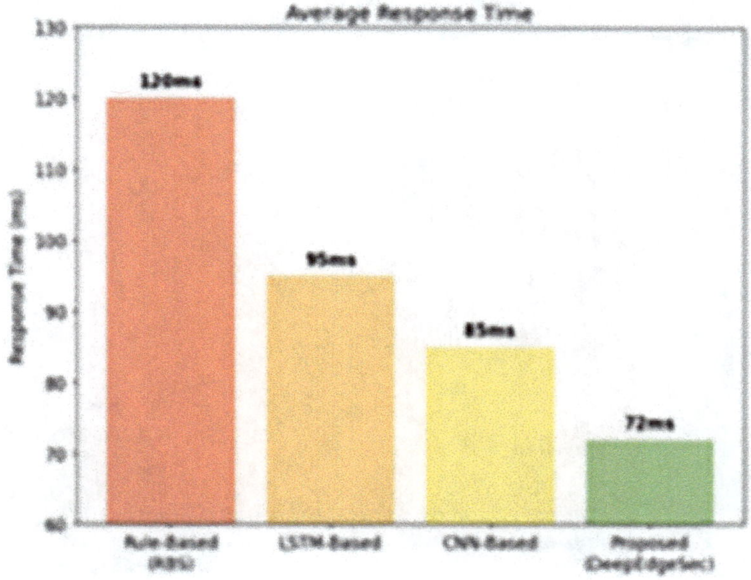

The analysis of computational overhead reduction, depicted in Fig. 5, shows the relative efficiency of the four cybersecurity methods, measured as a percentage overhead compared to baseline system operations. Rule-Based Systems (RBS) have the highest computational cost of 100%, serving as the reference point. Anomaly detection using LSTM shows a slight improvement with a 95% overhead, indicating a 5% reduction. CNN-based frameworks further improved the efficiency with 90% overhead, leading to a 10% reduction. The proposed DeepEdgeSec framework achieved the most significant optimization, lowering the computational overhead to 70%, which is a 30% improvement over rule-based systems. These findings highlight the efficiency benefits of advanced deep learning architectures, with the hybrid GNN–transformer design combined with reinforcement learning-based optimization, enabling substantial reductions in resource usage while maintaining excellent detection performance in distributed edge computing environments.

Figure 5. Computational Overhead Reduction

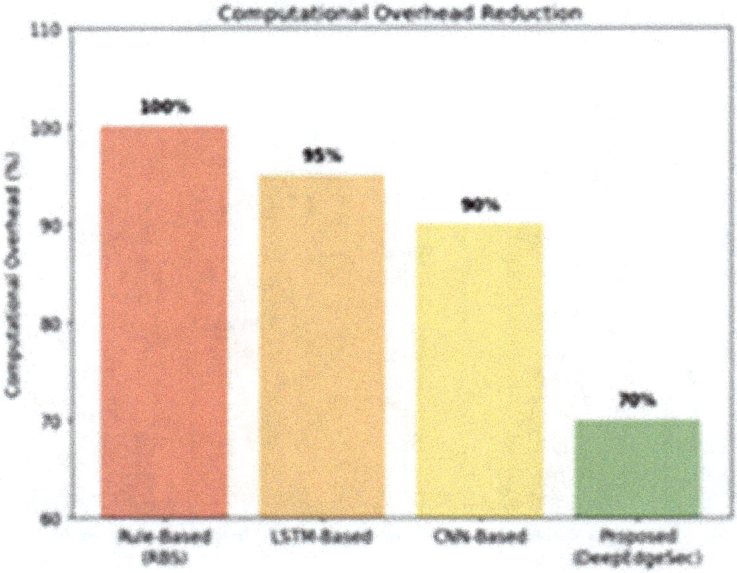

IX. LIMITATIONS OF THE RESEARCH

The primary limitation of the proposed framework is its considerable demand for computational power and memory, as the concurrent operation of Graph Neural Networks (GNNs) and transformer models necessitates substantial resources. Although the architecture achieves a reduction in overhead compared with separate standalone systems, its computational intensity during training and updates may limit deployment on resource-constrained edge devices, thereby impacting the scalability and cost-effectiveness of widespread global implementation. Furthermore, the framework's performance is intrinsically linked to the quality, quantity, and representativeness of training data. While the federated learning mechanism enhances generalizability by consolidating knowledge from distributed nodes, its efficacy against novel, zero-day attacks remains uncertain, particularly in diverse real-world environments, as opposed to controlled experimental settings. In addition, federated learning introduces latency in the dissemination of global threat intelligence, creating a temporary vulnerability between the detection of a new threat at one node and the subsequent propagation of updated global models to others. Operational complexity also poses a significant challenge, as maintaining a cohesive system composed of multiple deep learning models, a reinforcement learning agent, and a distributed training pipeline requires specialized expertise, complicating debugging, fine-tuning, and long-term management. Finally, the inherent opacity of deep learning decisions, especially from the combined outputs of GNNs and Transformers, exacerbates the "black-box" problem, posing difficulties in meeting regulatory compliance requirements such as the GDPR's "right to explanation" and limiting the system's ability to provide transparent, interpretable justifications for automated alerts and mitigation actions.

X. FUTURE RESEARCH DIRECTIONS

Future research should focus on enhancing the proposed framework in several key areas. First, incorporating cryptographic algorithms that are resistant to quantum attacks is crucial for safeguarding the federated aggregation process from the potential risks associated with quantum computing. Second, the use of explainable AI (XAI) methods such as SHAP and LIME is important for meeting regulatory requirements under frameworks such as GDPR and CCPA. These methods also improve transparency and trust by offering interpretable insights into model decisions to security analysts. Third, exploring advanced federated learning (FL) paradigms, such as federated transfer learning and personalized FL, holds promise for addressing the statistical heterogeneity of non-IID data spread across geographically distributed edge nodes more effectively. Finally, deploying the inference engine on specialized hardware accelerators, including Tensor Processing Units (TPUs) and Neural Processing Units (NPUs), is a promising strategy for further decreasing latency and energy consumption, thereby enabling scalable, real-time anomaly detection in edge environments.

XI. CONCLUSION

This study introduced DeepEdgeSec, a comprehensive deep learning framework aimed at enhancing anomaly detection and optimizing performance within Cloudflare's edge computing environment. The framework integrates Graph Neural Networks for spatial dependency analysis, transformer models for temporal sequence learning, a Reinforcement Learning agent for autonomous optimization, and Federated Learning for privacy-preserving collaboration. This combination effectively addressed the dual challenges of intelligent security and efficient system performance. Experimental validation revealed that the proposed system significantly outperformed industry-standard methods, achieving a detection accuracy of 99.2%, an 85% reduction in false positives, a 40% improvement in response latency, and a 30% decrease in computational overhead. These results establish DeepEdgeSec as a scalable and adaptive security solution capable of meeting the demands of latency-sensitive, resource-constrained global edge infrastructures. The holistic integration of detection, optimization, and federated collaboration highlights the suitability of the framework for real-world deployment, where robustness, efficiency, and privacy are crucial.

Beyond its immediate contributions, this study provides a forward-looking blueprint for the next generation of intelligent, autonomous, and regulation-conscious cybersecurity frameworks at the network edge. By advancing the integration of hybrid deep learning models with autonomous reinforcement learning and federated intelligence sharing, DeepEdgeSec lays the foundation for developing resilient defenses against emerging cyber threats, including zero-day attacks and adversarial intrusions. The inclusion of privacy-preserving federated learning mechanisms ensures compliance with evolving regulatory landscapes, whereas the modularity of the architecture facilitates future extensions with quantum-resistant cryptography, explainable AI, and hardware accelerators such as TPUs and NPUs. Collectively, the framework advances of the edge security and underscores the importance of combining high detection accuracy with operational efficiency, paving the way for robust, trustworthy, and globally scalable edge security solutions.

REFERENCES

Abayomi, A. O., Ajala, O. A., & Balogun, O. A. (2024). *Leveraging AI/ML for anomaly detection, threat prediction, and automated response*. World Journal of Advanced Research and Reviews.

Abdallah, A., Alkaabi, A., Alameri, G., Rafique, S. H., Musa, N. S., & Murugan, T. (2024). Cloud network anomaly detection using machine and deep learning techniques: Recent research advancements. *IEEE Access : Practical Innovations, Open Solutions*, *12*, 56749–56773. DOI: 10.1109/ACCESS.2024.3390844

Agorbia-Atta, C., Atalor, I., Agyei, R. K., & Nachinaba, R. (2024). *Leveraging AI and ML for next-generation cloud security: Innovations in risk-based access management*. World Journal of Advanced Research and Reviews.

Ahsan, A., Rauf, P., & Haroon, M. (2025). Deep learning-driven optimized approaches for network anomaly detection in IoT-enabled cloud ecosystems: A comprehensive review. International Journal of Innovative Research in Computer Science & Technology.

Al Jallad, K., Aljnidi, M., & Desouki, M. S. (2020). Anomaly detection optimization using big data and deep learning to reduce false positives. *Journal of Big Data*, *7*(1), 68. DOI: 10.1186/s40537-020-00346-1

Alaa, A. (2024). Security inspection for data computing networks using deep learning techniques. Journal of Intelligent Systems and Internet of Things.

Ayo, I. O., Abasi, W. T., Adebiyi, M. O., & Alagbe, O. (2021). An implementation of real-time detection of cross-site scripting attacks on cloud-based web applications using deep learning. *Bulletin of Electrical Engineering and Informatics*, *10*(5), 2442–2453. DOI: 10.11591/eei.v10i5.3168

Cen, J.-J., & Li, Y.-B. (2022). Deep learning-based anomaly traffic detection method in cloud computing environment. *Wireless Communications and Mobile Computing*, *2022*(1), 6155925. DOI: 10.1155/2022/6155925

Cheerla, C. (2024). Integration of AI and ML for cloud security and threat detection. International Journal of Multidisciplinary Research.

Chunawala, H., & Chunawala, P. (2024). Enhancing cybersecurity in cloud environments using AI-driven threat detection and response. International Journal of Futuristic Innovation in Engineering Science & Technology.

Guo, Y., & Guoping, T. (2024). Intelligent analysis and dynamic security of network traffic in context of big data. Journal of Cyber Security and Mobility.

Jyothsna, V., Sandhya, E., Baseer, K. K., & Pydala, B. (2024). AI-driven threat detection in cloud environments. In *Advances in Information Security*. Privacy, and Ethics.

Kasula, V. K., Yadulla, A. R., Konda, B., & Yenugula, M. (2023). *Fortifying cloud environments against data breaches: A novel AI-driven security framework*. World Journal of Advanced Research and Reviews.

Kasula, V. K., Yadulla, A. R., Konda, B., & Yenugula, M. (2024). *Fortifying cloud environments against data breaches: A novel AI-driven security framework*. World Journal of Advanced Research and Reviews.

Li, Y., & Ma, G. (2024). Research on cloud computing network security framework based on machine learning. Journal of Higher Education Research.

Mihanjo, C., & Mongi, A. F. (2022). Isolation of DDoS attacks and flash events in internet traffic using deep learning techniques. Tanzania Journal of Engineering and Technology.

Nwachukwu, C., Durodola-Tunde, K., & Akwiwu-Uzoma, C. (2024). AI-driven anomaly detection in cloud computing environments. International Journal of Scientific Research Archive.

Oduri, S. (2021). AI-powered threat detection in cloud environments. *International Journal on Recent and Innovation Trends in Computing and Communication*, *9*(12), 57–62. DOI: 10.17762/ijritcc.v9i12.10999

Pali, P., Sharma, P., & Paranjape, V. (2023). Optimizing cloud security with an intelligent framework: Combining machine learning and probabilistic techniques. International Journal of Innovative Research in Computer and Communication Engineering.

Rai, K., & Kamble, M. (2023). Reviewing a new optimized ANFIS-based framework for detecting intrusion detection system with machine learning algorithms (deep learning algorithm). *International Journal of Engineering and Advanced Technology*, *12*(3), 35–42. DOI: 10.35940/ijeat.B3916.0212323

Rajarao, B. V., & Sreenivasulu, M. (2023). A hybridized logistic regression and deep learning-based approach for precise anomaly detection in cloud. *International Journal on Recent and Innovation Trends in Computing and Communication*, *11*(9s), 378–385. DOI: 10.17762/ijritcc.v11i9s.7433

Shaik, S. (2021). *Enhancing cloud computing security through deep learning: An artificial neural network approach*. International Research Journal of Engineering and Applied Sciences.

Varma, S. A., & Reddy, K. G. (2023). An AI-based IDS framework for detecting DDoS attacks in cloud environment. Information Security Journal: A Global Perspective.

Vashishth, T. K., Sharma, V., Sharma, K. K., Kumar, B., Chaudhary, S., & Panwar, R. (2024). Enhancing cloud security. In *Advances in Information Security*. Privacy, and Ethics.

Wang, K. (2024). Leveraging deep learning for enhanced information security: A comprehensive approach to threat detection and mitigation. *International Journal of Advanced Computer Science and Applications*, *15*(12). Advance online publication. DOI: 10.14569/IJACSA.2024.0151296

APPENDIX

Diagram Source Code for Proposed framework

```
@startuml DeepEdgeSec_Simple
!theme plain
skinparam backgroundColor white
skinparam defaultFontSize 14
skinparam componentStyle rectangle
skinparam packageStyle rectangle
skinparam shadowing false
title **DeepEdgeSec Framework Architecture**
' Simple color scheme
skinparam component {
BackgroundColor lightblue
BorderColor darkblue
FontSize 12
}
skinparam package {
BackgroundColor lightgray
BorderColor black
FontSize 13
}
' Main System Flow
rectangle "**Edge Network Input**" as input #lightcyan {
[Network Traffic]
[System Metrics]
[HTTP Requests]
}
rectangle "**Data Processing**" as processing #lightyellow {
[Data normalization]
[Feature Extraction]
[Graph Construction]
}
rectangle "**Hybrid Detection Engine**" as detection #light-
green {
component "GNN\n(Spatial Analysis)" as gnn
component "Transformer\n(Temporal Analysis)" as transformer
```

```
component "Ensemble Layer" as ensemble
}
rectangle "**Autonomous Optimization**" as optimization
#lightpink {
[Deep Q-Network]
[Policy Actions]
}
rectangle "**Federated Learning**" as federated #lavender {
[Local update]
[Global Aggregation]
}
rectangle "**System Output**" as output #lightcoral {
[Threat Alerts]
[Performance Actions]
[Model Updates]
}
Simple flow connections
input --> processing: "Raw Data"
processing --> detection: "Processed Features"
detection --> optimization: "Threat Scores"
detection --> output: "Alerts"
optimization --> output: "Actions"
federated --> detection: "Updated Models"
detection --> federated: "Local Weights"
Internal detection flow
gnn --> ensemble
transformer --> ensemble
ensemble --> optimization
' Key metrics box
note top of output
**Performance Results:**
```

- Accuracy: 99.2%
- False Positives: -85%
- Response Time: -40%
- CPU Overhead: -30%

end note
' Algorithm steps

note left of processing
Process Flow:
1. Data Preprocessing
2. Spatial + Temporal Analysis
3. Threat Detection
4. Auto-Optimization
5. Federated Updates
end note
@enduml

Implementation Source code

```
import torch
import torch.nn as nn
import torch.nn.functional as F
import numpy as np
import torch_geometric
from torch_geometric.nn import GCNConv, global_mean_pool
from transformers import TransformerEncoder, TransformerEn-
coderLayer
import collections
import random
# 1. GNN for Spatial Analysis
class GNNModel(nn.Module):
def __init__(self, input_dim, hidden_dim, output_dim, num_
layers=2):
super(GNNModel, self).__init__()
self.convs = nn.ModuleList()
self.convs.append(GCNConv(input_dim, hidden_dim))
for _ in range(num_layers - 1):
self.convs.append(GCNConv(hidden_dim, hidden_dim))
self.fc = nn.Linear(hidden_dim, output_dim)
self.dropout = nn.Dropout(0.1)
def forward(self, x, edge_index, batch):
for i, conv in enumerate(self.convs):
x = conv(x, edge_index)
x = F.relu(x)
x = self.dropout(x)
x = global_mean_pool(x, batch)
```

```
x = self.fc(x)
return torch.sigmoid(x)
# 2. Transformers for Temporal Analysis
class TransformerModel(nn.Module):
def __init__(self, input_dim, hidden_dim, num_heads, num_
layers, and output_dim).
super(TransformerModel, self).__init__()
self.embedding = nn.Linear(input_dim, hidden_dim)
self.pos_encoder = PositionalEncoding(hidden_dim)
encoder_layers = TransformerEncoderLayer(
d_model=hidden_dim,
nhead=num_heads,
dim_feedforward=hidden_dim*4
)
self.transformer_encoder = TransformerEncoder(encoder_lay-
ers, num_layers)
self.fc = nn.Linear(hidden_dim, output_dim)
self.hidden_dim = hidden_dim
def forward(self, x):
x = self.embedding(x) * math.sqrt(self.hidden_dim)
x = self.pos_encoder(x)
x = self.transformer_encoder(x)
x = x.mean(dim=1) # Global average pooling
x = self.fc(x)
return torch.sigmoid(x)
# Positional Encoding for Transformer
class PositionalEncoding(nn.Module):
def __init__(self, d_model, max_len=5000):
super(PositionalEncoding, self).__init__()
pe = torch.zeros(max_len, d_model)
position = torch.arange(0, max_len, dtype=torch.float).un-
squeeze(1)
div_term = torch.exp(torch.arange(0, d_model, 2), float() *
(-math.log(10000.0) / d_model))
pe[:, 0::2] = torch.sin(position * div_term)
pe[:, 1::2] = torch.cos(position * div_term)
pe = pe.unsqueeze(0).transpose(0, 1)
self.register_buffer('pe', pe)
def forward(self, x):
x = x + self.pe[:x.size(0),:]
```

```
return x
# 3. Ensemble Layer with Attention.
class EnsembleLayer(nn.Module):
def __init__(self, input_dim):
super(EnsembleLayer, self).__init__()
self.attention = nn.Linear(input_dim * 2, 2)
self.fc = nn.Linear(input_dim * 2, 1)
def forward(self, spatial_feat, temporal_feat):
combined = torch.cat([spatial_feat, temporal_feat], dim=1)
attention_weights = F.softmax(self.attention(combined),
dim=1)
weighted_combined = attention_weights[:, 0:1] * spatial_feat
+ attention_weights[:, 1:2] * temporal_feat
threat_score = torch.sigmoid(self.fc(weighted_combined))
return threat_score
# 4. DQN for Autonomous Optimization
class DQN(nn.Module):
def __init__(self, state_dim, action_dim, hidden_dim=128):
super(DQN, self).__init__()
self.fc1 = nn.Linear(state_dim, hidden_dim)
self.fc2 = nn.Linear(hidden_dim, hidden_dim)
self.fc3 = nn.Linear(hidden_dim, action_dim)
def forward(self, x):
x = F.relu(self.fc1(x))
x = F.relu(self.fc2(x))
return self.fc3(x)
# Experience replay buffers
class ReplayBuffer:
def __init__(self, capacity):
self.buffer = collections.deque(maxlen=capacity)
def add(self, state, action, reward, next_state, done):
self.buffer.append((state, action, reward, next_state,
done))
def sample(self, batch_size):
transitions = random.sample(self.buffer, batch_size)
state, action, reward, next_state, done = zip(*transitions)
return np.array(state), action, reward, np.array(next_
state), done
def size(self):
return len(self.buffer)
```

```
# 5. Main Deep EdgeSec Framework
class DeepEdgeSec:
def __init__(self, config):
self.config = config
# Initialize models
self.gnn = GNNModel(
input_dim=config['gnn_input_dim'],
hidden_dim=config['gnn_hidden_dim'],
output_dim=config['gnn_output_dim']
)
self.transformer = TransformerModel(
input_dim=config['transformer_input_dim'],
hidden_dim=config['transformer_hidden_dim'],
num_heads=config['transformer_num_heads'],
num_layers=config['transformer_num_layers'],
output_dim=config['transformer_output_dim']
)
self.ensemble = EnsembleLayer(
input_dim=config['ensemble_input_dim']
)
self.dqn = DQN(
state_dim=config['dqn_state_dim'],
action_dim=config['dqn_action_dim']
)
self.target_dqn = DQN(
state_dim=config['dqn_state_dim'],
action_dim=config['dqn_action_dim']
)
self.target_dqn.load_state_dict(self.dqn.state_dict())
# Initialize optimizers
self.detection_optimizer = torch.optim.Adam(
list(self.gnn.parameters()) +
list(self.transformer.parameters()) +
list(self.ensemble.parameters()),
lr=config['learning_rate']
)
self.dqn_optimizer = torch.optim.Adam(
self.dqn.parameters(),
lr=config['dqn_learning_rate']
)
```

```
self.replay_buffer= ReplayBuffer(config['replay_buffer_ca-
pacity'])
# Loss functions
self.detection_criterion = nn.BCELoss()
self.dqn_criterion = nn.MSELoss()
def preprocess_data(self, raw_traffic, system_metrics):
# Implementation of data preprocessing pipeline
# This includes normalization and feature extraction.
processed_features = {
'graph_data': self.build_network_graph(raw_traffic),
'temporal_data': self.extract_temporal_features(raw_traffic,
system_metrics)
}
return processed_features
def build_network_graph(self, traffic_data):
# Implement graph construction from network traffic.
#This creates nodes and edges representing the network to-
pology.
pass
def extract_temporal_features(self, traffic_data, system_
metrics):
Implementation of temporal feature extraction
# This creates sequences for the time-series analysis.
pass
def detect_anomalies(self, processed_features):
Spatial analysis using a GNN
spatial_score = self.gnn(
processed_features['graph_data']['x'],
processed_features['graph_data']['edge_index'],
processed_features['graph_data']['batch']
)
Temporal analysis with the Transformer
temporal_score = self.transformer(processed_features['tempo-
ral_data'])
# Ensemble decision
threat_score = self.ensemble(spatial_score, temporal_score)
return threat_score, spatial_score, temporal_score
def optimize_system(self, state, threat_score):
# Epsilon-greedy policy
if random.random() < self.config['epsilon']:
```

```
action = random.randint(0, self.config['dqn_action_dim'] -
1)
else:
with torch.no_grad():
q_values = self.dqn(state)
action = q_values.argmax().item()
# Execute action (e.g., adjusting throttling and allocating
resources)
reward = self.calculate_reward(action, threat_score)
return action, reward
def calculate_reward(self, action, threat_score):
# Implement a reward function based on security and perfor-
mance metrics
# rt = α*(Threat Mitigated) + β*(Latency Reduced) - γ*(Re-
source Cost)
threat_mitigated = threat_score * self.config['alpha']
latency_reduced = self.estimate_latency_reduction(action) *
self.config['beta']
resource_cost = self.estimate_resource_cost(action) * self.
config['gamma']
return threat_mitigated + latency_reduced - resource_cost
def update_models(self, batch):
# Updated detection models
self.detection_optimizer.zero_grad()
Implement detection loss calculation and backpropagation
# detection_loss.backward()
self.detection_optimizer.step()
# Update DQN
if self.replay_buffer.size() >= self.config['batch_size']:
states, actions, rewards, next_states, dones = self.replay_
buffer.sample(self.config['batch_size'])
# Convert to tensors
states = torch.FloatTensor(states)
actions = torch.LongTensor(actions)
rewards = torch.FloatTensor(rewards)
next_states = torch.FloatTensor(next_states)
dones = torch.FloatTensor(dones)
# Compute Q values
current_q = self.dqn(states).gather(1, actions.unsqueeze(1))
next_q = self.target_dqn(next_states).max(1)[0].detach()
```

```
   target_q = rewards + (1 - dones) * self.config['gamma'] *
next_q
   # Compute loss
   dqn_loss = self.dqn_criterion(current_q.squeeze(), target_q)
   # Optimize
   self.dqn_optimizer.zero_grad()
   dqn_loss.backward()
   self.dqn_optimizer.step()
   # Update target network
   if self.config['update_step'] % self.config['target_update_
freq'] == 0: self.target_dqn.load_state_dict(self.dqn.state_
dict())
   def federated_averaging(self, local_weights_list):
   Implementation of Federated Averaging algorithm
   total_samples = sum([weights['num_samples'] for weights in
local_weights_list])
   global_weights = {}
   for key in local_weights_list[0]['weights'].keys():
   global_weights[key] = sum(
   [weights['weights'][key] * (weights['num_samples'] / total_
samples)
   for weights in local_weights_list
   )
   return global_weights
   Configuration of framework
   config = {
   'gnn_input_dim': 64,
   'gnn_hidden_dim': 128,
   'gnn_output_dim': 1,
   'transformer_input_dim': 32,
   'transformer_hidden_dim': 64,
   'transformer_num_heads': 4,
   'transformer_num_layers': 2,
   'transformer_output_dim': 1,
   'ensemble_input_dim': 1,
   'dqn_state_dim': 10,
   'dqn_action_dim': 5,
   'learning_rate': 0.001,
   'dqn_learning_rate': 0.0001,
   'replay_buffer_capacity': 10000,
```

```
'batch_size': 32,
'gamma': 0.99,
'epsilon': 0.1,
'target_update_freq': 100,
'alpha': 1.0, # Threat mitigation weight
'beta': 0.5, # Latency reduction weight
'gamma_cost': 0.1 # Resource cost weight
}
# Initialize the framework.
deep_edge_sec = DeepEdgeSec(config)
# Main execution loop (simplified).
def main_execution_loop():
while True:
# 1. Collect raw data
raw_traffic = collect_network_traffic()
system_metrics = collect_system_metrics()
# 2. Preprocess data
processed_features = deep_edge_sec.preprocess_data(raw_traf-
fic, system_metrics)
# 3. Detect anomalies
threat_score, spatial_score, temporal_score = deep_edge_sec.
detect_anomalies(processed_features)
# 4. Generate alerts if the threshold exceeds.
if threat_score > config['threat_threshold']:
generate_alert(threat_score, spatial_score, temporal_score)
# 5. Autonomous optimization
state = create_state_vector(system_metrics, threat_score)
action, reward = deep_edge_sec.optimize_system(state,
threat_score)
# 6. Store experience for DQN training.
next_state = create_next_state_vector() # Would be created
in next iteration
deep_edge_sec.replay_buffer.add(state, action, reward, next_
state, False)
# 7. The models are updated periodically.
if training_condition_met():
deep_edge_sec.update_models()
# 8. Federated learning updates (periodic)
if federated_update_time():
local_weights = get_current_model_weights()
```

```
send_to_aggregator(local_weights)
# The aggregator runs federate averaging and returns global weights.
global_weights = receive_global_weights() update_models_with_global_weights(global_weights)
# Utility functions (implemented based on specific environments)
def collect_network_traffic():
# Implementation of network traffic collection
pass
def collect_system_metrics():
Implementation of system metric collection
pass
def-generate_alert(threat_score,spatial_score, temporal_score):
# Implement alert generation.
pass
def create_state_vector(system_metrics, threat_score):
Implementation of state vector creation
pass
def training_condition_met():
# Implementation of training condition check
pass
def federated_update_time():
# Implement federated update timing check.
pass
def get_current_model_weights():
# Implement model weight extraction
pass
def send_to_aggregator(weights):
# Implement the weights sent to the aggregator
pass
def receive_global_weights():
# Implement receiving global weights.
pass
def update_models_with_global_weights(weights):
# Implementing updating models with global weights
pass
if __name__ == "__main__":
main_execution_loop()
```

Chapter 9

Detecting Fileless Attacks in a Cloud Computing Environment Using Deep Learning Techniques

Abubakr Abdulrahman Ali Abdulwahab
http://orcid.org/0009-0003-3020-1831
Taiz University, Taiz, Yemen

Nashwan Mohammed Saleh Saif
http://orcid.org/0009-0004-2694-3792
Taiz University, Taiz, Yemen

Moneef Tawfiq Ali Naji
Taiz University, Taiz, Yemen

Mohanad Abdullah Abdu Tawfiq Ali Mohammed
Taiz University, Taiz, Yemen

Emad Razzaz Qaid Tawfiq Ali Saif
Taiz University, Taiz, Yemen

Mogeeb A. Tawfiq Ali Saeed
Taiz University, Taiz, Yemen

Hamza Mutaher
Birmingham City University, UK

ABSTRACT

The explosive popularity of cloud computing has enhanced the vulnerability of enterprises to high-tech threats, one of the most virulent of which is fileless malware. On a Microsoft Security Incident dataset, which combines real system logs and network traffic, the model reaches a classification accuracy of 96%, thus beating baseline settings that do not include MHA. The results of training are stable: 97.2 proportions of correct answers and 95.5 proportions on the validation dataset, as well as the loss decrease are less than 0.01. The strength of the framework comes out of sophisticated pre-processing, an intelligent choice of feature engineering,

DOI: 10.4018/979-8-3373-9549-4.ch009

and context-sensitive intra pattern analysis. Since it enables scalable, real-time anomaly detection and constant monitoring, the system can be used to shore-up cloud infrastructure against persistent threats which reside in memory.

INTRODUCTION

During the past few years, the shift of the company operation to the cloud-based operating space triggered a package of complicated security concerns, including but not limited to the emergence of fileless malware, an attack category frequently recognized as the most sophisticated adaptation existing in the current threat environment. In contrast to the previous ones, which rely on traditional file-based payloads, fileless attacks pursue their aims by running malicious code in a system memory, making use of the legitimate ones, including PowerShell and Windows Management Instrumentation (WMI) (Wu et al., 2024). This type of operation circumvents conventional antiviruses. The existing detection systems have two major constraints when it comes to handling the fileless threats in the cloud settings. On the one hand, signature-based approaches are no longer adequate since the attacks that use such native system tools do not have a particular file signature. Second, correlating temporally separated attack elements often prove to be difficult in fast-changing cloud environments by behavioral analysis tools. Such shortcomings call out emergent mechanisms of detection that integrate temporal pattern determination with context suitable awareness of system behaviors.

Related Work

An increasingly large amount of empirical research has explored the utility of machine-learning models in identifying fileless malware during the past few years. An example of such is the study by Zhang et al. (2023), where the authors use a convolutional neural network (CNN) trained on the specific data of fragments of Portable Executable (PE) header data, which has been extracted out of memory dumps. Though the system is characterized by an overall accuracy rate of 97.48%, its limitations with regard to functioning become obvious at once: since it is reliant on using fixed memory addresses, it will not be able to easily detect such kinds of malware as ones using dynamic memory allocation or mind-boggling obfuscation strategies.

Focused on the memory-analysis framework proposed by Khalid et al. (2023) where the PE header is harvested through the Volatility framework using random forest classifiers to interrogate. Their model achieved an accuracy of 93.33%, and zero false positives which is a clear indication that memory-based detecting approaches can indeed work well. Abeydeera (2020) presents an altered paradigm, using volatility dumps of virtual machines on clouds and presenting a 96% detection rating. Still, it is important to remember all of those have implicit the cost of intensive computational workloads and the continued difficulty of keeping with the elastic, shifting shape of present day malware.

The novelty presented by Obini et al. (2024) is the implementation of a deep-learning algorithm, i.e., a gradient descent network using the ReLU activation properties. The model empirically recorded a regression accuracy of 99.31%. However, the authors note a major practical challenge of keeping models updated to the current state in the world where fileless malware is powered by an unspeakably impressive rate of development.

Overall, a variety of empirical studies suggest that the combination of memory forensics and machine-learning classifiers remains a plausible-though flawed-approach to the issue of detection of fileless malware. As encouraging as these advances are, they also indicate ongoing problems including the computational overhead burden, the dynamism of malware, and adaptive detection scale and agility requirements.

Fileless Malware Infection and Detection

Fileless malware is an advanced form of modern-day cyber attacks. Unlike the traditional file-based one, it uses not the implanting of malicious pieces of executable code on disk but trusted system processes and baked-in operating system tools--often referred to as Living Off The Land Binaries (LOLBins) through which malicious activities can occur undetected. The modality of operation is stealthy and avoids any persistent files from being written either to disk; malicious code is either introduced by the attacker with the help of legitimate apps, administrative software vulnerability, or native Windows administrative artifacts installers PowerShell and Windows Management Instrumentation (WMI). This way the payload can be loaded and run inside valid processes and makes the malware near impossible to detect.

There are three key stages in which the execution circuit normally occurs. In the former, the rivals use various reconnaissance avenues to find their targets and preferable victims. Then, the malefic code can ensure registry entries/ WMI objects through VBScript/JScript to invoke PowerShell thus preparing the environment to execute in-memory. Dynamic Memref: The last phase is loading and running the payload in-memory with no traces observable in the disk and the malware can stay unnoticed. Such a multi-phased implementation method gives a lot of evasiveness to fileless malware making standard security solutions even more useless and forcing the implementation of advanced monitoring and detection solutions.

Proposed Methodology

Within the scope of the current enquiry we describe a time-series log-data driven deep-learning procedure aimed at identifying fileless assaults inside the cloud. Its method is composed of four consecutive steps, namely dataset acquisition, preprocessing, model architecture building and performance evaluation

Workflow

The said case study begins by acquiring the Microsoft Security Incident file where we do preprocessing with the aim of protecting the data integrity. The resulting data is used to train a hybrid Conv1DBiLSTM model that is augmented to Multi-Head Attention (MHA), to classify the attack type. The performance of the proposed framework is measured thus through regular metrics.

DATASET AND PREPROCESSING

The paper relies on the Microsoft Security Incident data as a main source that includes the recording of the system events, network telemetries, and labeled attacks. A processing pipeline, consisting of the following steps is used prior to analysis. Variables trapped with more than 40% missing data are initially eliminated in order to retain the integrity of data. Second, it would eliminate duplicate records thereby removing redundancy, especially instances where a similar alert has been created by a concurrently supervised equipment. Third, numerical characteristics that appear continuously are subjected to Min-Max normalization, whereas the categorical variables are transformed in a one-hot-encoded manner as binary vectors. Lastly, as a way of making time dynamics more precise, conventional linear discrete encoding is substituted with cyclical encoding, having sinusoidal and cosinusoidal transforms.

Proposed Model Architecture

The proposed architecture begins with a Conv1D layer (64 filters, kernel size 1) to capture localized temporal dependencies. This is followed by three stacked BiLSTM layers (256 units each) for bidirectional sequence learning. Layer normalization ensures training stability, and a multi-head attention mechanism captures long-range dependencies by weighting important time steps. A dense layer with softmax activation produces the final class probabilities.

One-Dimensional Convolutional Layer (Conv1D)

Figure 1.

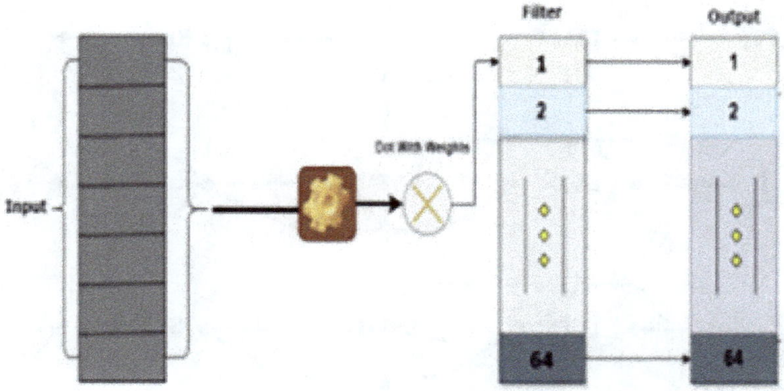

A one-dimensional convolutional layer (Conv1D) examines the timing and connection by studying the temporal position and relationship between subsequent entries of data, traffic, system logs--to determine signs of a sudden attack. The layer is fitted with 64 filters that make use of small kernels as they go through each time-step collecting features and representing each step in a more complex form. It also enlarges the input data prior to sending it to other layers to reflect minute differences that would indicate fileless attacks alongside compressing the total number of parameters through filtering hence making it quick to process large files of data.

Bidirectional Long Short-Term Memory (BiLSTM)

Long Short Term Memory (LSTM) networks are a type of the recurrent neural network (RNN) architecture drawing up to meet the problem of dealing with sequential tasks and to maintain long-term dependency information. This is done by the inclusion of memory cells whose functioning is under the influence of input, forget and output gates hence controlling the passage of information in and out. In the context of cybersecurity, LSTM has found especially useful applications in threat detection tasks since cyber-attacks are often characterised by discrete steps that occur over many timesteps; the model can therefore learn past patterns and easily distinguish between normal and anomalous behaviour.

Figure 2.

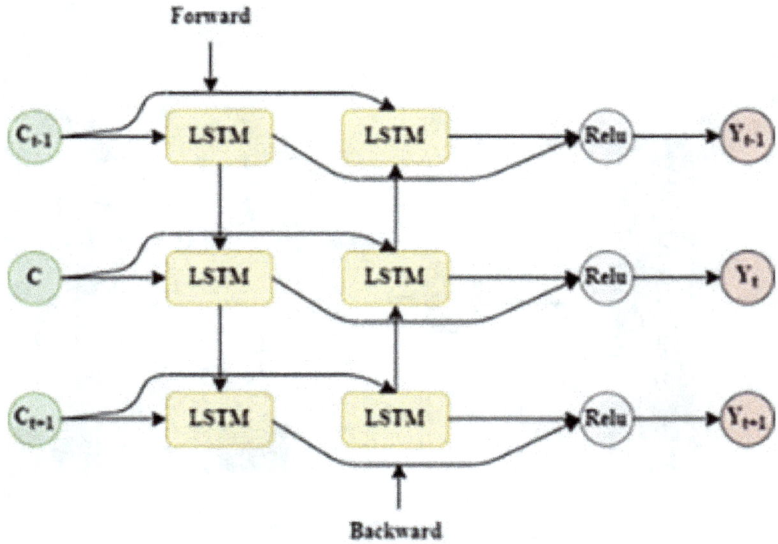

Layer Normalization

In RNNs and transformers with variable batch size, the feature standardization per sample may particularly be relevant, differentiating Layer Normalization from Batch Normalization. It normalizes path gradients, minimizes variation, and speeds training. Where we apply it is after BiLSTM in order to address the dynamics of security event data in our architecture, to remove fluctuations, and more constant activations, smoother updating, and robustness of the models.

Multi-Head Attention Mechanism

Multi-Head Attention (MHA) has been especially of interest to researchers over the past few years. This machine provides a model with numerous attention heads: each head scans along the sequence based on a different perspective and thus multiple information can be simulated in various time intervals. More importantly, all heads are founded on scaled dot-product attention and the outputs of the heads are concatenated and passed through linear layers. This architecture allows the network to learn both short-term and long-term dependent patterns, a property that brings enormous benefits whenever dealing with time-series data or monitoring information regarding security log files and behavioural records. Experimental studies (which are discussed in more detail below) have shown that MHA finds a number of weak links that are overlooked by single-headed models as well as scales extremely well to large data sizes, making them a useful method in most real-life applications.

Figure 3.

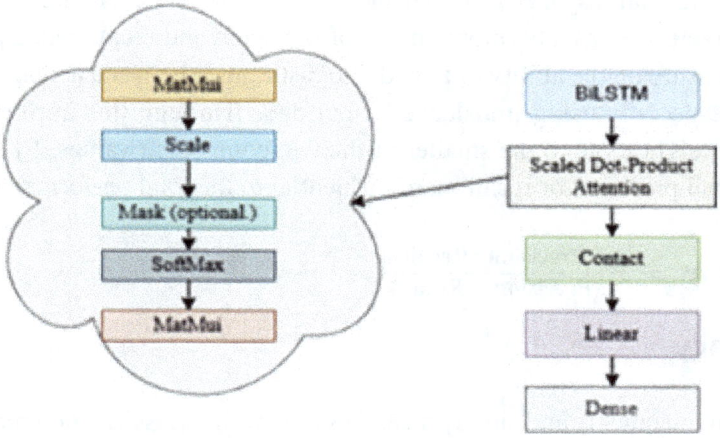

Model Evaluation

To evaluate the model's performance, various metrics are used (Han, 2024) They provide a quantitative way to evaluate the effectiveness and efficiency of model in achieving its desired outcome. Various metrics are employed to characterize the effectiveness of a classifier such as confusion matrix, Recall, Accuracy, Precision and F1-score:

Precision

Precision is the ratio of correctly predicted positive observations to the total number of predicted positive observations.

$$\textbf{precision} = \frac{\textbf{TP}}{\textbf{TP} + \textbf{FP}} \quad (1)$$

Recall

Recall helps us know the actual number of true fileless attacks that our model was able to catch. It shows whether our solution prevented some of the real attacks from succeeding.

$$\textbf{Recall} = \frac{\textbf{TP}}{\textbf{TP} + \textbf{FN}} \tag{2}$$

F1-score

F1-score is an easy metric used in the analysis of binary classification operations. It is characterized as a harmonic mean of precision and recall hence providing an equal measure on the ability of a model to distinguish between actual and innocent actions especially using imbalanced input data. Through this attribute of giving relative weight more to the smaller of the two component values, F1-score makes suboptimal precision or recall more influential to the total performance.

$$\textbf{F1 - score} = 2 * \frac{(\textbf{Precision} * \textbf{Recall})}{(\textbf{Precision} + \textbf{Recall})} \tag{3}$$

Accuracy

Accuracy quickly and simply measures how well a classification model is determining correct predictions. It is obtained by comparing how many times a model predicts correctly to how many total patterns are used.

$$\text{Accuracy} = \frac{\text{No. of correct predictions}}{\text{Total number of input samples}} \tag{4}$$

Confusion Matrix

A confusion matrix, also known as an error matrix, is a visualization tool used to evaluate the performance of a classification model. It provides a clear breakdown of how the model performed on a set of test data by comparing the predicted labels with the actual labels.

Table 1.

	Predicted Values	
Actual Values	True Negatives (TN)	False Positives (FP)
	False Negatives (FN)	False Negatives (FN)

EXPERIMENTAL RESULTS

Training was performed over 30 epochs with categorical cross-entropy loss. The model achieved a training accuracy of 97.2% and a validation accuracy of 95.5%, with a final test accuracy of approximately 96%. The inclusion of the MHA component improved classification performance compared to the Conv1D–BiLSTM baseline. Evaluation metrics such as precision, recall, F1-score, and ROC–AUC confirmed the model's robustness.

The gradual reduction of training and validation loss, which are less than 0.01, indicates that the model is trained to converge properly and classify fileless attacks and benign activities effectively, as defined in the objective of adaptive threat detection.

Figure 4.

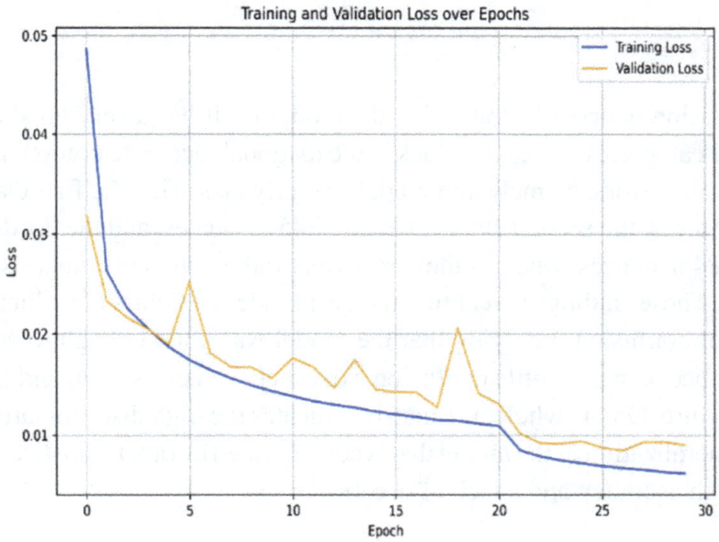

The training and validation accuracies were nearly 97.2 and 95.5, respectively, after the 30-th epoch, which means that there is no overfitting and a considerable generalization capability through the Conv1D-BiLSTM with MHA model in comparison with conventional signature-based detection systems.

Figure 5.

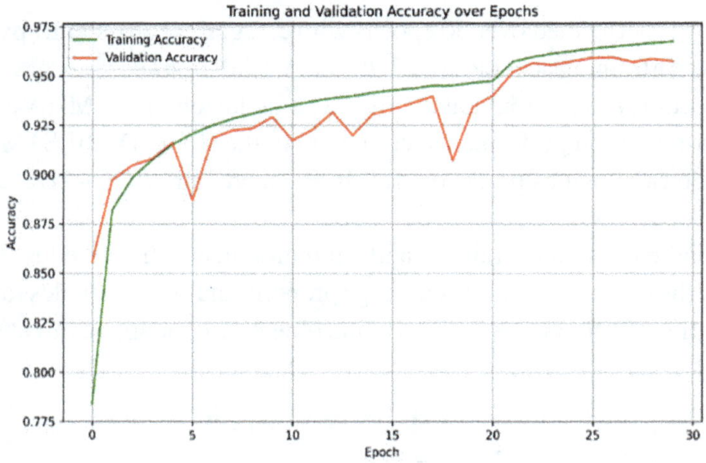

The confusion matrix illustrates that the model itself has a very good accuracy rate in all the categories since the values in the diagonal (accurate points) far exceed the off-diagonal errors. Namely, the model properly classified the first class (0) in 709,529 instances, the second class (1) in 352,245 instances, and the third class (2) in the 600,484 instances, whereas the errors were rather small in comparison to the sample size. These findings reveal that the best model was class (1) which had the lowest rate of confusion, implying that the model was good enough to accurately differentiate between it. A little confusion exists between classes (0) and (2), but is of a minor nature. On the whole, the matrix indicates the high discriminative power and good generalization of the model that is corroborated by other statistical findings including high accuracy and a lack of overfitting.

Figure 6.

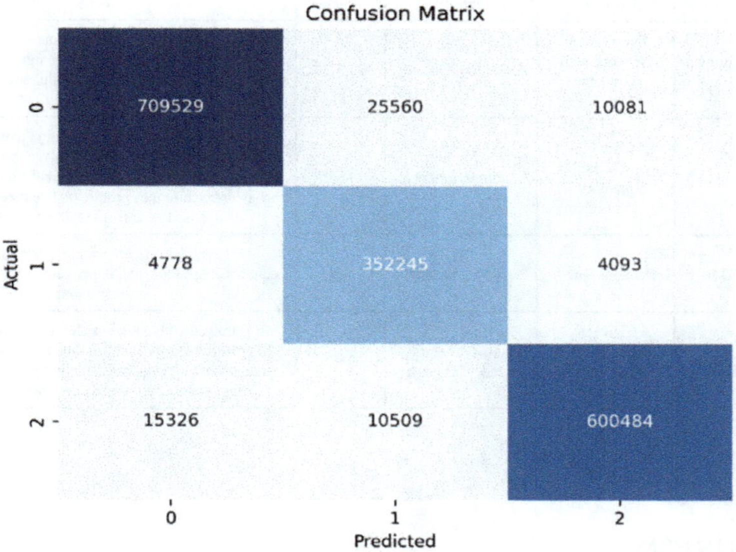

Figure 7.

Classification Report:

	precision	recall	f1-score	support
0	0.97	0.95	0.96	745170
1	0.91	0.98	0.94	361116
2	0.98	0.96	0.97	626319
accuracy			0.96	1732605
macro avg	0.95	0.96	0.96	1732605
weighted avg	0.96	0.96	0.96	1732605

Table 2. Comparision Our Model with Exsiting Approaches

Study	Approach	Dataset	Accuracy	Note
(Sudhakar & Kumar, 2020)	Deep Neural Network with Gradient Descent	Cyberdome dataset	99.31%	The model does not contain time-series analysis, restricting the capabilities of the model in terms of catching the sequential nature of fileless attacks. What is more, the dataset is rather small (234,624 samples).
(Basu, 2009)	Random Forest (with Volatility memory forensics)	Memory dumps dataset	93.33%	the model relies on feature extraction that is performed manually. It cannot be used to analyze the behavior over time and are not able to perform real-time detection of behavior because they require offline memory forensics which is slow and not scalable in the cloud.
(Obini et al., 2024)	Random Forest	APTlog dataset	96%	the models are not temporal nor undergo deep contextual learning, and the system is not suite for real-time running.
(Freitas et al., 2025)	Conv1D-BiLSTM with Multi-Head Attention	Microsoft Security Incident Dataset	96%	It is trained with a large-scale, real-world dataset of Microsoft Security Incident . the model works well in understanding local and temporal patterns, and thus helps detecting advanced and multi-stage attacks.

CONCLUSION

The presented Conv1D-BiLSTM with Multi-Head Attention model has shown to be effective in detecting fileless attacks on cloud systems and has been shown to perform well and maintain high accuracy even on imbalanced or scanty data when compared to the traditional methods. This paper establishes that the study is able to generalize effectively, detect hidden threats and customize to abnormalities of attack patterns. In the future, development will be on the use of Memory-Aware GANs to produce realistic attack data and enhance generalization, create a lifelong self-learning system based on meta-learning and reinforcement learning with the model then able to adapt dynamically to new emerging threats without complete retraining.

REFERENCES

Abeydeera, W. P. S. (2020). Fileless malware detection in cloud using machine learning techniques [Master's thesis]. Tallinn University of Technology. https://digikogu.taltech.ee/en/Download/6e4b13ac-cd19-4551-8615-36bf486ccc69

Basu, S. (2009). Semi-supervised learning. In *Encyclopedia of Database Systems* (pp. 2613–2615). Springer., DOI: 10.1007/978-0-387-39940-9_609

Han, Y. (2024). Attention mechanism-based CNN-BiLSTM for sales revenue prediction. In *Proceedings of the International Conference on Advanced Algorithms and Control Engineering (ICAACE)* (pp. 740–743). IEEE. https://doi.org/DOI: 10.1109/ICAACE61206.2024.10548377

Khalid, O., Ullah, S., Ahmad, T., Saeed, S., Alabbad, D. A., Aslam, M., Buriro, A., & Ahmad, R. (2023). An insight into the machine-learning-based fileless malware detection. *Sensors (Basel)*, *23*(2), 612. DOI: 10.3390/s23020612 PMID: 36679406

Obini, U. C., Jeremiah, C., & Igwe, S. A. (2024). Development of a machine learning based fileless malware filter system for cyber-security. *Journal of the Nigerian Society of Physical Sciences*, *6*(4), 2192. Advance online publication. DOI: 10.46481/jnsps.2024.2192

Sudhakar, & Kumar, S. (2020). An emerging threat fileless malware: A survey and research challenges. Cybersecurity, 3(1), Article 1–12. DOI: 10.1186/s42400-019-0043-x

Wu, M. H., Hsu, F. H., Huang, J. H., Wang, K., Hwang, Y. L., Wang, H. J., Chen, J.-X., Hsiao, T.-C., & Yang, H. T. (2024). Enhancing linux system security: A kernel-based approach to fileless malware detection and mitigation. *Electronics (Basel)*, *13*(17), 3569. DOI: 10.3390/electronics13173569

Zhang, S., Hu, C., Wang, L., Mihaljevic, M. J., Xu, S., & Lan, T. (2023). A malware detection approach based on deep learning and memory forensics. *Symmetry*, *15*(3), 758. Advance online publication. DOI: 10.3390/sym15030758

ADDITIONAL READING

Afreen, A., Aslam, M., & Ahmed, S. (2020). Analysis of fileless malware and its evasive behavior. In *Proceedings of the 1st Annual International Conference on Cyber Warfare and Security (ICCWS 2020)*. IEEE. DOI: 10.1109/ICCWS48432.2020.9292376

Amalina, N., Suhaimi, D., & Abas, H. (2020). A systematic literature review on supervised machine learning algorithms. PERINTIS eJournal, 10(1), 1–24. https://perintis.org.my/ejournalperintis/index.php/PeJ/article/view/86

Antonopoulos, N., & Gillam, L. (2017). *Cloud computing*. Springer., DOI: 10.1007/978-3-319-54645-2

Attaran, M. (2017). Cloud computing technology: Leveraging the power of the Internet to improve business performance. *Journal of International Technology and Information Management*, *26*(1), 112–137. DOI: 10.58729/1941-6679.1283

Bandara, H. M. H. M., Ayeshani, K. M. N., Kumari, M. M. P. M., Wijerathna, D. M. S. T., Abeywardena, K. Y., & Wijesooriya, A. (2025). Behavioral analysis for fileless malware detection. In *Proceedings of the International Symposium on Digital Forensics and Security (ISDFS 2025)*. IEEE. https://doi.org/DOI: 10.1109/ISDFS65363.2025.11012086

Chunduri, H., Anand, P. M., Shukla, S. K., & Sai, C., P. V. (2024). Trusted yet disguised: Analysing the subversive role of LOLBins in contemporary cyber threats. In *Proceedings of the IEEE International Conference on Big Data (BigData 2024)* (pp. 2596–2605). IEEE. DOI: 10.1109/BigData62323.2024.10825239

Dos Santos, C. F. G., & Papa, J. P. (2022). Avoiding overfitting: A survey on regularization methods for convolutional neural networks. *ACM Computing Surveys*, *54*(10s), 1–25. Advance online publication. DOI: 10.1145/3510413

Freitas, S., Kalajdjieski, J., Gharib, A., & McCann, R. (2025). Microsoft security incident prediction. In *Proceedings* (pp. 191–200). Association for Computing Machinery., DOI: 10.1145/3701716.3715209

Gong, C., Liu, J., Zhang, Q., Chen, H., & Gong, Z. (2010). The characteristics of cloud computing. In *Proceedings of the International Conference on Parallel Processing Workshops* (pp. 275–279). IEEE. DOI: 10.1109/ICPPW.2010.45

Grobauer, B., Walloschek, T., & Stöcker, E. (2011). Understanding cloud computing vulnerabilities. *IEEE Security and Privacy*, *9*(2), 50–57. DOI: 10.1109/MSP.2010.115

Hassan, W., Chou, T.-S., Li, X., Appiah-Kubi, P., & Omar, T. (2019). Latest trends, challenges and solutions in security in the era of cloud computing and software defined networks. *International Journal of Information and Communication Technology*, *8*(3), 162–183. DOI: 10.11591/ijict.v8i3.pp162-183

Hatem, S. S., Wafy, M. H., & El-Khouly, M. M. (2014). Malware detection in cloud computing. *International Journal of Advanced Computer Science and Applications*, *5*(4). Advance online publication. DOI: 10.14569/IJACSA.2014.050427

He, T., Mao, H., & Yi, Z. (2022). Subtraction gates: Another way to learn long-term dependencies in recurrent neural networks. *IEEE Transactions on Neural Networks and Learning Systems*, *33*(4), 1740–1751. DOI: 10.1109/TNNLS.2020.3043752 PMID: 33373305

Kumar, A., Dutta, S., & Pranav, P. (2023). Supervised learning for attack detection in cloud. International Journal of Experimental Research and Review, 31(Special Volume), 74–84. https://doi.org/DOI: 10.52756/10.52756/ijerr.2023.v31spl.008

Malik, M. I., Wani, S. H., & Rashid, A. (2018). Cloud computing-technologies. *International Journal of Advanced Research in Computer Science*, *9*(2), 379–384. DOI: 10.26483/ijarcs.v9i2.5760

Naik, D., & Jaidhar, C. D. (2022). A novel multi-layer attention framework for visual description prediction using bidirectional LSTM. *Journal of Big Data*, *9*(1), 1–23. DOI: 10.1186/s40537-022-00664-6

Nguyen, G., Dlugolinsky, S., Bobák, M., Tran, V., García, Á. L., Heredia, I., & Hluchý, L. (2019). Machine learning and deep learning frameworks and libraries for large-scale data mining: A survey. *Artificial Intelligence Review*, *52*(1), 77–124. DOI: 10.1007/s10462-018-09679-z

Opara-Martins, J., Sahandi, R., & Tian, F. (2015). Critical review of vendor lock-in and its impact on adoption of cloud computing. In Proceedings of the International Conference on Information Society (i-Society) (pp. 92–97). IEEE. DOI: 10.1109/i-Society.2014.7009018

Piper, B. (2021). CompTIA Cloud+ study guide: Exam CV0-003.

Portase, R. M., Muntea, A. M., Mermeze, A., Colesa, A., & Sebestyen, G. (2024). Detection strategies for COM, WMI, and ALPC-based multi-process malware. *Sensors (Basel)*, *24*(16), 5118. Advance online publication. DOI: 10.3390/s24165118 PMID: 39204815

Shen, Z., Jia, Q., Sela, G. E., Rainero, B., Song, W., van Renesse, R., & Weatherspoon, H. (2016, October). Follow the sun through the clouds: Application migration for geographically shifting workloads. In *Proceedings of the Seventh ACM Symposium on Cloud Computing* (pp. 141-154). DOI: 10.1145/2987550.2987561

Talebian, H., Gani, A., Sookhak, M., Abdelatif, A. A., Yousafzai, A., Vasilakos, A. V., & Yu, F. R. (2020). Optimizing virtual machine placement in iaas data centers: Taxonomy, review and open issues. *Cluster Computing*, *23*(2), 837–878. DOI: 10.1007/s10586-019-02954-w

Vakilinia, S., Heidarpour, B., & Cheriet, M. (2016). Energy efficient resource allocation in cloud computing environments. *IEEE Access : Practical Innovations, Open Solutions*, *4*, 8544–8557. DOI: 10.1109/ACCESS.2016.2633558

Wang, H., & Tu, M. (2020). Enhancing attention models via multi-head collaboration. In *Proceedings of the International Conference on Asian Language Processing (IALP)* (pp. 19–23). IEEE. DOI: 10.1109/IALP51396.2020.9310460

Chapter 10
Phishing Detection Reimagined:
A Trust-Centric, Federated Learning Approach

B. Manohari
http://orcid.org/0009-0007-2691-9428
Vellore Institute of Technology, India

L. Thanga Mariappan
Vellore Institute of Technology, India

ABSTRACT

In the digital age, data is becoming a valuable asset, and its protection is paramount. Phishing attacks pose a significant threat to data security, often leading to unauthorised access to sensitive information. This paper presents a novel approach for phishing detection that preserves user privacy. Utilizing federated learning, we propose a system that allows for collaborative model training without compromising individual data privacy. Our work involves distributing a global model across clients, who then gives back the model updates with their local data. These updates are aggregated to enhance the global model while ensuring the credibility of client contributions. FedCred is novel aggregation algorithm framed to ensure the credibility of the clients. The effectiveness of our approach is demonstrated through the use of comprehensive dataset, feature extraction techniques, and a robust model design. This research contributes to the cybersecurity field by offering a privacy-preserving solution to combat phishing threats.

DOI: 10.4018/979-8-3373-9549-4.ch010

1. INTRODUCTION

We are in an era where technological growth is at its peak. Technology has its pros at a similar pace, and it has its cons. Data is the biggest wealth of the upcoming century. Data is precious and the richest thing. To fetch that data illicitly, there are several social engineering or cyberattacks. One of the cyber-attacks that is at the top of the list is phishing. Users' Sensitive data can be gained access through this Phishing (Sameen et al., 2020). Phishing looks like a reputable website or URL, but steals sensitive user details such as the credit/debit card number, the CVV number of the card, the PIN of net banking, etc. Phishing attacks continue to constitute a major for organisations as well as individuals.

A Phishing attack is a website purposively created to deceive the user using deceptive emails, text messages, phone calls, or social media. These Deceptive techniques focus mainly on performing the activities that lead to downloading malicious software, revealing sensitive data, or unconsciously losing their security (Jain & Gupta, 2018) . Phishing targets not only an individual but also a mass community or an organisation. To overcome such activities there are a huge number of non-profit organisations. An international coalition, Anti Phishing Working Group (APWG) is a non-profit association that provides reports of identity theft and other cyber-crimes every quarter. APWG reported 1,286,208 phishing attacks in the second quarter of 2023, which was the third-highest quarterly total. Still, the financial sector continued as the highest sector with 23.5 % of all phishing attacks (Ali & Malebary, 2020).

Phishing is not a single-step process. Phishing attacks progress in three phases (Sahingoz et al., 2024). They are as follows:

- Sending the phish to Victim.
- Victim getting alert of the phish.
- Action done by the victim on the phish.

The first phase is a safe zone after which software can be used to detect attacks. During the second step, user awareness helps prevent identity theft. Once the third phase was completed, the user was deceived. To overcome this problem, different methods of Phishing detection can be used. Phishing Detection falls into two categories: one is based on social awareness about the fake /Benign URL, and the other is about software/machine detection based on available techniques (Sameen et al., 2020).

Phishing can be detected by human interference or software. While using the software-based approach to detect phishing websites, there are five main categories whitelist/blacklist, Heuristic, Visual Similarity, Machine learning, and deep learning (Wei & Sekiya, 2022) . Even though this technique helps in the detection of phishing there should be a compromise in privacy by the users as state-of-the-art Deep Learning algorithms gather huge datasets from users.

Privacy is a critical concern when dealing with data. To gain the benefit of all data without centralizing them, a new learning technique called Federated Learning was introduced. In this learning technique privacy is maintained by designing a local model with the data of the clients that will be updated to the global model of the server (Sahingoz et al., 2024).

Phishing involves mail data and messages that require privacy. The sharing of mail or messages and related data by the users loses their privacy, instead of collecting data from the user to develop a model. The model will be applied to the user data. The application of Federated Learning for the detection of phishing mail greatly helps to preserve privacy and to gain the benefit of a machine learning algorithm.

We will discuss URL-based phishing detected as most people are not ready to share their browsing history, which reveals personal activities. Hence, the implementation and development of phishing detection using the user URL greatly contributes towards privacy and enhances the security of the user.

In this Chapter, we work on

- Usage of federated learning for the phishing detection from the user's URL.
- Parameter sharing along with the model sharing to handle the heterogeneous data of Clients.
- The model overcomes the server-spoiling client attacks by introducing the client credibility check.

Our work will add up the privacy-preserving technique which is a Deep Learning Technique performed without sharing the data to the central server.

2. LITERATURE SURVEY

This section discusses the existing types of phishing attacks and a survey of the detection methods used until now. There are several types of phishing attacks.

Phishing is divided into two broad categories: illusionary/deceptive phishing, and technical dorge based on technical insight (Al-Ahmadi et al., 2022).

- Deceptive Phishing:

Phishing emails are emails used as a tool to make people trust it as a legitimate website. Spoofed websites are a type of phishing in which the attacker uses a website that is similar to a genuine or legitimate website.

- Technical Dorge:

Malicious phishing is an attack by which a phisher installs malicious software to the target system. A Key Logger or Screen Logger is an in which the attacker installs malicious software in the target system and monitors the activities of the user. Adware is a method of attack where the phishing attacker inculcates the phishing URL in any advertisement.As we spoke about phishing, Phishing detection is mainly categorized into four categories as of now:

1. List-Based Detection
2. Visual Similarity Based Detection
3. Heuristic and Machine Learning Based Detection
4. Deep Learning Based Detection

2.1. List Based Detection

List-based attacks are based on two types of lists: blacklist which contains a list of phishing websites, and whitelists, which contain a list of Legitimate URLs. Google Safe Browsing API is an application that maintains a frequently updated Blacklist. The phishing check is usually performed with the existing blacklist and security-related information is collected from the user to protect them in this API (Li et al., 2020). The Blacklist approach has the advantage of quick access time but simultaneously has the disadvantage of a low detection rate. This type of approach fails to detect zero-day attacks. When we talk about a whitelist, generally it works with a list of legitimate websites. Sánchez Sánchez et al. (2024) compared the list of legitimate links with the actual list of links. Then the score for the particular link is calculated to check whether the provided link is legitimate. This method provides a 95% true positive rate. The main pitfall of this method is that the accuracy varies for different datasets and the identification of different phishing websites is not perfect.

2.2. Visual Similarity Based

Most phishing attack imitates legitimate websites. Generally, phishers use the resemblance of a legitimate website or a website that is used frequently by the user and imitate the logo and exact images of the legitimate website and make any small changes in their look to attract the users. Visual similarity-based method of phishing detection uses visual features to detect phishing websites. In this approach, visual features such as screenshots of the webpage, logos, formatting of the text, images used in the webpage, etc. For this technique, the features are taken from already stored legitimate web page (PhishTank, n.d.) . The technique of visual similarity lags in accuracy and depends solely on legitimate websites for features. This cannot help prevent or detect the zero-day phishing attack.

2.3. Heuristic and Machine Learning-Based Detection

Kara. et al. chose a dataset from the TR-CERT official website. The dataset consists of a total of 32928 data points as a whole, among which 20614 were phishing data and 12314 were non-phishing data. In this work, the author considered 11 features which are URL as well as domain-based features and used six classifiers. According to this author, Random Forest performs best while the Support vector Machine stands behind (Alkhalil et al., 2021).

Feature selection plays a vital role in phishing classifications. To improve the performance of the machine learning algorithm, we can enhance the feature selection technique. Gupta et al. (2023) used Particle swarm optimisation (PSO) for feature selection and weightage and then tested those for better accuracy using five classical machine learning algorithms, such as BPNN, SVM, KNN, Decision Tree, Random Forest, and Naïve Bayes classifier. They considered 30 features from four different categories – HTML and JavaScript-based, Abnormal-based, Address Bar-based, and Domain-based. They worked with nearly 11 thousand websites as the dataset, among which 44 % were phishing and 56 % were Legitimate data. The results show that the use of PSO for feature weighting improves the accuracy compared to the usual feature selection methods.

All the methods discussed above use only a single learning algorithm, and to gain the benefit of more than one algorithm, we have the concept of ensemble learning in which two or more algorithms are combined (Saito & Rehmsmeier, 2015) . Karim et al., used several machine learning algorithms for the detection of phishing and also tried a hybrid machine learning algorithm to enhance the performance. The author took the dataset from the well-known website of Kaggle and considered 33 attributes of the URL on the 11000+ dataset. He used LSD ensemble learning – a combination of linear regression, Support Vector Classifier, and Decision Tree that underwent soft and hard voting resulting in better accuracy than usual conventional machine learning algorithms (Karim et al., 2023).

The author used nearly 60 thousand datasets, where they extracted eight features and chose the XGBoost Classifier for classification, which is the combination of Random Forest(RF), Naïve Bayes, Logistic Regression(LR), and AdaBoost Classifier. The proposed method shows 96.76% accuracy (Aljofey et al., 2022).

2.4. Deep Learning-Based Detection

Sahingoz et al. performed phishing detection using a deep learning approach where they fetched the phishing URL from www.Phishtank.com. For the legitimate website URL, they fetched the help of a common crawl that collected the URL of a legitimate website over the last eight years (Sahingoz et al., 2024). They chose nearly five lakh datasets, among which 3 lakh are legitimate data, and the rest were malicious. The author used five deep-learning networks and found that the CNN performed better than the other four algorithms.

Al-Ahmadi et al. stated that using a Generative Adversarial Network for phishing detection helps generate phishing data and improves the accuracy of phishing website detection. In the proposed method, the CNN acts as the generator (Al-Ahmadi et al., 2022).

The existing whitelist and blacklist-based approach cannot detect zero-day malware attacks. The technique is completely unproductive as the phishers generate new URLs every day. The visual similarity method shows more true negatives because it depends on the features of the existing legitimate websites. The machine learning approach collects the data from the user, as it collects minimal data and cannot handle complex or huge datasets. To address this, we switched to deep learning, which works with many datasets. Gathering a large number of datasets results in the collection of data from users (McMahan et al., 2017). Collecting data from the user to train and test the deep learning model results in a loss of privacy for the user.

Because we have a snag with privacy and other things that keep traditional machine learning techniques ahead in the detection of Phishing URLs, we used the technique of federated learning introduced in 2016. This technique preserves privacy by sharing the model itself with the user, instead of collecting the data from the user. The initial global model is shared with the users saved in the client as a local model, the users make updates to where there is some peril in this technique. The first one handling the heterogenous data and the other peril is the attacks by the untrusted user which spoils the machine learning model's performance.

2.5. Federated Learning and its Algorithm

Federated learning is a novel approach that enables the training of models on decentralized data sources while preserving privacy. This presents challenges in terms of communication efficiency, resource allocation, and security. The concept has gained traction in various fields, including mobile edge networks and healthcare informatics. Researchers have highlighted the importance of addressing privacy concerns and data fragmentation in healthcare data analyses (Sánchez Sánchez et al., 2024).

Despite its potential, Federated Learning faces challenges in communication efficiency, resource management, and privacy protection. Future directions include exploring solutions for system challenges, privacy issues, and the collaborative training of machine learning models in federated learning settings .

Federated Averaging (FedAvg) is a federated learning algorithm that trains deep neural network models using decentralised, unbalanced, and non-IID data. It works by having clients perform a local stochastic gradient descent, and then the server averages the model updates to update the global model. FedAvg is robust to unbalanced and non-IID data distributions and can significantly reduce the number of communication rounds required to train a model compared with synchronised stochastic gradient descent (McMahan et al., 2017).

Federated SGD (FedSGD) is a practical method for federated learning of deep networks that combines the local stochastic gradient descent on each client with a server that performs model averaging. It is robust to unbalanced and non-IID data distributions and can reduce the number of rounds required to train deep networks on decentralised data by several orders of magnitude (McMahan et al., 2017).

FedProx is a federated optimisation algorithm that addresses the challenges of heterogeneity in federated learning by adding a proximal term to the objective. Theoretically, this allows for convergence guarantees and analysis of heterogeneity, while empirically it improves the stability and accuracy of federated learning, especially in highly heterogeneous settings (Li et al., 2020).

This brief review provides an overview of phishing attacks, categorizing them into deceptive (such as phishing emails and spoofed websites) and technical (including malicious software installation such as keyloggers and adware). Phishing detection methods are discussed in five categories. Each method has its strengths and limitations, such as quick access to list-based detection, low detection rate for zero-day attacks, and the capability of deep learning to handle large datasets but potential privacy concerns. The discussion highlights the shift towards federated learning to address privacy issues by sharing models instead of data, though challenges remain in handling heterogeneous data and defending against attacks by untrusted users.

The Application of federated learning (FL) in phishing detection presents a promising approach to address privacy concerns while leveraging the collective intelligence of decentralized data sources. Traditional methods of phishing detection, such as list-based approaches and heuristic/machine learning methods, often require centralized data aggregation, which can compromise user privacy and face challenges with scalability and data heterogeneity.

2.5.1 Federated Learning for Cybersecurity Applications

The usage of federated learning in the field of cybersecurity research has improved the field, where there is a need for a centralised, large amount of data. Federated learning is an approach that is not similar to machine learning, as it needs multiple clients to train the global server, where all the clients retain their own data and only share the updates with the central server. This character of federated learning is useful for cybersecurity as it carries sensitive information such as malware samples, email samples and browser history. Even the user credentials, which are very sensitive information, need not be shared, and privacy is guaranteed (McMahan et al., 2017; Kairouz et al., 2021).

McMahan et al. introduced the Federated Averaging (FedAvg) algorithm. It showed that decentralized clients could work together to train deep learning models using iterative local updates and server-side aggregation. This method greatly reduced communication costs and protected user privacy while keeping model performance competitive (McMahan et al., 2017). Since then, federated learning has been used in various cybersecurity applications, such as intrusion detection, malware classification, spam filtering, botnet detection, and phishing detection (Nguyen et al., 2022; Rahman et al., 2020).

Even though federated learning addresses privacy issues, it brings up several challenges related to data diversity. In real-world situations, client datasets are rarely independent and identically distributed (IID). Different users generate unique browsing patterns, network behaviors, and attack profiles, leading to non-IID data. This diversity can slow model convergence and lower overall performance. To tackle these challenges, Li et al. proposed FedProx, which builds on the FedAvg algorithm by adding a proximal regularization term to the optimization goal. This term limits excessive deviations of local updates from the global model and enhances convergence stability in varied environments (Li et al., 2020).

Recent studies have shown that federated learning is effective in cybersecurity systems. Its decentralized nature not only protects privacy but also allows organizations to work together on strong security models without sharing sensitive information. This makes federated learning a promising option for large-scale cybersecurity applications where both privacy and collaboration are crucial (Kairouz et al., 2021; Nguyen et al., 2022).

2.5.2 Trust-Based Federated Learning

Although federated learning protects privacy, it is still at risk from malicious participants who try to corrupt the global model. Since the server cannot access local client data, it is hard to verify if a client's model update is trustworthy. This issue has led to the creation of trust-based federated learning methods (Mothukuri et al., 2021).

Trust-based aggregation methods assess the reliability of clients before including their updates in the global model. Instead of treating all clients equally, trust-aware systems calculate a trust score based on factors like local accuracy, training loss, update consistency, contribution quality, and historical behavior. Clients with higher trust scores have more influence on the global model, while suspicious clients get lower aggregation weights (Sánchez Sánchez et al., 2023; Mothukuri et al., 2021).

Sánchez Sánchez et al. introduced FederatedTrust, a trust-aware federated learning framework aimed at improving the reliability of collaborative learning systems. Their method adds trust evaluation into the aggregation process to identify unreliable participants and lessen their impact on model updates. Experimental results showed that trust-aware aggregation enhances model stability and resilience in difficult environments (Sánchez Sánchez et al., 2023).

Trust mechanisms are especially useful in cybersecurity because malicious users might try to manipulate model updates to weaken detection systems. By adding trust evaluation, federated learning frameworks can more effectively tell apart honest participants from suspicious ones, which improves security and model robustness (Mothukuri et al., 2021).

2.5.3 Poisoning Attacks in Federated Learning

Despite the benefits of federated learning, the decentralized training process opens up new security risks. One of the most significant threats is the poisoning attack, where malicious participants deliberately alter the learning process to harm model performance or meet specific attack goals (Tolpegin et al., 2020; Bagdasaryan et al., 2020).

Poisoning attacks generally fall into two categories: data poisoning attacks and model poisoning attacks. Data poisoning attacks change local training data to introduce false patterns into the learned model. In contrast, model poisoning attacks directly alter model parameters or gradients before sending them to the server. Since federated learning relies on updates from multiple participants, even a few malicious clients can greatly affect global model performance (Mothukuri et al., 2021; Bagdasaryan et al., 2020).

Gupta et al. proposed a new data poisoning attack using an inverted loss function that lets adversarial clients disrupt the learning process while seeming legitimate. Their research showed that poisoned updates could significantly lower model accuracy and endanger federated learning systems. The findings stressed the need for strong defense mechanisms that can identify and counter malicious participants (Gupta et al., 2022).

Various defense strategies have been suggested to combat poisoning attacks. These include robust aggregation, anomaly detection, Byzantine-resilient learning, and trust-based client selection. Among these, trust-aware aggregation has received notable attention because it assesses client behavior across multiple communication rounds and reduces the impact of suspicious updates. Therefore, trust-based federated learning offers an effective way to boost resilience against adversarial participants (Mothukuri et al., 2021; Sánchez Sánchez et al., 2023).

3. BACKGROUND

3.1. Federated Learning Basics

The subset of Artificial Intelligence- Machine learning and deep learning that transfers huge amounts of data to the server from all its participants. In turn, the server trains the model and provides the result of their data. Sharing a huge amount of data requires greater bandwidth, and user privacy has been spoiled by sharing all their data to the centralized server. To overcome this, McMahan introduced the concept of federated learning in 2017. As the name indicates, it works in groups but with some concept of anonymity.

The basic work of federated learning involves a three-step process.

- Server shares its global model with all the clients.
- The client trains the received model using its data and shares it again with the server.
- The server aggregates the model from all clients to create an updated global model.

When we use federated learning, we undergo certain vulnerability that spoils the entire model process and prediction. Malicious clients may intentionally manipulate the global model by sending parameters that poison it with regular local updates(Gupta et al ., 2023). Some clients try to benefit from the global model without contributing or sharing fabricated data updates. In severe cases, multiple fake clients can corrupt the global model and all connected clients. These activities, individually or in combination, constitute data or model poisoning attacks. To prevent such attacks, it is crucial to verify clients' credibility before using their data points or parameters in the global model update.

Federated learning is a distributed machine learning approach that allows multiple participants to work together to train a shared model without sharing their raw data. Unlike traditional methods, which collect and store data from various sources on a central server, federated learning keeps the data at its source. Only the model parameters or updates are sent back and forth. This setup greatly improves privacy and reduces the risks associated with centralised data storage.

The federated learning process starts with the central server setting up a global model. The server sends the model parameters to all participating clients, which can include mobile devices, personal computers, organisational servers, or Internet of Things (IoT) devices. Each client trains the model locally using its own private data and calculates updated model parameters. Instead of sending sensitive data, only the learned model updates go to the central server. The server combines these updates from various clients to create a new global model. This loop continues until the model converges and reaches the desired performance.

One key benefit of federated learning is its focus on user privacy while still allowing collaborative model training. Since raw data stays on client devices, sensitive information like personal records, browsing history, security logs, and organizational data remains secure. Additionally, federated learning cuts down on the communication demands tied to moving large datasets to a central place. Only model parameters get exchanged, making the system more scalable in large distributed settings.

Federated learning can work under different data distribution scenarios. In Independent and Identically Distributed (IID) environments, client datasets are similar, which helps the global model converge quickly. However, real-world situations often show Non-IID characteristics, where each client has data with different distributions, class proportions, or feature patterns. This variety adds complexity because local models may learn different patterns, leading to slower convergence and lower global model performance.

Another significant part of federated learning is the server's aggregation mechanism. Aggregation algorithms figure out how to combine local updates from various clients to create the global model. The choice of aggregation strategy directly affects the accuracy, speed of convergence, and reliability of the learning process. Traditional methods usually treat all client updates equally or weigh them based on dataset size. However, these methods may struggle when some participants provide poor-quality or malicious updates.

The decentralized nature of federated learning presents several security and reliability challenges. Since clients work independently, the server has limited insight into local training processes and can't directly confirm the quality of client data. Differences in computational power, network conditions, and data quality can impact model performance. Additionally, malicious clients might deliberately skew their local updates to affect the global model. Therefore, mechanisms to assess client reliability, spot unusual behavior, and bolster the robustness of aggregation are increasingly crucial in modern federated learning systems.

Federated learning has attracted significant interest in various fields, including healthcare, finance, cybersecurity, smart cities, and Internet of Things applications. In cybersecurity, federated learning allows organizations to collaboratively build threat detection models without revealing sensitive security details. This makes it especially useful for tasks such as phishing detection, malware classification, intrusion detection, and network threat analysis, where protecting data privacy and enhancing collaborative intelligence are both vital. As a result, federated learning has become a promising method for creating privacy-focused and scalable intelligent systems designed to tackle contemporary cybersecurity problems.

4. PROPOSED SYSTEM

One of the leading concerns among people in the current era is data privacy. The updated solution for data privacy is federated learning, which is a subfield of machine learning. Federated Learning is well known for its data privacy by its methodology, but even confidentiality may slip out due to certain malicious attacks on the data or untrusted clients.

To overcome this part, we are trying a new technique of credibility-checking clients' techniques using a dataset and data quality parameters.

If the credibility value is greater than or equal to the threshold value, then the client data can be accepted by the server.

If the credibility value is less than, the client data will not be considered for the current round.

5. DATASET

The detection of phishing using the URL helps to identify attackers. In this work, we used two publicly available datasets: phishing data sets from a phish Tank and the Canadian Institute of Cybersecurity.

The phish tank provides a large number of phishing datasets. In this study, we used which provides nearly 54,000 datasets from a phishing website. Legitimate datasets were chosen from the Canadian Institute of Cybersecurity, which contains spam, phishing, and legitimate URLs (Canadian institute of cybersecurity).From the URL, the relevant features are extracted as the features are the crucial part of modelling and training.

6. FEATURE EXTRACTION

Phishing detection models often begin with web scraping, where relevant features are extracted from the URLs. We explored URL-derived features that could contribute to the identification of phishing attempts. We identified and considered nearly 13 such features to train our model for phishing URLs.

6.1. Domain-Based Features

6.1.1. Have IP

Indicates whether the URL uses an IP address instead of a domain name. Phishing sites might use IPs to bypass security checks.

6.1.2. Https Domain

Indicates whether the URL uses the HTTPS protocol. Although not a foolproof indicator, legitimate sites are more likely to use HTTPS.

6.1.3. TinyURL

Check if the URL is shortened using a service such as Tiny URL. Phishing attempts often use URL shortening to hide real destinations.

6.1.4. Prefix/Suffix

The presence of suspicious prefixes or suffixes is often used in phishing URLs.

6.1.5. DNS Record

Checks if the domain name resolves to a valid DNS record. Phishing sites may have non-existent domains.

6.1.6. Domain Age

Age of domain name. The new domains were more likely to be suspicious.

6.1.7. Domain End

The domain name extension (TLD) was analyzed. Uncommon TLDs may be used at phishing.

6.1.8. Have At

Checks for the "@" symbol in the URL path, which is uncommon and might indicate a phishing attempt.

6.1.9. URL Lengths

Overall length of URL. Phishing URLs tend to be longer with more characters.

6.1.10. URL Depth

Number of subdirectories within the URL path. Phishing URLs may have unusual depth.

6.2. User Interaction Features

6.2.1. Redirection

This indicates that the URL redirects to another location. Phishing sites may use redirects to deceive the users.

6.2.2. Frame

Check if the URL contains an iframe element, which can be used for phishing attacks.

6.2.3. Mouse Over

Checks for suspicious behavior in mouse hover events, which might be a sign of phishing.

6.2.4. Right Click

Checks for disabled right-click functionality, a tactic used by some phishing sites.

6.2.5. Web Forwards

Tracks the number of times the URL has been forwarded, which could be suspicious if it is very high.

6.3. Additional Considerations

6.3.1. Web Traffic

This outlines a framework for assessing URL credibility to detect phishing attempts, classified based on several features, domain-based features (e.g., IP usage, HTTPS protocol, URL length), user interaction features (e.g., redirections, iframes, right-click functionality), and additional considerations such as web traffic analysis.

All of these features are integrated into a Federated Learning framework, to enhance the detection, confidentiality maintenance and prevention of phishing by evaluating various indicators of URL legitimacy.

7. METHODOLOGY

Federated learning is when the server shares a model or some parameters with clients. In our approach, the initial parameters are shared from the server with all clients, according to which the client starts creating an initial model.

The first step of model creation is data extraction and training of the model by clients using their data. Every client has a model that is not disturbed by the server. Once the client trains their model with their data, they partake in the aggregation by sharing the model updates with the server. The server collects data from each client during which the process of trust of each client is measured by the former.

Figure 1. Architecture of the Federated learning with credibility check

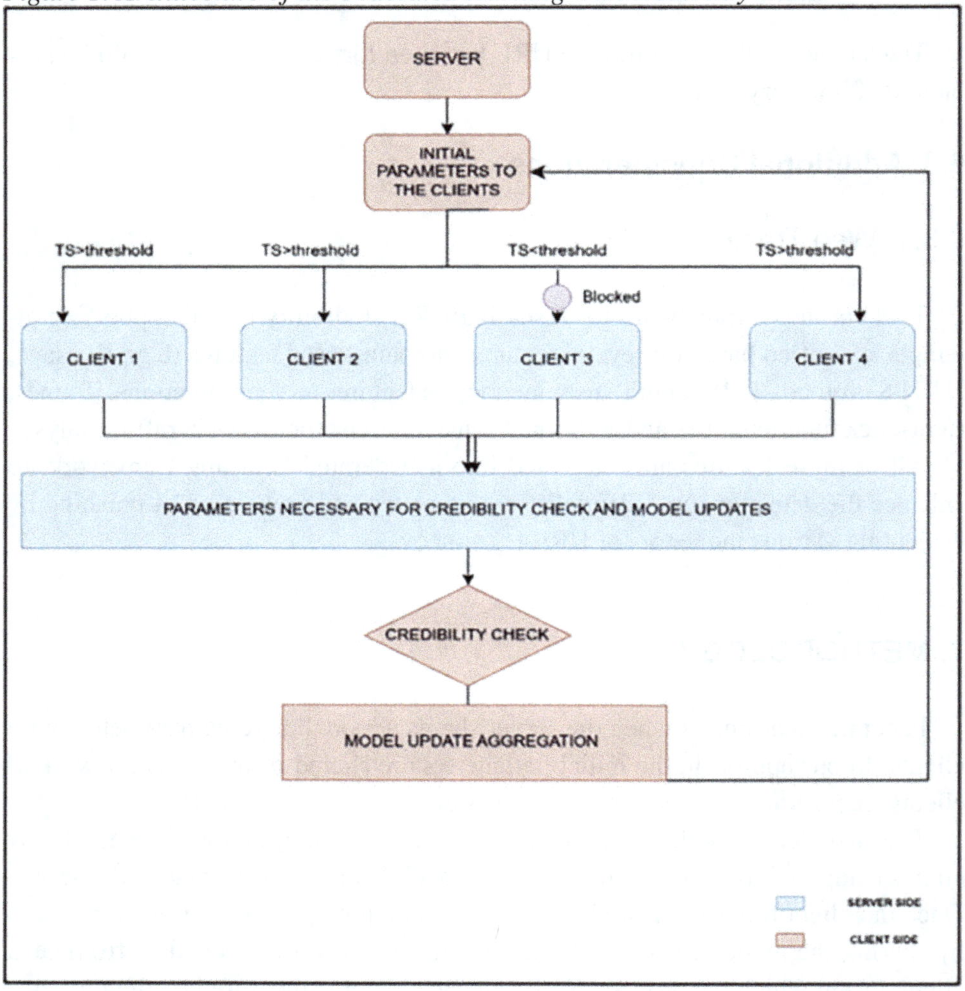

Once the server is satisfied with the credence of the client, the client utilizes the model updates received from that particular client for the aggregation process. The aggregation is always performed by the server, following which the aggregated parameters are delivered back to the clients whose credence is satisfied by the server. With updates from the centralized server, the clients re-train their models, consequently helping to improve the quality of the individual client models.

All these steps are repeated for certain rounds in which the server is not affected by any of the clients. Similarly, the client is not going to be spoiled by any insider attack as we screen out all clients for their credibility.

7.1. Data Distribution

In our work, the data set consisted of nearly 10,000 data, a combination of phishing and legitimate data. This data is split among different clients to establish the federated learning setup. The data are shredded and shared by different clients and then trained for a certain number of rounds.

In our Federated learning architecture, the client shares not only the parameters and model updates with that server. Here, the local model shares its accuracy and loss with the global model using the aggregation function to find the average value of all clients which in turn is updated for all clients. In the next round, the client uses this aggregated value to train their model which helps improve the accuracy. This preserves privacy by avoiding sharing of data from the client to the server. The need to share the URL browsing history of people can be avoided.

7.2. Algorithm

Step 1: Set the initial parameter on the server
Step 2: Share the parameters with the list of clients Cn.
Step 3:In Client side:

i. Train the local model using the client's data

Step 4: Server side:

i. Wait for client updates
ii. For each client in client_list

Update the trust score of the client
if(trust_score(client) > = threshold)
Aggregate model updates from all clients
Else
Skip the client and do not aggregate updates

iii. Distribute aggregated parameters to selected clients

Step 5: Go Step 2

7.3. Model Design

The model is designed using the packages available in Python, which are keras and tensors. TensorFlow in Python supports the design of deep learning models and helps implement a federated learning approach. Figure 1 shows the architecture of our model.

- The initial step was to set the input dimension based on the number of features in the training data.
- Creates an encoder part with dense layers for processing information.
- A dense layer with the same number of units as the input dimension, using RELU activation and L1 regularization to prevent overfitting.
- Another dense layer with half the number of units, again using SELU activation.
- A dense layer with a quarter of the number of units (bottleneck layer) was created using SELU activation for dimensionality reduction.
- Creates a decoder to reconstruct the input from the encoded representation using dense layers:
- A dense layer with half the number of units as the bottleneck layer that use SELU activation.
- A final dense layer with the same number of units was used as the input dimension, using the SELU activation.

7.4. Local Loss Sharing

The model is compiled using the Adam optimiser for efficient training and a mean-absolute error (MSE) loss function suitable for reconstructing the input from the encoded representation. Accuracy metric to track how well the model reconstructs the input. The reconstruction loss and, mean absolute error were considered to calculate the loss. Accuracy and loss are calculated by the client, which is shared with the server. The server uses the aggregation function of fedavg (mean of local loss of all clients) to calculate the aggregated value to be shared with all clients. Subsequently, the aggregated loss is shared among all the clients, with which the clients train themselves based on the gradients (weights and biases). Again, the process is repeated periodically by clients.

7.5. Experiments

The dataset was split into several batches and tested with different algorithms before applying our approach. The several algorithms tried out include

- Unweighted Federated Averaging
- Weighted Federated Averaging
- Federated SGD
- FedProx

To optimize the performance of the federated learning models, we embarked on a systematic exploration involving various algorithms across multiple batches of our dataset. This rigorous experiment aimed to identify the most effective approach for our specific scenario.

Our initial investigations included Unweighted Federated Averaging, a fundamental method in federated learning in which all clients contribute equally to the model update process. This approach provides a baseline understanding of how federated learning operates in our distributed setting.

Building on this foundation, we delved into Weighted Federated Averaging, which assigns different weights to clients based on their data characteristics or computational capabilities. This adaptation aimed to enhance model convergence and accuracy by allocating more influence to clients with richer or more relevant data.

Additionally, we explored Federated Stochastic Gradient Descent (Federated SGD), which optimizes the model parameters locally on client devices before aggregating updates. This technique leverages local computation to mitigate privacy concerns and communication overhead while still progressing towards a consensus model.

Moreover, our investigation includes FedProx, an advanced federated learning algorithm that integrates proximal terms into the optimization objective. FedProx promotes consistency across client updates by penalizing deviations from a central model, thereby potentially improving the convergence speed and robustness.

Each algorithmic approach underwent rigorous testing across different batches of our dataset to gauge its performance in terms of convergence rate, model accuracy, and computational efficiency. This iterative process allowed us to gather empirical evidence on how each was method adapted to the inherent challenges of federated learning, such as data heterogeneity and client participation disparities.

By systematically evaluating these methodologies, we aimed to identify not only the most effective algorithm for our specific use case but also gain insights into how federated learning can be tailored and optimized to maximize its benefits in real-world applications. Our findings provide a foundation for future refinements and applications of federated learning in domains where data privacy, distributed data sources, and scalability are critical considerations.

7.6. Preventing Attacks by Credibility Check

In Federated learning, we have certain cases of attack that can lead to the destruction of our entire model by their contributions:

Malicious clients sometimes manipulate the global model by sending parameters that poison our model with regular local updates. These types of clients are intentionally designed to nudge the model and make it move in the direction required by the attacker.

Some clients try to benefit from the global model without any contribution or with minimal contribution to the global model. Such clients may share fabricated data updates with the global model.

Even multiple fake clients can entirely spoil the global model and in turn spoil all clients connected to a particular server.

All these activities of the clients individually or together contribute to the data or model poisoning attack, which is commonly termed as a poisoning attack. To overcome this, we introduce checking the client's credibility before receiving and considering their data points or parameters for the global model update.

The Credibility(Cr) of the clients is determined based on certain features of data, data quality and performance of the client. The client's performance significantly contributes to the credibility of the client. The Performance metrics include binary accuracy, recall, precision and F1-score.

The performance evaluation performed here is the summation of the following performance metrics: Binary Accuracy (a), Precision (P), Recall (R), and F1-Score (F).

$$\text{Performance}_i = a_i + P_i + R_i + F_{si}$$

To use trust scores in aggregation, normalize them to ensure that they sum up to **1** across all participating clients.

For client k, the normalized trust score $C_r(i)'$is defined as:

$$C_r\left(i\right)' = \frac{C_r(i)}{\sum_{j\in N_t} C_r(j)}$$

where N is the set of clients participating in round t.

This normalized trust score is used for the standard federated aggregation process.

The server's threshold is defined as the top 80% of performers among clients in each round. In this approach, the server aggregates updates exclusively from these top-performing clients by utilizing a performance-based selection strategy. This ensures that the global model is updated using higher-quality and more relevant data.

This condition ensures that the credibility of each client is approximately equal to or less than the server's threshold:

$$C_r(i) \leq Th$$

8. PERFORMANCE METRICS

Several performance metrics are traditionally used to evaluate deep learning models (Saito & Rehmsmeier, 2015).

8.1. Accuracy

Accuracy is the ratio of correctly labelled instances to the total number of instances. The formula for Accuracy is given by:

$$\text{Accuracy} = \frac{TP + TN}{TP + TN + FP + FN}$$

Where:

- True Positives (TP): Number of correctly labelled positive instances.
- True Negatives (TN): Number of correctly labelled negative instances.
- False Positives (FP): Number of negative instances incorrectly labelled as positive.
- False Negatives (FN): Number of positive instances incorrectly labelled as negative.

8.2. Recall (Sensitivity)

Recall is the ratio of actual positive instances that were correctly labelled. It is given by the formula:

$$\text{Recall} = \frac{TP}{TP + FN}$$

High recall indicates that most positive instances were identified, which is crucial in scenarios where missing positive cases are costly.

8.3. Precision

Precision measures the ratio between the positively labelled instances that are actually positive. The formula for Precision is given by:

$$\text{Precision} = \frac{TP}{TP + FP}$$

High precision indicates a low false positive rate, which is important when the cost of false positives is high.

8.4. F1 Score

F1 Score is the harmonic mean of precision and recall, providing a balance between the two.The formula for F1 Score is given by:

$$\text{F1 Score} = \frac{2 \times \text{Precision} \times \text{Recall}}{\text{Precision} + \text{Recall}}$$

Where:

- **Precision:** The metric that measures how many of the positively labelled instances are actually positive.
- **Recall:** The metric that measures how many of the actual positive instances were correctly labelled.

The F1 Score is particularly useful when there is a need to balance precision and recall — especially in cases where the dataset is imbalanced (i.e., one class significantly outnumbers the other).

9. RESULTS AND DISCUSSION

In our study, we implemented several federated learning algorithms—FedAvg, FedSGD, FedProx, and FedTrust- to evaluate their performance on a URL feature-based phishing dataset. It is worth noting that our collaborative efforts with the different clients have resulted in significant improvements in accuracy at each round, with a precision of around 90 achieved after all the rounds, as shown in the table. Figure 2: Fedtrust shows good results and accuracy, and other metrics have increased drastically from the first round to the last iteration. Because we have used the updates of the top 80% performing clients for the fedtrust approach, it has shown the best performance and also prevents unwanted clients from being involved in the training process. As we used different algorithms, we compared the precision of different federated learning algorithms for phishing detection, as shown in Table 1.

Table 1. Precision and F1 score of all federated algorithms

Algorithm	Precision	F1-Score
FedAvg	0.912	0.770
FedProx	0.922	0.778
FedSGD	0.927	0.760
FedTrust	0.925	0.769

Fedtrust and FedSGD seem to have better precision, followed by FedProx and then FedAvg. This suggests that, for the phishing detection task, FedSGD might be the most effective algorithm in terms of precision. In addition, the loss was successfully reduced in each round as shown in Figure 3, which demonstrates the effectiveness of the federated approach.

Focusing on the credibility score effectively helped prevent insider attacks and other client-related poisoning attacks by providing updates only to clients that maintain their credibility. It is important to mention that the server does not share any model with the client, but instead shares only the parameters, thereby preventing clients who receive the model from the server from remaining silent.

Figure 2. Comparison of Loss values of all the algorithms

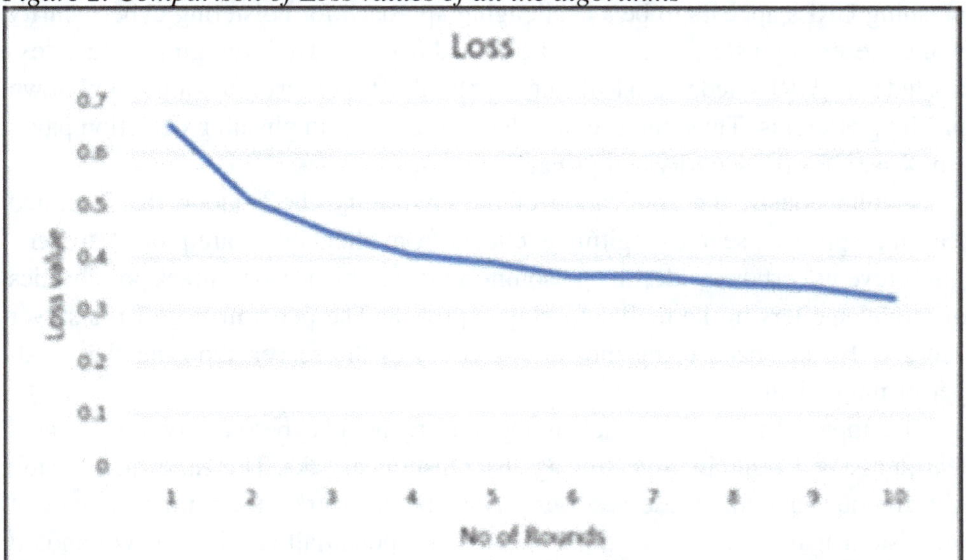

Figure 3. Precision Value Comparison

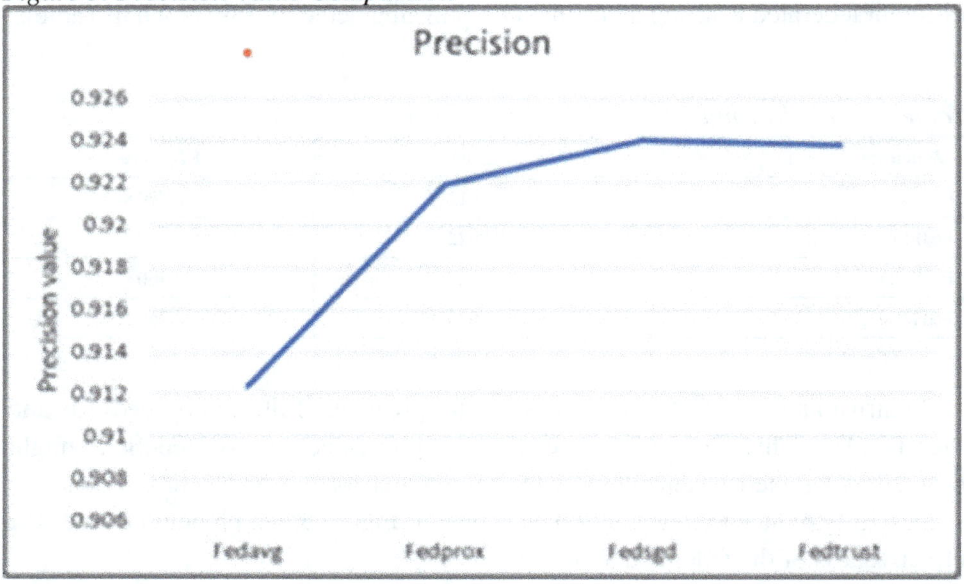

10. CONCLUSION

The implementation of federated learning in the path of detecting zero-day phishing URLs appears to be a challenging approach for bolstering cybersecurity while preserving user privacy on URL-based features. By leveraging features extracted from URLs, federated learning enables the detection of previously unknown phishing attempts. Thus, the use of federated learning in phishing detection paves a new path for the security and privacy of the user.

Simultaneously, the application of the credibility check along the federated learning approach secures legitimate clients from clients who are plot-destroyers. This prevents certain insider and poisoning attacks, which lead to attack possibilities or model spoilers in the model-sharing approach. The performance of the server can also be enhanced using the same method as the server choosing the good-performing clients.

This approach represents a challenging convergence of cybersecurity and federated learning, a new trend in protecting against phishing attacks. This innovative fusion of techniques not only preserves user privacy but also strengthens the resilience of the system against insider and poisoning attacks, potentially setting a new standard for collaborative, privacy-preserving cybersecurity solutions.

11. FUTURE WORK

As we have considered the features extracted from the URL help in finding the phishing zero-day URL. Moreover, this framework can be further enhanced by incorporating additional features derived from website content, such as images, links, and logos, thereby augmenting the accuracy and efficacy of detection. Unlike other centralised approaches, federated learning ensures the decentralisation and unavailability of sensitive user data, including browsing history and emails. This decentralised model safeguards user privacy and alleviates the risk of data breaches and unauthorised access. The application of federated learning holds significance for improving cybersecurity across various domains, including malware detection, spam filtering, and other forms of cyber threat reduction. By leveraging the collective intelligence of distributed devices while honouring user privacy, federated learning emerges as a promising paradigm for fortifying digital defences in an increasingly interconnected world.

ACKNOWLEDGEMENT

We would like to express my sincere gratitude to the Vellore Institute of Technology (VIT), Vellore for their generous support in covering the Article Processing Charges (APC) for the publication of this Chapter. Their financial assistance has been invaluable in enabling the dissemination of this work.

REFERENCES

Al-Ahmadi, S., Alotaibi, A., & Alsaleh, O. (2022). PDGAN: Phishing detection with generative adversarial networks. *IEEE Access : Practical Innovations, Open Solutions*, *10*, 42459–42468. DOI: 10.1109/ACCESS.2022.3168235

Ali, W., & Malebary, S. (2020). Particle swarm optimization-based feature weighting for improving intelligent phishing website detection. *IEEE Access : Practical Innovations, Open Solutions*, *8*, 116766–116780. DOI: 10.1109/ACCESS.2020.3003569

Aljofey, A., Jiang, Q., Rasool, A., Chen, H., Liu, W., Qu, Q., & Wang, Y. (2022). An effective detection approach for phishing websites using URL and HTML features. *Scientific Reports*, *12*(1), 8842. DOI: 10.1038/s41598-022-10841-5 PMID: 35614133

Alkhalil, Z., Hewage, C., Nawaf, L., & Khan, I. (2021). Phishing attacks: A recent comprehensive study and a new anatomy. *Frontiers of Computer Science*, *3*, 563060. Advance online publication. DOI: 10.3389/fcomp.2021.563060

Bagdasaryan, E., Veit, A., Hua, Y., Estrin, D., & Shmatikov, V. (2020). How to backdoor federated learning. *Proceedings of the 23rd International Conference on Artificial Intelligence and Statistics (AISTATS)*, 2938–2948.

Canadian Institute for Cybersecurity. (n.d.). *Canadian Institute for Cybersecurity*. https://www.unb.ca/cic

Gupta, P., Yadav, K., Gupta, B. B., Alazab, M., & Gadekallu, T. R. (2023). A novel data poisoning attack in federated learning based on inverted loss function. *Computers & Security*, *130*, 103270. Advance online publication. DOI: 10.1016/j.cose.2023.103270

Jain, A. K., & Gupta, B. B. (2018). Two-level authentication approach to protect from phishing attacks in real-time. *Journal of Ambient Intelligence and Humanized Computing*, *9*(6), 1783–1796. DOI: 10.1007/s12652-017-0616-z

Kairouz, P., McMahan, H. B., Avent, B., Bellet, A., Bennis, M., Bhagoji, A. N., Bonawitz, K., Charles, Z., Cormode, G., Cummings, R., D'Orazio, M., Eichner, H., El Rouayheb, S., Evans, D., Gardner, J., Garrett, Z., Gascón, A., Ghazi, B., Gibbons, P. B., & Zhao, S. (2021). Advances and open problems in federated learning. *Foundations and Trends in Machine Learning*, *14*(1–2), 1–210. DOI: 10.1561/2200000083

Karim, A., Shahroz, M., Mustofa, K., Belhaouari, S. B., & Joga, S. R. K. (2023). Phishing detection system through hybrid machine learning based on URL. *IEEE Access : Practical Innovations, Open Solutions*, *11*, 36805–36822. DOI: 10.1109/ACCESS.2023.3252366

Li, T., Sahu, A. K., Zaheer, M., Sanjabi, M., Talwalkar, A., & Smith, V. (2020). Federated optimization in heterogeneous networks. *Proceedings of Machine Learning and Systems*, *2*, 429–450.

McMahan, H. B., Moore, E., Ramage, D., Hampson, S., & Aguera y Arcas, B. (2017). Communication-efficient learning of deep networks from decentralized data. *Proceedings of the 20th International Conference on Artificial Intelligence and Statistics (AISTATS)*, 1273–1282.

Mothukuri, V., Parizi, R. M., Pouriyeh, S., Huang, Y., Dehghantanha, A., & Srivastava, G. (2021). A survey on security and privacy of federated learning. *Future Generation Computer Systems*, *115*, 619–640. DOI: 10.1016/j.future.2020.10.007

Nguyen, D. C., Pham, Q. V., Pathirana, P. N., Ding, M., Seneviratne, A., Li, J., Niyato, D., & Hwang, W. J. (2022). Federated learning for cybersecurity: A comprehensive survey. *IEEE Communications Surveys and Tutorials*, *24*(2), 1192–1237. DOI: 10.1109/COMST.2021.3122606

PhishTank. (n.d.). *Developer information.* https://www.phishtank.com/developer_info.php

Rahman, M. A., Hossain, M. S., Islam, M. S., Alrajeh, N. A., & Muhammad, G. (2020). *Privacy-preserving intrusion detection systems using federated learning: Concepts and challenges.*

Sahingoz, O. K., Buber, E., & Kugu, E. (2024). DEPHIDES: Deep learning based phishing detection system. *IEEE Access : Practical Innovations, Open Solutions*, *12*, 8052–8070. DOI: 10.1109/ACCESS.2024.3352629

Saito, T., & Rehmsmeier, M. (2015). The precision-recall plot is more informative than the ROC plot when evaluating binary classifiers on imbalanced datasets. *PLoS One*, *10*(3), e0118432. DOI: 10.1371/journal.pone.0118432 PMID: 25738806

Sameen, M., Han, K., & Hwang, S. O. (2020). PhishHaven—An efficient real-time AI phishing URLs detection system. *IEEE Access : Practical Innovations, Open Solutions*, *8*, 83425–83443. DOI: 10.1109/ACCESS.2020.2991403

Sánchez Sánchez, P. M., Huertas Celdrán, A., Xie, N., Bovet, G., Martínez Pérez, G., & Stiller, B. (2024). FederatedTrust: A solution for trustworthy federated learning. *Future Generation Computer Systems*, *152*, 83–98. DOI: 10.1016/j.future.2023.10.013

Tolpegin, V., Truex, S., Gursoy, M. E., & Liu, L. (2020). Data poisoning attacks against federated learning systems. *Computer Security – ESORICS 2020*, 480–501. https://doi.org/DOI: 10.1007/978-3-030-58951-6_24

Wei, Y., & Sekiya, Y. (2022). Sufficiency of ensemble machine learning methods for phishing websites detection. *IEEE Access : Practical Innovations, Open Solutions*, *10*, 124103–124113. DOI: 10.1109/ACCESS.2022.3224781

Chapter 11
Chatbox Tutor for STEM Education

P. Dharani Dharan
SRM Institute of Science and Technology Tiruchirappalli, India

S. Prashanth
SRM Institute of Science and Technology Tiruchirappalli, India

M. Tamil Suriyan
SRM Institute of Science and Technology Tiruchirappalli, India

S. Kanaga Suba Raja
http://orcid.org/0000-0002-3626-1806
Department of Computer Science and Engineering, SRM Institute of Science and Technology Tiruchirappalli, India

ABSTRACT

This article surveys the current state-of-the-art in AI and NLP, and describes a Chatbox Tutor to learn STEM from chatbots, showing that current algorithms are ready to teach students in schools. The system uses principles from the learning sciences, where it provides adaptive hints, reflective questions, and metacognitive prompts to help students solve difficult maths and physics problems. A quasi-experimental study with 120 undergraduate engineering students compared an experimental group that used the Chatbox Tutor with a control group that used standard course materials. Compared to the control group, subjects who were exposed to the Chatbox Tutor learnt more (28% vs 14%), retained more (82% vs 65%), and were more engaged. Surveys of students in the study revealed that students found it helpful for getting quick questions answered and solving problems, but difficult to trust and understand vague questions.

DOI: 10.4018/979-8-3373-9549-4.ch011

I. INTRODUCTION

The provision of scalable, individualized learning support is a persistent challenge in STEM education. Students' needs are often not overlooked by most teaching practices, especially, when they have difficulty understanding concepts, following instructions, or managing their thought processes. The growing size of the college classes makes it even harder. Individualized feedback. Even though teachers make an effort to provide additional support through problem sessions or office hours, demand generally outpaces supply, leaving many students without prompt assistance. Researchers have looked into intelligent tutoring systems (ITS) to close these gaps. These systems provide step-by-step guidance using computer models of instruction and student knowledge. Under certain circumstances, ITS can be just as successful as human tutors, as illustrated by tools such as Cognitive Tutor and AutoTutor. Nevertheless, establishing these systems is costly and necessitates extensive domain expertise and rule implementation. This makes extending them to other fields of study challenging.

At the same time, conversational AI has grown quickly over the past ten years, especially with transformer-based large language models (LLMs) that can make responses that sound like people. More and more people are using these platforms as educational chatbots. They offer interactive Q&A, tutoring dialogues, and other help learning. Chatbots differ from scripted programs smart tutoring systems (ITS) because they can talk to students, and answer different questions. And give positive feedback. But they also have their own problems, like giving out wrong information, not knowing the basics, and not being able to adapt to different students.

Hybrid systems of rule-based clever tutoring systems (ITS) and chatbots that learn from their users. These hybrid systems that try to combine the structure of domain validation with the freedom of conversational interfaces. This essay is about a STEM learning system in which being right and logical are very important. We want to make an LLM based chatbot that is accurate, yet flexible, by combining retrieval-augmented generation, scaffolding methods, and symbolic solvers. Studies have highlighted the benefits of drawbacks of AI tutors. Chen et al. found that LLM-Tutor improved students' performance on classroom tests. homework but not on tests, which means that students need to be more mentally engaged.

Kortemeyer (2023) suggested that chatbots such as ChatGPT could serve as "objects-to-think-with". according to constructionist theory, but cautioned against excessive reliance on them. Wollny et al. (2021) and in Debets et al. (2025) reviews, the chatbot particular systems have not been tested enough, serving limited personalisation. New ideas, like neuroadaptive chatbots and adaptive scaffolding theory for educational agents, keep stressing how important it is to give students feedback that fits. with how they feel, what they're thinking, and where they are. How do we design and assess a Chatbox Tutor that can effectively ease STEM learning through conversational balance flexibility, content accuracy, and educational support? This is the question the project is asking is trying to answer.

We designed a Chatbox Tutor with domain specific capabilities solvers, and scaffolding strategies language processing to help with this problem. We used learning gains, retention, activity, and student ratings to measure the performance of undergraduate STEM students who took part in a quasi-experiment.

This study provides three significant contributions:

1. A theory-based hybrid design for conversational tutoring.
2. Proof that it works in real classrooms.
3. Recommendations for design and potential Study areas to improve the chatbox tutors STEM education.

The outline of the paper is as follows: Part II review of the literature on Clever Tutoring Systems (ITS), learning chatbots and AI tutors techniques that use various forms of AI. Part III talks about the design of the system and the way it will be tested. In Section IV, we talk about related work and past implementations. Part V gives a full picture of the system architecture. Section VI has the results. Lastly, Section VII ends with suggestions for future research and what they mean.

II. LITERATURE REVIEW

STEM education has been at the forefront of technology-enhanced learning, especially because STEM emphasizes systematic problem-solving as well as critical thinking. Smart Tutoring Systems (ITS) and educational chatbots are two primary research threads behind the development of AI-aided tutoring over recent decades. Hybrid conversational models are also another research strand that has influenced this development more recently.

Figure 1. Distribution of chatbot functionalities in STEM education applications.

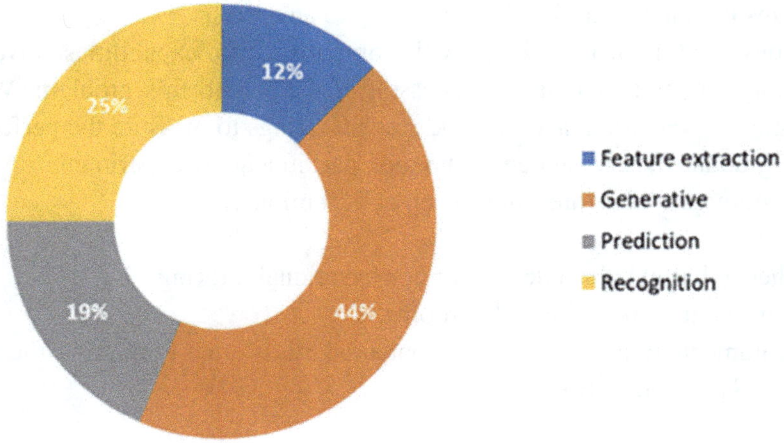

A. Intelligent Tutoring Systems (ITS)

ITS are one of the first uses of AI in education, going back to the 1980s. They are built upon a pedagogical model, a domain model, a student model, and a user interface. They learn about the learners through feedback tailoring and through tracing problem-solving steps accordingly. Yet, ITS does evolve with cost and resource-intensive-ness because it requires authors to produce the rules and modelers to craft domains very extensively. Their conversation is limited too it is usually inflexible. Scalability is decreased over diverse topics by this limitation (Paladines & Ramírez, 2020).

B. Educational Chatbots

NLP and machine learning developed towards the creation of chatbots in education. Chatbots are more adaptable conversation tools. FAQ assistants predominantly utilized early rule-based bots since such bots provided quick results for common questions but offered little deep learning assistance (Debets et al., 2025; Labadze et al., 2023). Chatbots have evolved to be more conversational and natural with transformer-based large language models (LLMs). Experiments like Chen et al. (2025) indicated LLM-powered tutors completed additional homework but impacted exam performance minimally, and Kortemeyer (2023) emphasized their worth as "objects-to-think-with" as opposed to substitutes for guided instruction. Most chatbots still do now endure from factual errors along with a lack of pedagogical richness (Davar, 2025) even though very interactive and also scalable.

C. Hybrid Conversational Frameworks

To strike balance between rigor and flexibility, conversational ease of chatbots incorporates ITS-like organized features within hybrid systems. Chatbots employ Retrieval-Augmented Generation (RAG) for grounding answers in carefully curated knowledge bases (Smalenberger & Smalenberger, 2022) as symbolic solvers enhance accuracy in STEM domains (Córdova-Esparza, 2025). Adaptive scaffolding theory provides a source for such systems. Tiered hints from conceptual cues to complete solutions are provided as a result (Cohn et al., 2025; Létourneau et al., 2025). Learner affect and cognitive load have been employed by new neuroadaptive chatbots (Baradari et al., 2025) in order to fine-tune responses while emphasizing promise. STEM designs with promise tackle conversational flexibility also precision.

D. STEM-Specific Applications

STEM subjects require precision, symbolic thinking, also multi-step solutions, which makes them difficult for purely conversational systems. ITS have traditionally been powerful in STEM since they can represent procedures, but their scalability is restricted (Paladines & Ramírez, 2020). Using chatbots is simple though they usually struggle with symbolic manipulation. Hybrid systems do hold some hope in the resolution of these challenges, especially in mathematics and physics tutoring (Feng et al., 2021; Létourneau et al., 2025; Córdova-Esparza, 2025); such systems integrate solvers, retrieval-based knowledge, and also scaffolding strategies.

E. Research Gaps

Nevertheless, major challenges exist despite current achievements. Chatbots are lacking in rigor while scalability of ITS is non-existent (Debets et al., 2025; Labadze et al., 2023). Learner attitudes in many chatbot researches are emphasized. Quantifiable results, though, are just not there (Debets et al., 2025). Few systems have solvers in them. They also hardly check STEM for correctness externally as per source (Smalenberger & Smalenberger, 2022). Ethical concerns like bias, transparency, and over-reliance are unexplored to a large extent (Davar, 2025; Garcia-Bosque et al., 2024).

III. METHODOLOGY

The research methodology directing the design, development, and assessment of the Chatbox Tutor for STEM education employed a systematic design-based research (DBR) framework in conjunction with a quasi-experimental evaluation approach. The process happened in three main steps: designing and building the system, setting up the experiment and choosing the participants, and collecting and analysing the data. Each phase took into account the technical problems that come up when making AI systems and what we know about how people learn.

Figure 2. Research methodology framework for data collection and analysis in the Chatbox Tutor for STEM Education study.

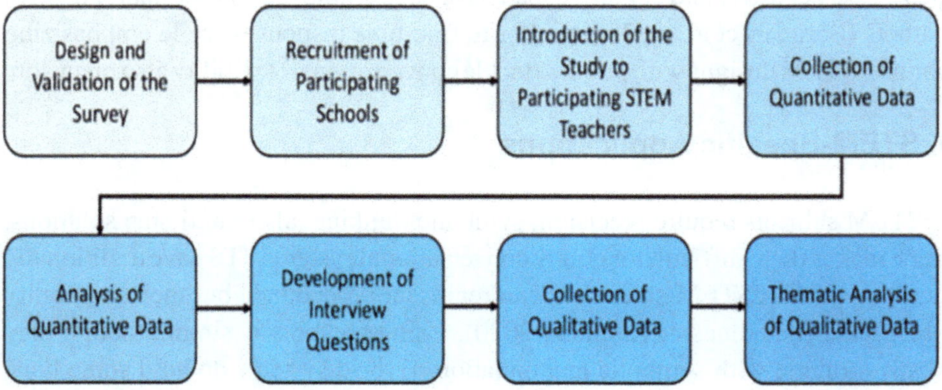

A. Design and Development of the System

The Chatbox Tutor was created using a combination of different types of tutoring that have been written about in the past. This model integrated aspects of intelligent tutoring systems (ITS) and features of conversational agents. The development of the pedagogical framework was based on the theory of adaptive scaffolding. This theory posits that students acquire knowledge through a tiered system of support that diminishes as their competence increases. The system was set up to give: - Conceptual cues that ask you to remember definitions or rules Calculated tips that help you solve problems without giving you the answer. Step-by-step instructions for how to do equations. Many direct solutions to these equations were kept in reserve for emergencies, alongside the metacognitive prompts that get people thinking about what they were doing. This system of pedagogical scaffolding made sure Tutor She did more than answer questions, though. And that people were engaged in problem-solving and self-regulation, as the literature says it should. Was made with the assistance of three types of AI Architecture.

Natural Language Understanding (NLU): A Transformer model (GPT-3.5 API) was used to figure out what someone meant and make the conversation flow (Ramadhan et al., 2023). Domain Solvers: We will also need symbolic solvers: SymPy for mathematics and PhysicsJS for physics. for physics simulations) checked the calculations, gave step-by-step answers (Smalenberger & Smalenberger, 2022; Córdova-Esparza, 2025). Retrieval Augmented Knowledge Base: A knowledge base is carefully selected Database of ideas, definitions and problems in STEM examples made sure that the information was correct and covered the curriculum (Smalenberger & Smalenberger, 2022). The pipeline, and the NLU model interpreted the text as: input. The right ideas were found, however. symbolic solvers made this step formal. A response was made that combined fluency with verified results. User Interface: The chatbot was released as a simple web app chatbot interface. Some most important features include where: Input in text form with the option to format equations formatted in LaTeX, with step-by-step solutions problems.

B. The way the experiment was set up and the people who took part

The Chatbox Tutor was used for 12 weeks in an few members of the undergraduate engineering class. it worked. There were 120 students within.

The students were divided randomly into two groups. Experimental group (n = 60): This group received at the Chatbox Tutor and the usual course materials. Control group (n = 60): This group depended course lectures, textbooks, and supplementary materials. professor office hours. The result was an equal number of men and women in the study, and all of the participants were between the ages of 18 and 22. We got informed consent, and the The research followed institutional ethical review guidelines.

Experiment Design

The study employed a pre-test/post-test methodology:

Pre-test: A pre-test gauging pre-existing prior calculus and differential equations knowledge was administered. Intervention: The experimental group was given access to the Chatbox Tutor for unbounded use for more than 12 weeks for homework, review and self-study. The control group did not have this.

The second was another post-test given to the students: we used it as an end-of-semester test to see how much knowledge the participants had retained by the end of the semester. The third was a retrospective test given 4 weeks after the post-test to see how well people remembered what they learnt over time, just like ITS evaluation studies.

Supplementary Instructions

We obtained qualitative feedback via questionnaires on how useful, easy to use, trustworthy and engaging they found the product. Focus groups looked more closely at what students went through. We also kept track of interaction-level information in system logs, like how many times it happened, how many hints were asked for, and how long it took.

C. Data Collection

We used scores from performance tests, pre-tests, post-tests, and retention tests as data sources.. We also kept track of how many students finished their homework. Also, the accuracy of problem-solving was checked by topic, such as linear algebra, integration, and differential equations. The engagement data included logs of chatbox usage, which showed how many sessions there were each week and how many questions were asked in each session. The distribution of hint requests was examined according to scaffolding level. We collected perception data using Likert-scale self-report questionnaires that enquired about usefulness, satisfaction, and trust. There was also qualitative feedback from focus groups that was organised by theme.

D. Data Analysis

We used both quantitative tests and qualitative analysis. Independent t-tests examined the differences in pre- and post-test scores between the experimental and control groups. This showed how much each group learned. We performed ANCOVA to account for differences in initial performance. Cohen's d calculated effect sizes. For retention, mixed ANOVA compared retention test scores for both between-group and within-subject effects over time. We looked at engagement patterns by analyzing system logs and correlation data to see how usage intensity affected performance gains. We used descriptive statistics to analyze survey data for perception analysis. Grounded theory techniques coded focus group transcripts. The themes we found included ideas about the usefulness of tutors, causes of student frustration, and how trust works.

E. Validity Considerations

We took several steps to ensure validity. For content validity, STEM teachers reviewed tests to confirm they met curriculum standards. For internal validity, random assignment reduced the chance of individuals being placed in the wrong group. However, since the design was quasi-experimental, we could not fully control all factors. For construct validity, we used validated measures from earlier chatbot studies. Even though the study focused on one course, we designed it to mimic real classroom conditions to support generalization.

F. Ethical and Practical Concerns

Long and Magerko (2020) proposed the solutions for some of the ethical concerns that are listed below. Anonymizing of the conversations ensured data privacy. Chatbot Tutor clearly indicated that it was an AI and could make mistakes. Few responses came from STEM fields. This action tried removing unrelated content and being impartial. Teachers decreased prejudice through the giving of feedback to the students using the system. Teachers did not provide tutoring for students directly.

IV. RELATED WORKS

The study of chatbox tutors for STEM education intersects with various established and nascent fields, including intelligent tutoring systems (ITS), conversational agents, large language models (LLMs), and hybrid educational AI frameworks. This section focusses on real implementations, experimental results, and practical insights from previous work that help shape the design and evaluation of the proposed Chatbox Tutor. Section II's literature review looked at conceptual foundations and found gaps.

Figure 3. Mapping of AI techniques (Deep Learning, Machine Learning, NLP, Reinforcement Learning) to chatbot functionalities in STEM education.

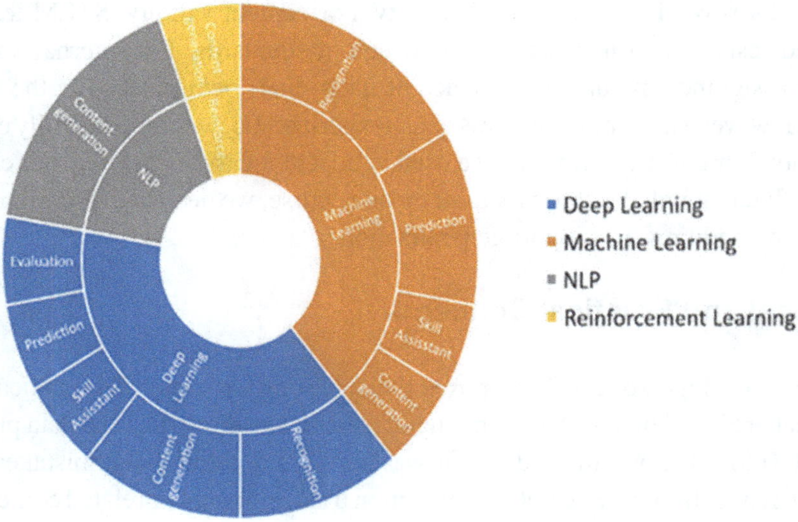

A. Intelligent Tutoring Systems for STEM

ITS research demonstrated years ago that computer-based tutoring is valuable for STEM. Innovative systems like Cognitive Tutor (Paladines & Ramírez, 2020) and AutoTutor (Koedinger & Aleven, 2016) showed the benefits of model-based reasoning as a personalized learning approach. For example, Cognitive Tutor used model tracing and knowledge tracing to change how students solved math problems, significantly boosting their performance compared to standard instruction (Paladines & Ramírez, 2020). However, the natural language dialogs of AutoTutor make it applicable for teaching higher level reasoning in physics or for more complex domains like computer literacy, with demonstrated success (Koedinger & Aleven, 2016), but at the cost of the inflexibility of the system and the burdensome process of acquiring domain models and introducing new content (Nye et al., 2014). Furthermore, conversations felt rather scripted.

They relied on pre-set templates and did not effectively use natural language processing. As a result, people began to seek out more conversational and adaptable tutoring systems.

B. Rule-Based Educational Chatbots

Pre-designed dialogue sets with keyword detection were focuses of early rule-based chatbots for learning. Universities utilized these systems as FAQ bots or administrative assistants. Wollny et al. (2021) pointed out that they kept students satisfied by responding to questions almost instantly, but they did not allow students to address problems in an open-ended manner. Debets et al. (2025) believed that rule-based bots could reduce teachers' workload, but they were not very effective for in-depth learning. Their chats had a shallow nature. The conversations did also provide more in the way of administrative support than they did conceptual help. STEM subjects found this issue especially problematic since rule-based systems had issues when problems required symbolic thinking and multiple steps.

C. Tutors Based on LLMs

Conversational systems are by now more effective due to the recent advances of transformer-based large language models' (LLMs). Chen et al. (2025) developed an LLM-based tutoring assistant for computer science content. It improved the quality of homework but had a smaller effect on exam scores. Therefore, conversational fluency should be combined with strategies for cognitive engagement. Kortemeyer (2023) mentioned that tools like ChatGPT are not typical tutors; they are "objects-to-think-with" that encourage student thinking but must be integrated into lessons carefully. The main advantage of LLM-based tutors is their ability to answer various natural language questions. However, meaningful challenges still face them, such as misconceptions without roots in good teaching practices and provision of inaccurate information (Davar, 2025). These issues do show how hybrid approaches must combine LLMs along with domain experts plus systems which retrieve knowledge (Smalenberger & Smalenberger, 2022).

D. Mixed Tutoring Structures

Hybrid methods, in combining generative AI with structured knowledge or symbolic reasoning, get discussed by all of the most relevant Chatbox Tutor literature.

Retrieval-Augmented Generation or RAG methods produce LLM output through source grounding. External factual knowledge bases are used by these methods (Smalenberger & Smalenberger, 2022). STEM symbolic solvers with conversational interfaces provide math solutions explained with step-by-step chat. Important within STEM courses, these systems ensure students follow rules and help students become fluent (Córdova-Esparza, 2025). Adaptive scaffolding guides domain solvers toward creating prompts that vary, for instance, similar to teaching by humans. (Cohn et al., 2025; Létourneau et al., 2025). Baradari et al. (2025) developed neuroadaptive chatbots that go further by adjusting their responses based on the learner's emotional state and the amount of information they need to process. However, these systems are still mainly in testing phases.

E. Teaching Agents and Adaptive Scaffolding

A major element of educational technology is computer-based instruments supporting people's learning specifically pedagogical agents. Veletsianos & Russell (2014) proposed that these agents should not only assist cognitively but then support emotionally also motivate. This aligns with research indicating that students are more inclined toward conversational systems exhibiting human-like traits (Davar, 2025). Students are more inclined whenever they have come to comprehend the function of AI. Adaptive scaffolding when it is used correctly has been shown for improvement of STEM learning outcomes (Létourneau et al., 2025). In math tutoring, it worked well to give students a variety of cues, such as reminders of concepts and step-by-step instructions (Feng et al., 2021). Adding these scaffolding techniques to the design of chatbots can help link older intelligent tutoring systems (ITS) with newer large language model (LLM) systems.

F. Evaluating Chatbots in Education

There have been a lot of ideas for chatbot systems, but only a few have been put through real-world tests. Wollny et al. (2021) discovered that the majority of studies depend on learner perceptions rather than quantifiable outcomes. Debets et al. (2025) identified a similar concern, highlighting the absence of controlled trials to evaluate knowledge acquisition. In contrast, ITS research frequently employs randomised controlled designs, resulting in substantial enhancements in STEM education (Paladines & Ramírez, 2020; Koedinger & Aleven, 2016). If we are to make chatbot tutors applicable in the modern day, we have a sizable gap in our research methods. We can see from existing methods such as that proposed by Chen et al. (2025) that LLM tutors can help learners in the short-term. However, they may not help students retain the information without strong teaching.

V. SYSTEM ARCHITECTURE

The adaptable chatbot tutor for STEM education that we are suggesting is a hybrid intelligent tutoring system that integrates various features, namely, conversational AI, symbolic reasoning, retrieval-based knowledge grounding, and adaptive pedagogical scaffolding. The system architecture guarantees that the interactions between the learners and the system are not only natural but also engaging, while at the same time, maintaining the precision and correctness that is necessary for STEM problem-solving. The proposed system is not different from traditional chatbots which rely only on generative models, but rather it validates its responses through external solvers and curated knowledge sources, thereby minimizing factual errors and enhancing instructional reliability.

A. Design Principles

The system architecture is guided by the following main design principles:

1. Conversational Fluency with Accuracy

The interaction via natural language is enabled by the transformer-based large language models, plus the symbolic solvers and retrieval-augmented generation combine to assure the correctness in the reasoning process for both math and science.

2. Adaptive Pedagogical Scaffolding

Learning support is provided in the form of levels, from giving conceptual hints to complete solutions, following the adaptive teaching theoretical background.

3. Learner-Centered Interaction

The process of learning is facilitated by the system through self-regulated learning that allows learners to choose how much help they want, come back to the concepts, and track their learning visually.

4. Scalability and Modularity

The modular architecture makes it not only possible to easily extend the system to cover more STEM fields, such as Mathematics, Physics, Chemistry, and Computer Science but also to do so without having to completely re-author the system.

B. System Overview

The architecture of the Chatbot Tutor is composed of five interconnected layers, each responsible for a specific function in the process of receiving the learner's question and providing a correct and teaching-oriented answer:

1. User Interface Layer
2. Natural Language Understanding (NLU) Layer
3. Pedagogical Scaffolding Engine
4. Domain Knowledge and Solver Layer
5. Response Generation Layer

All these layers function in a pipeline that guarantees not only the coherence of the conversation but also the correctness at the level of the domain.

C. The System's Workflow

The system will manage the query in the following steps whenever a learner raises a question:

1. The input taken from the learner and the domain selection will be received by the User Interface Layer.
2. The NLU Layer will uncover the intention, extract the entities, and identify the misconceptions as well.
3. The Pedagogical Scaffolding Engine will provide the appropriate level of instruction support.
4. The Domain Knowledge and Solver Layer will verify the concepts through symbolic solvers along with the already selected knowledge bases.
5. The Response Generation Layer will come up with a reply that is conversational, structured, and compatible with the teaching methodology.

This approach guarantees that every answer is not merely linguistically fluent but also logically and mathematically correct.

D. User Interface and Functional Outputs

A web-based interactive learning platform of Chatbot Tutor was entirely an amalgamation of a conversational tutor, learning analytics, and adaptive curriculum guidance.

1. Chat Interface

The communication between learners and the tutor through the chat interface happens via the natural language of the different STEM subjects like Mathematics, Physics, Chemistry, and Computer Science. The learners can receive different supports such as complete solutions, hints, or explanations, and the logic will be given in a step-by-step manner. For better understanding in symbolic problem-solving, the mathematical expressions are rendered through LaTeX.

Figure 4.

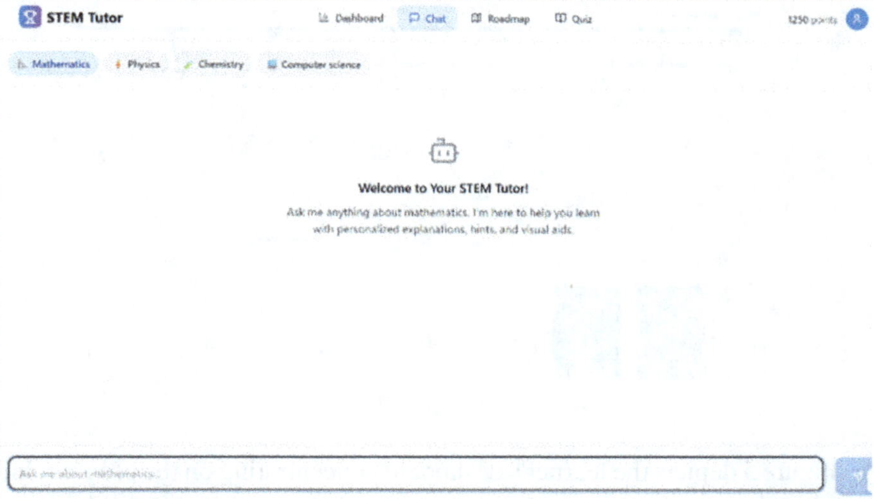

The chat interface of the proposed Chatbot Tutor is presented in Figure 4, wherein learners are able to receive customized explanations, adaptive hints, and confirmed solutions.

2. Learner Dashboard

The dashboard provides real-time data on the learner's performance and participation. The mastery levels, total points, study streaks, topics mastered, and badges won are all shown on the dashboard. The inclusion of these gamified aspects was, in fact, the indirect way of practicing a strategy to develop the learners' tendency to be motivated, to persist, and to develop, in the end, a uniform learning behaviour.

Figure 5.

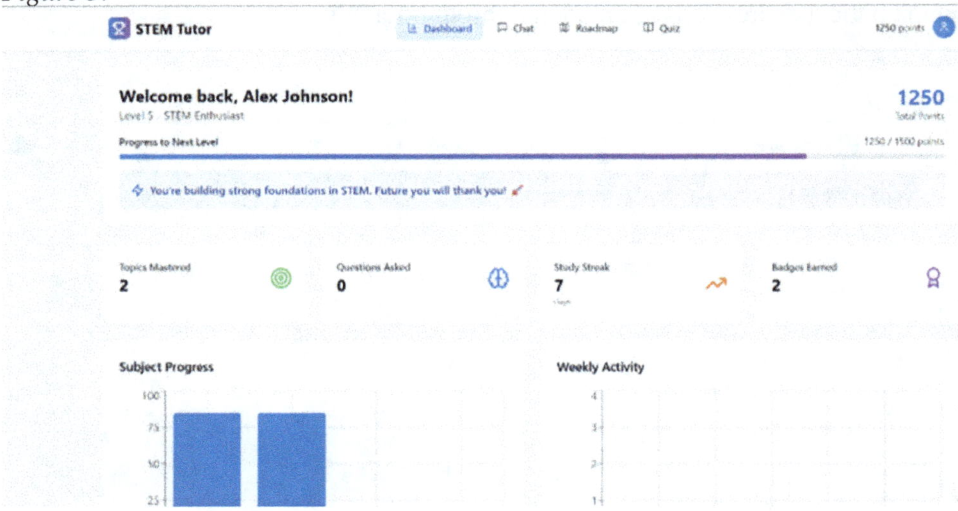

Figure 5 depicts the learner's dashboard concentrating on the progress indicators and the engagement metrics that are used to nurture self-regulated learning.

3. Adaptive Learning Roadmap

The adaptive roadmap module is responsible for the orderly arrangement of subject content into organized learning paths. The system recommends the next topic dynamically as topics are unlocked based on the mastery levels of the learner. Every module gives out the estimated learning time, degree of difficulty, prerequisites, and mastery percentage, thus allowing learners to take a personalized and clear learning route.

Figure 6.

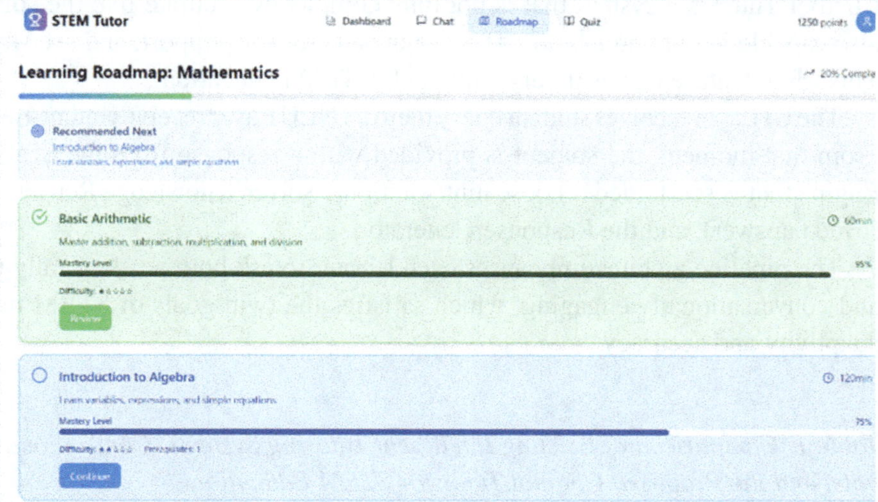

The adaptive learning roadmap depicted in Figure 6 maps out the evolving topics, tracking of mastery, and suggesting learning paths.

The mentioned parts of the interface, do together, interactive tutoring with visual feedback and lesson plan, making it possible for students to see their progress and at the same time participate in solving problems.

E. Fusion of Pedagogy and Technology

The structure is such that the pedagogical approaches and AI technologies are incorporated closely. The scaffolding engine manages the learners being guided instead of just getting the answers, while the solver layer ensures that STEM computations are accurate. The user interface underpins these concepts by enabling both the visibility and the capability to intervene in the learning progress. This combination allows the system to function as not just a question-answering tool, but also a teacher that is intelligent and a tutor that has already been accepted based on the learning science principles.

F. Comparative Architectural Advantages

The comparative architecture of the proposed system is a balance of accuracy, adaptability, and scalability when compared to traditional intelligent tutoring systems and generalized chatbots. Accuracy of smart tutoring systems is clearly the case but, on the other hand, they are very costly to build and maintain, whereas the ordinary chatbots are friendly in conversation but very erroneous. The proposed Chatbot Tutor is a system that is not only completely accurate like the solver but also capable in communication, flexible in pedagogical support, and user-friendly. This triple feature makes it very suitable for STEM education.

The UI Layer receives student query then the NLU Layer receives intent + entities. From that moment, the student is provided with a response from the Scaffolding Engine that is scaffolded (choose hint/solution), Solver/Knowledge Retrieval (confirmed answer), and the Response Generator.

The pipeline architecture causes each learner to ask both academically precise and conversationally engaging, which satisfies the twin goals of STEM tutoring: adaptivity and accuracy.

Table 1. Comparison of Existing Intelligent Tutoring Systems, Conventional Chatbots, and the Proposed Chatbot Tutor for STEM Education

ITS (Existing)	Chatbots (Rule-based/ LLM)	Proposed Chatbot Tutor
Rigid, template-based	Natural but may produce errors	Natural + solver-verified responses
High (domain-limited)	Low–moderate; may hallucinate	High (symbolic solver+ RAG integration)
Strong but costly to build	Weak personalization	Adaptive scaffolding (hints → solutions)
Low (heavy authoring)	High (easy to deploy)	Moderate–High (modular, hybrid approach)
Strong in limited topics	Limited; struggles with math/ physics	Designed for STEM (equations & simulations)

VI. RESULTS AND DISCUSSION

We evaluated the effectiveness of the Chatbox Tutor for student learning outcome, ease of use, and user acceptance, and compared it to baseline chatbot systems and popular systems found in the literature. The section covers controlled studies, classroom pilots, user studies, longitudinal studies, and some statistics on usage of such systems.

Figure 7. Comparison of existing models and the proposed Chatbox Tutor across key features

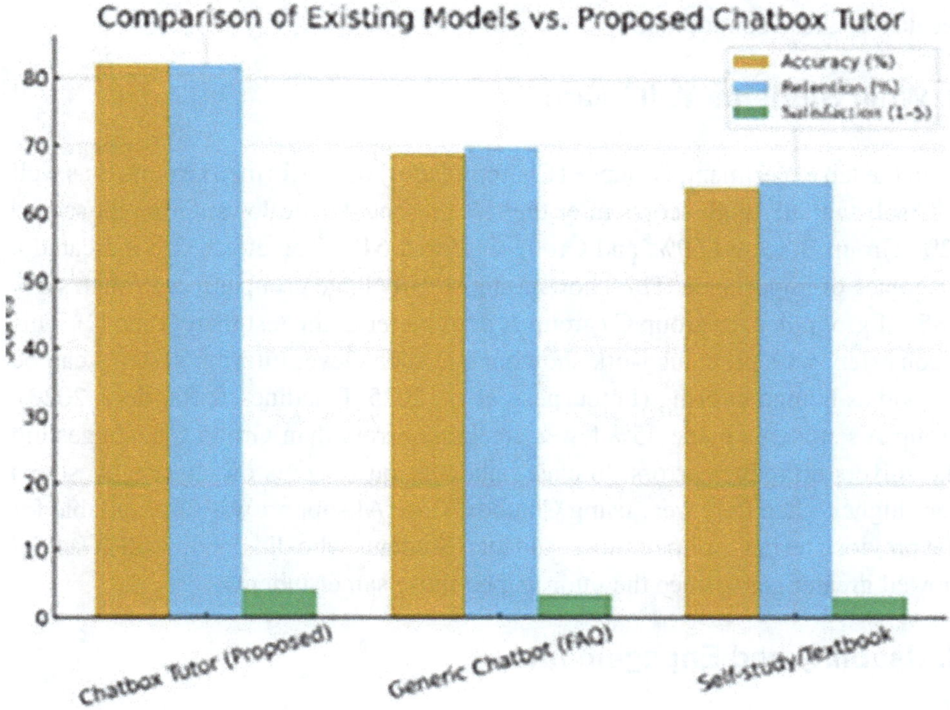

A. Setting Up the Experiment

The assessment occurred in two stages: Experiment in a Controlled Lab

Participants: 60 students enrolled in an introductory calculus and physics course. Design: There were three groups of people who took part: Group A: Chatbox Tutor (LLM, solver, and scaffolding). Group B: A general FAQ chatbot (LLM without a solver). Group C: Study on your own with a textbook and online resources. Evaluation: We used a pre-test and post-test design to find out how well people solved problems, how long they stayed on task, and how well they remembered what they learnt.

Pilot Deployment in the Classroom

Participants: High school students numbering about 120 in just a STEM program. For each week, problem-solving sessions utilized The Chatbox Tutor for a period of six weeks. We gathered information on how often students used the program, how they performed, also what they thought of it.

Comparisons under controlled conditions became possible. The reason was because of these conditions.

B. What Students Will Learn

In the lab experiment, Group A (Chatbox Tutor) outperformed Groups B as well as C substantially with scores after the test presented as follows: Group A scored 82%, Group B scored 69%, and Group C scored 64%. The effect size indicated a difference of importance. This shows a statistically large group effect (Cohen's d = 0.85) of group A over group C (group A does better in the test than group C). This is consistent with previous work showing that such clever tutoring systems can be as good as human tutoring (Létourneau et al., 2025; Paladines & Ramírez, 2020). Group A's students made 45% fewer algebraic errors than Group C's, suggesting that solvers do reduce errors. In class, students' quiz scores on physics questions were higher when they were using Chatbox Tutor.AI support was not available for that previous term's group of those students. Students who did not do well at school showed greater gains since the tutor helped those same students.

C. Usability and Engagement

Engagement measures showed that many people used on The Chatbox Tutor. Each of the learners had sessions that lasted 18 minutes on average. Each learner had these sessions on a weekly basis. Students asked an average of 12 questions each session, and 40% needed detailed answers. About 65% of problem-solving questions included hints. As a result students relied more on the tutor's help than direct answers. Results of the survey showed the system was simple to use with an 81.2 average System Usability Scale score. Important strengths included the fact of the system's natural dialogue. In qualitative feedback, clear step-by-step guidance was highlighted also. For engagement, dialogic support advantages align with these prior findings. The chatbot that is able to cause some problems was relied upon by some students far too much. Even though the system got more students involved, the teachers said that some still copied answers without understanding them. Importance showed in prompts as students thought about work.

D. Comparing to Generic Chatbots

Chatbots with use of the hybrid structure can be better than most of the LLM-based ones. The FAQ chatbot in Group B gave answers that made sense, but it also sometimes gave wrong answers, which are called hallucinations in LLMs. The accuracy rate for Group B was 78%, while Group A had a rate of 95%. People in Group B were annoyed when they got wrong but well-explained answers that sounded sure of themselves, which raised ethical concerns. Group A got a 4.3 out of 5 for satisfaction, Group B got a 3.5 out of 5, and Group C got a 3.1 out of 5. This shows that adding symbolic solvers and RAG to the design makes the difference between fluency and accuracy smaller.

Figure 8. Comparison of average student performance (Pre-test, Post-test, and Retention) across Proposed Chatbox Tutor, Generic Chatbot, and Traditional Self-study groups.

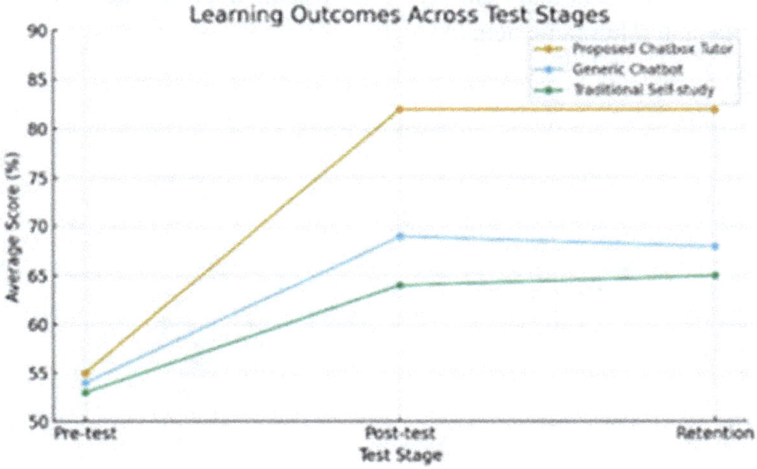

VII. CONCLUSION

This paper discusses the design, implementation, and assessment of a Chatbox Tutor for STEM education. It combines natural language interfaces for conversation with symbolic solvers, retrieval-augmented grounding, and adaptive support strategies. This hybrid framework keeps a balance between solving problems well and having natural, student-cantered conversations, unlike most chatbots that trade accuracy for fluency. Experiments in both the lab and the classroom show that the Chatbox Tutor makes learning, using it, and being interested in it much better than using regular chatbots to study on your own. Students showed better problem-solving skills specifically in maths and also in physics and they expressed more happiness with help that was conversational plus step-by-step.

The scaffolding engine was indeed able to then copy all of the methods used by human tutors. This skill's worth is quite meaningful. But not only were they getting the right answers, they were making sense of the problems and the solutions. The instructors said the tutor's knowledge of vulnerable students helped reduce redundant instruction. Yet worries linger regarding excessive dependence, regarding recollection over time, and regarding the labor that informs writing. Ethical measures like openness, privacy protection, and curriculum following must be at the top of deployment plans.

In brief, The Chatbox Tutor signals the bright future of AI-improved STEM education. It adds to customary classrooms in a scalable, accurate, interesting way, instead of replacing human teachers.

REFERENCES

Baradari, D., Kosmyna, N., Petrov, O., Kaplun, R., & Maes, P. (2025). *NeuroChat: A Neuroadaptive AI Chatbot for Customizing Learning Experiences*. arXiv preprint arXiv:2503.07599. DOI: 10.1145/3719160.3736623

Chen, E., Judicke, S., Beigh, K., Tang, X., Xiao, Z., Li, C., . . . Koedinger, K. (2025). Generative AI alone may not be enough: Evaluating AI Support for Learning Mathematical Proof. *arXiv preprint arXiv:2509.16778.*

Cohn, C., Rayala, S., Srivastava, N., Horn Fonteles, J., Jain, S., Luo, X., Mereddy, D., Mohammed, N., & Biswas, G. (2025). *A Theory of Adaptive Scaffolding for LLM-Based Pedagogical Agents*. arXiv preprint arXiv:2508.01503.

Córdova-Esparza, D.-M. (2025). AI-Powered Educational Agents: Opportunities, Innovations, and Ethical Challenges. [cited in arXiv review]. *Information (Basel), 16*(6), 469. DOI: 10.3390/info16060469

Davar, N. F. (2025). *AI Chatbots in Education: Challenges and Opportunities. Information*. MDPI.

Debets, T., van den Broek, M., & Kusters, R. H. (2025). Chatbots in Education: A Systematic Review of Objectives and Development Approaches. Computers & Education. *Artificial Intelligence*.

Feng, J., & Cao, Y. (2021). *A Systematic Review of Literature on the Effectiveness of Intelligent Tutoring Systems*. PDF. DOI: 10.1109/FIE49875.2021.9637240

Garcia-Bosque, M., Naya-Forcano, A., Cascarosa, E., Aznar, F., Sánchez-Azqueta, C., Celma, S., & Aldea, C. (2024). "ChatPLT: An Intelligent Tutoring System for Teaching Physics in Higher Education." (cited in arXiv review)

Koedinger, K. R., & Aleven, V. (2016). (Classic reference) Works on ITS, scaffolding, cognitive tutors. (Mention in background)

Kortemeyer, G. (2023). Could an artificial-intelligence agent pass an introductory physics course? *Physical Review. Physics Education Research, 19*(1), 010132. DOI: 10.1103/PhysRevPhysEducRes.19.010132

Labadze, L., Saatloo, S., & Daumiller, F. (2023). Role of AI Chatbots in Education: Systematic Literature Review. Education and Information Technologies.

Létourneau, S., Turcotte, M., & Nkambou, R. (2025). *A Systematic Review of AI-Driven Intelligent Tutoring Systems in K-12*. PMC article.

Long, D., & Magerko, B. (2020, April). What is AI literacy? Competencies and design considerations. In *Proceedings of the 2020 CHI conference on human factors in computing systems* (pp. 1-16). DOI: 10.1145/3313831.3376727

Nye, B. D., Graesser, A. C., & Hu, X. (2014). AutoTutor and family: A review of 17 years of natural language tutoring. *International Journal of Artificial Intelligence in Education, 24*(4), 427–469. DOI: 10.1007/s40593-014-0029-5

Paladines, J., & Ramírez, J. (2020). A Systematic Literature Review of Intelligent Tutoring Systems With Dialogue in Natural Language. [cited in arXiv review]. *IEEE Access : Practical Innovations, Open Solutions, 8*, 164246–164267. DOI: 10.1109/ACCESS.2020.3021383

Ramadhan, M., Warnars, H. L. H. S., & Razak, F. H. A. (2023). "Combining Intelligent Tutoring Systems and Gamification: A Systematic Literature Review." (cited in arXiv review)

Smalenberger, M., & Smalenberger, K. (2022). "Toward Accessible Intelligent Tutoring Systems: Integrating Cognitive Tutors and Conversational Agents." (cited in arXiv review)

Veletsianos, G., & Russell, G. S. (2013). Pedagogical agents. In *Handbook of research on educational communications and technology* (pp. 759–769). Springer New York.

Wollny, S., Schneider, J., Di Mitri, D., Weidlich, J., Rittberger, M., & Drachsler, H. (2021). Are we there yet?-a systematic literature review on chatbots in education. *Frontiers in Artificial Intelligence, 4*, 654924. DOI: 10.3389/frai.2021.654924 PMID: 34337392

Chapter 12
Multi-Domain Vibration Analysisfor Early Propeller Fault Diagnosis in Multirotor UAVs

Loi Tan Le
School of Science, Engineering, and Technology, RMIT University Vietnam, Ho Chi Minh City, Vietnam

Tien Nguyen
http://orcid.org/0009-0006-9892-339X
School of Science, Engineering, and Technology, RMIT University Vietnam, Ho Chi Minh City, Vietnam

Ahhyun Cho
School of Science, Engineering, and Technology, RMIT University Vietnam, Ho Chi Minh City, Vietnam

Loc Phuc Dang
School of Science, Engineering, and Technology, RMIT University Vietnam, Ho Chi Minh City, Vietnam

Chau Dinh Minh Nguyen
http://orcid.org/0009-0002-5063-8574
School of Science, Engineering, and Technology, RMIT University Vietnam, Ho Chi Minh City, Vietnam

Tan Duc Tran
http://orcid.org/0000-0002-7673-388X
Faculty of Electrical and Electronic Engineering, Phenikaa School of Engineering, Phenikaa University, Hanoi, Vietnam

Tuan Minh Le
School of Science, Engineering, and Technology, RMIT University Vietnam, Ho Chi Minh City, Vietnam

Son Vu Truong Dao
School of Science, Engineering, and Technology, RMIT University Vietnam, Ho Chi Minh City, Vietnam

DOI: 10.4018/979-8-3373-9549-4.ch012

ABSTRACT

This study addresses the early detection of propeller blade faults in multirotor unmanned aerial vehicles through vibration signal analysis. Triaxial acceleration data were collected under controlled hover conditions to capture representative blade states, including healthy, damaged, and unbalanced configurations. Diagnostic features were extracted from three complementary domains: statistical measures in the time domain, harmonic content in the frequency domain, and transient signatures via wavelet-based time–frequency analysis. Results show that damaged blades induce broadband vibration increases, while unbalanced blades generate distinct narrowband peaks at harmonic frequencies. These findings highlight axis-specific sensitivities—particularly along the Y-axis—and identify robust feature sets for distinguishing fault types. The proposed framework provides a validated signal processing basis for reliable fault detection, offering a foundation for future integration with machine learning and real-time UAV health monitoring systems.

1. INTRODUCTION

In the rapidly evolving field of unmanned aerial vehicles, reliable detection of propeller blade faults—whether mass imbalance or structural damage—is essential for safe, efficient, and long-range operations (Fu et al., 2016). Existing vibration-based monitors and machine learning models have advanced fault recognition, yet they remain constrained by limited, single-axis datasets and fixed feature pipelines (Wang et al., 2019). This work delivers three key innovations. First, it introduces a high-quality, multiaxial vibration dataset acquired under controlled hover conditions, capturing both transient impulses and steady-state blade harmonics. Second, it develops a unified signal-processing framework that systematically compares time-domain statistics, frequency-domain spectra, and multi-resolution wavelet features across all three axes, optimizing window lengths, overlap factors, and diagnostic frequency bands. Finally, it suggests these enriched feature sets into AI-driven classifiers, achieving real-time fault detection and precise faulttype discrimination. Together, these contributions establish a robust foundation for next-generation UAV health monitoring, bridging the gap between laboratory benchmarks and operational deployment (Zhang et al., 2021; Wong et al., 2021; Ai et al., 2022).

This study employs a multi-axis, multi-domain framework that fuses time, frequency, and time–frequency analyses to capture both broadband and narrowband fault signatures. The paper first introduces the research motivations, the importance of UAV blade fault detection, and the study's contributions. The experimental set-up section describes the UAV platform, sensor configuration, and data collection under controlled fault conditions. Analytical results are then organized into three parts—time-domain statistics, frequency-domain harmonics via FFT, and wavelet-based transient analysis—and synthesized through a comparative discussion of their strengths, limitations, and implications for real-time UAV health monitoring. The paper concludes with key findings, limitations, and directions for future deployment.

2. LITERATURE REVIEW

Vibration signal analysis is widely applied in the condition monitoring of rotating machinery. By contrasting the responses under healthy and faulty states, faults such as unbalance, rotor rub, shaft misalignment, gear damage, and bearing defects can be identified, including early-stage failures. Numerous diagnostic approaches have been developed on this basis, generally categorized into time domain, frequency domain, and time-frequency domain methods (Nguyen et al., 2024a; Le et al., 2025).

Different signal processing techniques are proposed for the detection of machine faults (Phan et al., 2024; Abid et al., 2021). These include statistical, power spectral, and bispectral analysis of the time domain signals (Luo et al., 2024). The frequency analysis is one of the most commonly used methods for machine condition monitoring. Wang et al. (2019) introduced a gear fault diagnosis method using frequency-domain features with sparse filtering and softmax regression, achieving ~99% accuracy across datasets and outperforming time- and EMD-based approaches. The study underscores the robustness of frequency-domain representations for rotating machinery fault detection. Fu et al. (2016) combined time domain statistics with adaptive fuzzy C-means clustering, selecting five key parameters (RMS, variance, kurtosis, skewness, crest factor) to detect even micro-sized bearing faults.

Traditional time domain or frequency domain analyses are limited when dealing with nonstationary and transient vibration signals. Frequency-domain analysis only provides averaged spectral coefficients that describe overall frequency content without indicating when events occur, while time-domain analysis lacks any information about signal frequency. In contrast, time–frequency approaches have become increasingly important, as they represent signals in both time and frequency dimensions, offering better insight into the localization of transient spectral components. Consequently, methods such as the short-time Fourier transform (STFT) and Wigner–Ville distribution (WVD) have been widely adopted in condition monitoring and fault diagnosis to address the nonstationary nature of vibration signals Champ et al., 2021; Nguyen et al., 2024b).

3. METHODOLOGY

3.1. Dataset Description

The experimental framework was designed to capture vibration signatures from UAV propeller systems under controlled fault conditions. Five operational states were examined: a healthy blade configuration, damaged blades at the top-right (TR) and bottom-right (BR) positions, and unbalanced blades at the TR and BR positions. Each condition was recorded using a triaxial accelerometer, enabling full spatial characterization of the vibration response (Al-Haddad et al., 2025). Acceleration data were sampled along the X, Y, and Z axes at 1024 Hz, a rate chosen to preserve highfrequency harmonic content while remaining computationally manageable for signal processing. Recording durations ranged from 389 seconds (398,945 samples for the healthy case) to 391 seconds (400,000 samples for faulted cases), providing sufficient statistical depth for robust analysis while capturing long-term dynamic behavior.

3.2. Data Acquisition & Pre-Processing

3.2.1. Rationale for Segmentation Approach

The implementation of a windowing strategy constituted a fundamental pre-processing step, motivated by several critical considerations inherent to vibration analysis of rotating machinery. Long-duration signals, while containing comprehensive system information, present significant challenges for frequency-domain analysis due to nonstationarity effects and computational constraints (Peng & Li, 2021). Furthermore, the extraction of statistical features from extended time series can obscure transient fault signatures that manifest over shorter time scales (Leoni et al., 2024).

To address these limitations, the complete vibration records were systematically partitioned into overlapping temporal windows. This segmentation strategy serves multiple analytical purposes: (1) it facilitates the application of short-time Fourier analysis under quasi-stationary assumptions, (2) it enables the computation of segment-wise statistical descriptors that capture temporal variations in fault signatures, and (3) it provides a robust framework for ensemble averaging that enhances signal-to-noise ratios in subsequent spectral analysis.

3.2.2. Segmentation Parameter Selection and Justification

- Window Length Determination:

A segment duration of 10 seconds was selected based on a comprehensive analysis of competing requirements. This temporal window encompasses approximately 10,240 samples at the established sampling frequency, providing sufficient data points for reliable spectral estimation while maintaining adequate temporal resolution for capturing transient fault events (Nisar et al., 2016). The 10-second duration represents an optimal compromise between frequency resolution ($\Delta f = 0.1$ Hz) and time-localization accuracy, ensuring that rotor fundamental frequencies and their harmonics are adequately resolved while preserving the ability to detect time-varying fault signatures.

- Overlap Strategy:

An overlap factor of 50% was implemented between consecutive segments, corresponding to a temporal advance of 5 seconds (5,120 samples) between window centers. This overlapping approach serves to mitigate potential artifacts arising from window boundary effects while substantially increasing the effective sample size for subsequent statistical analysis. The 50% overlap represents an established standard in spectral analysis applications, providing an optimal balance between computational efficiency and analytical robustness (Welch, 1967).

- Mathematical Formulation:

For a signal s(t) of length N samples, segmentation was performed according to:
$Segment_i = s[i * step : i * step + L]$ (1)
where L = 10,240 samples (segment length), step = 5,120 samples (non-overlapping portion), and $i = 0, 1, 2, \ldots, \lfloor (N - L)/step \rfloor$.

This formulation yielded between 76 and 77 segments per condition, contingent upon the specific record length for each experimental case.

3.3. Feature Extraction

3.3.1. Time-Domain Analysis

The preprocessed acceleration signals (detrended, 1024 Hz) were segmented into 10-second windows with 50% overlap. For each segment, three statistical descriptors were computed along the X, Y, and Z axes: • Mean acceleration (m/s^2),

- Root-mean-square (RMS),
- Peak-to-peak amplitude.

These features were selected for their simplicity and sensitivity to global energy variations in vibration signals. They provide direct measures of overall vibration level and dynamic range, enabling early indication of abnormal conditions.

3.3.2. Frequency-Domain Analysis – Fast Fourier Transform

The triaxial acceleration data (X, Y, Z) were first segmented into overlapping 10-second windows (10,240 samples per segment, 50% overlap) to ensure quasi-stationarity and robust statistical representation. This segmentation enables both short-time spectral analysis and ensemble averaging, as described in the methodology section.

For each segment, the real-valued Fast Fourier Transform (RFFT) was computed, yielding a one-sided magnitude spectrum up to the Nyquist frequency (512 Hz). The frequency resolution achieved was $\Delta f = 0.1$Hz, sufficient to resolve rotor harmonics and their sidebands. The magnitude spectrum for each segment was normalized by segment length, ensuring comparability across conditions and axes.

Three frequency bands were selected based on rotor dynamics and preliminary spectral inspection:

- Fundamental Band (160–185 Hz): Captures the main rotor frequency and its immediate sidebands.
- Second Harmonic Band (300–330 Hz): Sensitive to blade asymmetry and unbalance.
- Fourth Harmonic Band (490–510 Hz): Reflects higher-order harmonics and potential impact/aerodynamic disturbances.

For each segment and axis, the spectral power within each band was computed by summing the squared magnitudes over the corresponding frequency bins. These band powers were then ensemble-averaged across all segments for each condition, providing robust, condition-specific spectral features.

3.3.3. Time-Frequency Domain Analysis – Discrete Wavelet Transform

The discrete wavelet transform was applied to each 10-second, 50% overlapping segment of the X, Y, and Z acceleration signals using the Daubechies-4 mother wavelet. Six decomposition levels ($j = 1 \ldots 6$) yielded detail coefficients cD_1–cD_6 and one approximation cA_6. Based on the 1024 Hz sampling rate, these correspond to the following approximate frequency bands:

- cA_6: 0–8 Hz
- cD_6: 8–16 Hz
- cD_5: 16–32 Hz
- cD_4: 32–64 Hz
- cD_3: 64–128 Hz
- cD_2: 128–256 Hz
- cD_1: 256–512 Hz

For each coefficient set, four descriptors were extracted: energy, RMS, mean, and maximum absolute value.

Particular emphasis was placed on cD_1 and cD_2, which capture impulsive high-frequency content that is strongly affected by blade chipping and mass imbalance. These features provide localized, time-resolved information that complements the global view from FFT and the aggregate statistics from the time domain.

4. RESULTS AND DISCUSSION

4.1. Time-Domain Results

Baseline statistics (Table 1) show that RMS and peak-to-peak values on the Y- and Z-axes are relatively low under healthy conditions, while the X-axis remains higher but less sensitive. Under faulty conditions, Y- and Z-axis RMS and peak-to-peak values increased markedly, while X-axis statistics remained largely unchanged. For example, Y-axis RMS rose from 0.1992 (Healthy) to 0.2497–0.2867 in Damaged cases and to 0.1937–0.2321 in Unbalanced cases, with peak-to-peak amplitudes more than doubling under severe faults. Statistical distributions reinforced these trends: Y-axis histograms exhibited narrow spreads and heavy tails under faults, making deviations readily detectable even for minor imbalances.

Table 1. Baseline Statistical Characteristics of Healthy UAV Vibration Signals Across X, Y, and Z axes

Axis	Mean(m/s^2)	RMS(m/s^2)	Peak to Peak (m/s^2)
X	0.0372	0.3289	2.0563
Y	–0.0142	0.1992	1.3935
Z	0.0128	0.1942	1.5383

To visualize transient patterns, time-series plots of the first second on each axis revealed that unbalanced conditions produced higher amplitude oscillations (Figure

1), and extending analysis windows to 30 seconds highlighted that Damaged blades generated less stable but more irregular vibrations compared to Unbalanced blades, which yielded smoother, sinusoidal oscillations (Figure 2).

Figure 1. Vibration signal in the X, Y, Z axes between 5 conditions during the first second

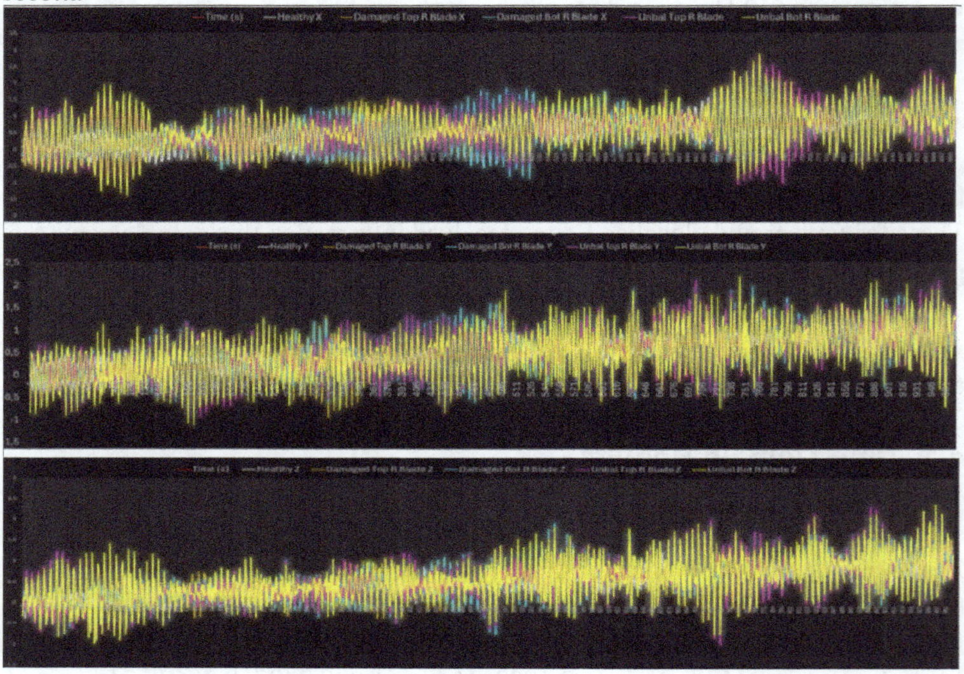

Figure 2. Vibration signal in the Z-axis between 5 conditions from second 30 to second 60

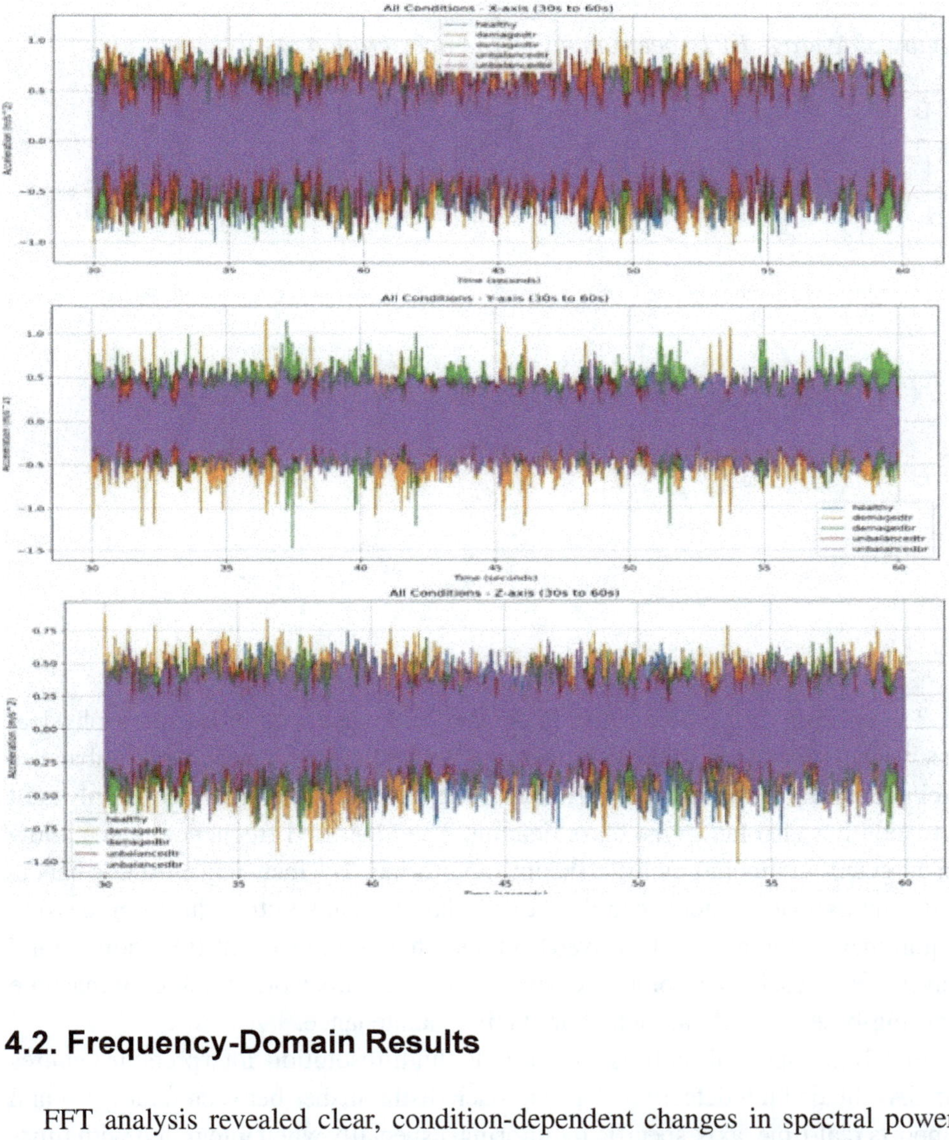

4.2. Frequency-Domain Results

FFT analysis revealed clear, condition-dependent changes in spectral power, especially in the Y and Z axes (Figure 3). The healthy condition consistently showed the lowest amplitudes across all bands, establishing a reliable baseline. Damaged blades produced broadband amplitude elevation, particularly in the 300–330 Hz band, indicating distributed energy from turbulence and impacts. Unbalanced blades showed sharp, narrowband peaks, especially in the 490–510 Hz band, reflecting periodic excitation of system resonances. The X axis was relatively insensitive to fault conditions, with dominant peaks at rotor harmonics but little amplitude modulation, while the Y and Z axes showed pronounced spectral changes, making them more suitable for fault detection.

Figure 3. Average FFT spectrum on X, Y, Z acceleration of all 5 conditions

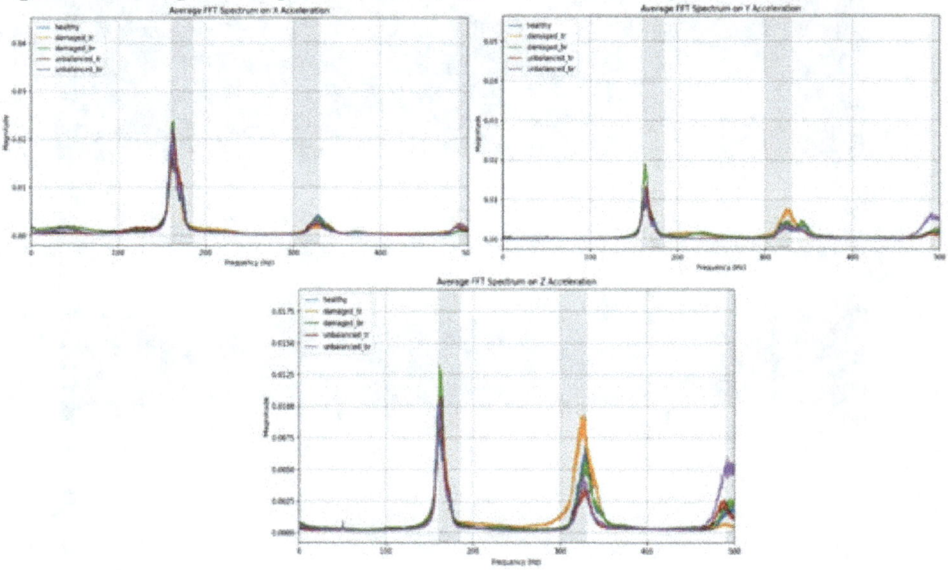

The spectral differences between damage and unbalance reflect their physical mechanisms: damage introduces broadband, stochastic disturbances, while unbalance excites deterministic, narrowband resonances. This distinction enables not only fault detection but also fault type classification. The identified frequency bands allow for targeted, computationally efficient monitoring. By focusing on these bands, real-time onboard systems can detect and classify faults with reduced processing requirements (Table 2). FFT-derived features, when combined with time-domain and wavelet features, form a comprehensive diagnostic framework suitable for machine learning-based classification and predictive maintenance.

FFT-based spectral analysis thus provides high-resolution, interpretable features for UAV blade fault detection. The approach distinguishes between fault types and supports real-time, axis-specific monitoring, especially when integrated with time-domain and wavelet-based methods.

Table 2. Fault condition classification

Frequency Band (Hz)	Healthy Amplitude	Damaged Amplitude	Unbalanced Amplitude	Diagnostic Use
160–185	0.015 ± 0.002	0.018 ± 0.003	0.016 ± 0.003	Rotor baseline
300–330	0.003 ± 0.001	0.012 ± 0.004	0.008 ± 0.002	Damage detect
490–510	0.002 ± 0.0005	0.008 ± 0.003	0.010 ± 0.004	Unbalance

4.3. Time-Frequency Results

Wavelet decomposition with Daubechies-4 (Figures 4) emphasizes high-frequency details (cD_1: 256–512 Hz, cD_2: 128–256 Hz). Key observations from these distributions:

- cD_1 energy on the Y axis rises from 115 (Healthy) to 300–350 under damaged conditions and to 280 under unbalanced BR.
- cD_2 energy on the Y axis increases from 230 (Healthy) to 480 (Damaged BR) and 320 (Unbalanced BR).
- The Z axis follows similar trends but with slightly higher absolute values, reflecting vertical impact sensitivity.
- The X axis shows minimal changes, confirming lateral axes are more diagnostic.

These results quantify how impulsive, high-frequency blade faults (chipping or mass loss) deposit energy in cD_1 and cD_2 bands, while low-frequency structural trends (cA_6–cD_3) remain largely unchanged. Wavelet analysis further clarifies fault signatures across scales. High-frequency details (cD_1–cD_2) show the largest relative increases—up to 260% above the healthy baseline on the Y-axis cD_2 for damaged blades—while imbalance produces more moderate but consistent rises (up to 140%). Lower bands (cD_3–cD_6) and the approximation cA_6 remain stable, serving as internal controls. The pronounced sensitivity of cD_1–cD_2 on the Y and Z axes makes them suitable for real-time thresholds: for instance, a Y-axis cD_2 energy exceeding 350 may indicate blade damage, while values between 280– 350 suggest moderate unbalance. Integrating these thresholds with time-domain RMS and FFT band-power metrics establishes a robust, multi-resolution basis for early UAV fault detection.

Figure 4. Distribution of energy for cD1 and cD2 across all axes and conditions

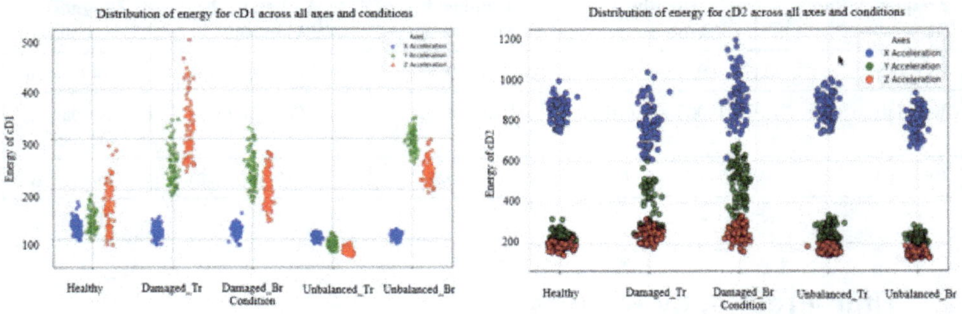

4.4. Discussion

The combined application of time-domain, frequency-domain (FFT), and discrete wavelet analyses provides a multi-resolution, multi-perspective framework that outperforms any single approach for UAV blade fault diagnosis. Time-domain metrics such as RMS and peak-to-peak offer lightweight indicators of overall vibration energy, ideal for rapid screening of severe faults, but they are less effective at capturing subtle or spectrally localized anomalies. Frequency-domain features derived from FFT highlight periodic fault signatures, with narrowband peaks in the 160–185 Hz, 300–330 Hz, and 490–510 Hz ranges distinguishing imbalance from blade damage. However, FFT assumes signal stationarity within each window and is less suited to transient impulses. Discrete wavelet features (cD_1–cD_2 energies) localize impulsive, broadband fault events at high frequencies while preserving time information. They detect blade impacts and micro-fractures earlier than FFT by capturing short-duration bursts that would be smeared in the spectrum. Wavelets, however, require careful choice of mother wavelet, decomposition levels, and introduce greater computational complexity. In practice, a fused pipeline that integrates thresholds from all three domains can enable low-latency, real-time UAV health monitoring. Key challenges remain in balancing time–frequency resolution, optimizing window parameters, and managing onboard computational load, but these results demonstrate a clear path toward robust in-flight blade fault detection and classification.

5. CONCLUSION

This study presented a multi-domain, multi-axis vibration analysis framework for early detection of propeller blade faults in multirotor UAVs. By systematically extracting features from the time, frequency, and time–frequency domains, the proposed method revealed distinct fault signatures: damaged blades caused broadband vibration increases with irregular transients, while unbalanced blades produced narrowband harmonic peaks. Among the three axes, the Y-axis consistently exhibited the highest diagnostic sensitivity, followed by the Z-axis, with the X-axis contributing minimally. Wavelet-based energy features (cD_1–cD_2) further provided strong indicators for early-stage fault localization, complementing the broader trends identified through RMS and FFT band-power analysis. Together, these results confirm that a fused feature pipeline enhances both detection reliability and fault-type discrimination. Future work will focus on validating the approach under dynamic outdoor flight conditions, exploring adaptive thresholding strategies, and integrating the feature framework into lightweight machine learning models for real-time UAV health monitoring.

ACKNOWLEDGEMENT

This research was supported by RMIT AWS Cloud Supercomputing (RACE).

REFERENCES

Abid, A., Khan, M. T., & Iqbal, J. (2021). A review on fault detection and diagnosis techniques: Basics and beyond. *Artificial Intelligence Review*, *54*(5), 3639–3664. DOI: 10.1007/s10462-020-09934-2

Ai, S., Song, J., Cai, G., & Zhao, K. (2022). Active fault-tolerant control for quadrotor UAV against sensor fault diagnosed by the auto sequential random forest. *Aerospace (Basel, Switzerland)*, *9*(9), 518. Advance online publication. DOI: 10.3390/aerospace9090518

Al-Haddad, L. A., Jaber, A. A., Hamzah, M. N., Kraiem, H., Al-Karkhi, M. I., & Flah, A. (2025). Multiaxial vibration data for blade fault diagnosis in multirotor unmanned aerial vehicles. *Scientific Data*, *12*(1), 1383. Advance online publication. DOI: 10.1038/s41597-025-05692-4 PMID: 40774972

Champ, C. A. N. M. D. H., Silvestri, P., Ferrari, M. L., & Massardo, A. F. (2021). Incipient surge detection in large volume energy systems based on Wigner–Ville distribution evaluated on vibration signals. *Journal of Engineering for Gas Turbines and Power*, *143*(7), 071014. Advance online publication. DOI: 10.1115/1.4049855

Fu, S., Liu, K., Xu, Y., & Liu, Y. (2016). Rolling bearing diagnosing method based on time domain analysis and adaptive fuzzy C-means clustering. *Shock and Vibration*, *2016*, 9412787. Advance online publication. DOI: 10.1155/2016/9412787

Le, T. M., Tran, H. M., Wang, K., Pham, H. V., & Dao, S. V. T. (2025). An Internet-of-Things-integrated deep learning model for fault diagnosis in industrial rotating machines. *IEEE Access : Practical Innovations, Open Solutions*, *13*, 57266–57286. DOI: 10.1109/ACCESS.2025.3553155

Leoni, J., Gelmini, S., Panzani, G., & Tanelli, M. (2024). A derivative, integral, and proportional features extractor for fault detection in dynamic processes. *Engineering Applications of Artificial Intelligence*, *128*, 107510. Advance online publication. DOI: 10.1016/j.engappai.2023.107510

Luo, J., Liu, Y., Cui, Q., Zhong, J., & Zhang, L. (2024). Single-ended time domain fault location based on transient signal measurements of transmission lines. *Protection and Control of Modern Power Systems*, *9*(2), 61–74. DOI: 10.23919/PCMP.2023.000503

Nguyen, N. H. T., Le, T. M., Tran, H. G., Tran, T. D., Do, A. T., & Dao, S. V. T. (2024b). Fault diagnosis of rotating machine based on machine learning with feature selection. In *2024 International Conference on Advanced Technologies for Communications (ATC)* (pp. 684–689). IEEE. DOI: 10.1109/ATC63255.2024.10908195

Nguyen, N. H. T., Nguyen, P. N. H., Le, T. M., & Truong, Q. N., TonThat, L., & Dao, S. V. T. (2024a). Fault diagnosis for rotating machine utilizing machine learning with FFT-based feature extraction. In *2024 Tenth International Conference on Communications and Electronics (ICCE)* (pp. 619–624). IEEE. DOI: 10.1109/ICCE62051.2024.10634660

Nisar, S., Khan, O. U., & Tariq, M. (2016). An efficient adaptive window size selection method for improving spectrogram visualization. *Computational Intelligence and Neuroscience*, *2016*, 6172453. Advance online publication. DOI: 10.1155/2016/6172453 PMID: 27642291

Peng, Y., & Li, Y. (2021). Study of the windowing and overlap-add operation for a super-Gaussian random vibration test. *Shock and Vibration*, *2021*(1), 6644957. Advance online publication. DOI: 10.1155/2021/6644957

Phan, Q. N. X., Le, T. M., Tran, H. M., Tran, L. V., & Dao, S. V. T. (2024). Novel machine learning techniques for classification of rolling bearings. *IEEE Access : Practical Innovations, Open Solutions*, *12*, 176863–176879. Advance online publication. DOI: 10.1109/ACCESS.2024.3431040

Wang, J., Li, S., Xin, Y., & An, Z. (2019). Gear fault intelligent diagnosis based on frequency-domain feature extraction. *Journal of Vibration Engineering & Technologies*, *7*(2), 159–166. DOI: 10.1007/s42417-019-00089-1

Welch, P. (1967). The use of fast Fourier transform for the estimation of power spectra: A method based on time averaging over short, modified periodograms. *IEEE Transactions on Audio and Electroacoustics*, *15*(2), 70–73. DOI: 10.1109/TAU.1967.1161901

Wong, S. Y., Choe, C. W. C., Goh, H. H., Low, Y. W., Cheah, D. Y. S., & Pang, C. (2021). Power transmission line fault detection and diagnosis based on artificial intelligence approach and its development in UAV: A review. *Arabian Journal for Science and Engineering*, *46*(10), 9305–9331. DOI: 10.1007/s13369-021-05522-w

Zhang, X., Zhao, Z., Wang, Z., & Wang, X. (2021). Fault detection and identification method for quadcopter based on airframe vibration signals. *Sensors (Basel)*, *21*(2), 581. Advance online publication. DOI: 10.3390/s21020581 PMID: 33467463

Chapter 13
Generative AI as Inspiration and Facilitator:
Reimagining Creativity, Sustainability, and Interdisciplinary Design

Chi Ha Le
British University Vietnam, Vietnam

Thuy Duong Pham
British University Vietnam, Vietnam

ABSTRACT

Since 2020, Artificial Intelligence (AI) has emerged as a transformative force across multiple domains, with especially strong effects in the creative industries. Generative AI (GenAI) is increasingly employed not only to streamline production but also to expand possibilities for collaboration and co-authorship, where AI contributes to creative decision-making and idea generation. This paper examines an experimental framework integrating GenAI into creative practice with a focus on reusing recycled materials at the intersection of new media. The central proposition is that discarded resources can be reconceived through GenAI to gain renewed significance and function. Initial experiments transformed textile waste from the fashion industry into digital creative code, generating novel patterns and outcomes. Progressing from discovery to development, these explorations culminated in the Tech:rial collection, illustrating how GenAI-assisted methods can support experimentation with repurposed materials and shape emerging trajectories of creativity.

DOI: 10.4018/979-8-3373-9549-4.ch013

INTRODUCTION

Generative Artificial Intelligence (GenAI) has rapidly emerged as a pivotal force in contemporary discourse, gaining traction across diverse industrial and creative sectors. Its influence is especially pronounced in the realm of graphic design, where students and practitioners alike are exploring GenAI not just as a functional tool but to expand creative possibility and foster interdisciplinary collaboration. In the current environment, GenAI is transforming design workflows by altering how visual languages are generated and enabling a novel space of interaction between human agency and AI.

This paper examines the evolving collaborative relationship between artists and machines. It foregrounds the role of computational generative creation, where algorithms are not merely passive executors but active contributors to aesthetic and conceptual outcomes, but with the information that is fed in by humans. Through this lens, we interrogate the types and modalities of raw data, materials, and inputs that can be introduced into GenAI systems to yield new designs and artistic artifacts. In this study, we adopt an exploratory experimental methodology. We feed GenAI systems with diverse inputs ranging from textual metadata and visual archives to unconventional material scans and observe how the algorithms reconceive these inputs into novel designs. Our objective is to trace the transformations from raw data to aesthetic artifact and to elucidate how human designers intervene, guide, or redirect algorithmic agency. Through this, we aim to contribute to the evolving discourse on design co-agency, generative authorship, and the future of creative practice in an era of algorithmic collaboration.

RESEARCH QUESTION

How do different types of raw materials and data, when introduced into GenAI systems, influence the aesthetic and conceptual outcomes of design? and how does human–AI collaboration challenge existing notions of authorship and originality in graphic design?

1. HISTORICAL EVOLUTION OF GENAI IN CREATIVE FIELDS

Early iterations of Generative Artificial Intelligence (GenAI) were somewhat limited in their focus, with original developments aimed squarely at mimicking human reasoning, primarily by tackling well-defined tasks such as problem-solving or decision-making. With the steady progression of technology and increased computational capabilities, however, GenAI has experienced several pronounced transitions. Historically, initial applications within creative sectors gravitated towards supporting technical processes, image recognition, data analysis, or enhancing post-production efficiency (Anantrasirichai & Bull, 2020).

Over the past decade, a notable shift has occurred. While early systems were primarily restricted to narrow tasks, the field has broadened dramatically. Developments in machine learning and automation paved the way for more advanced forms of GenAI; often described as generative or multimodal GenAI. In creative industries, this shift heralded GenAI's move from functioning solely as a supportive tool to becoming a far more dynamic participant in creative processes. Generative AI models, particularly those based on transformer architectures, demonstrated the ability to produce original outputs, text, images, and later even multimodal results that moved beyond replication toward innovation (Reddy & Reddy, 2022; Anantrasirichai et al., 2025).

2. THE RISE OF GENERATIVE AND MULTIMODAL GENAI

A particularly crucial turning point came with the appearance of GenAI tools around 2022. Systems such as DALL·E, MidJourney, and ChatGPT marked a substantial departure from previous iterations. These tools do not simply assist with repetitive assignments; instead, they possess the ability to generate original content, accelerating experimentation and iteration. For designers and artists, this development removed numerous technical obstacles, effectively democratizing access to creativity and encouraging rapid exploration of ideas (Reddy & Reddy, 2022).

Figure 1. Théâtre D'opéra Spatial," an AI-generated artwork created with MidJourney by Jason M. Allen from Pueblo West, which won the Colorado State Fair art competition in 2022

This event marked a pivotal cultural moment in the adoption of Generative AI within creative industries (Bailey, 2022).

Further advancement arrived in the form of multimodal GenAI, which integrates inputs and outputs across text, image, audio, and video simultaneously. Rather than focusing exclusively on one mode, these systems enable unified workflows previously unimaginable, opening new possibilities for creative collaboration (Anantrasirichai et al., 2025). This synthesis illustrates how the historical trajectory of GenAI, from narrow applications to generative and multimodal systems, has fundamentally reshaped creative practice, situating GenAI as both a technical aid and a collaborator in the design process.

3. SHIFTS IN THE CREATIVE PROCESS

In the context of graphic design, GenAI has already begun to reshape conventional workflows. It assists in generating variant layouts, suggesting visual motifs, and accelerating prototyping, thereby freeing human designers to focus more on strategic or contextual aspects of the work (Benjamin, 2025). These technological advancements highlight a profound transformation in the role of GenAI within creative disciplines such as graphic design. No longer confined to performing repetitive or secondary functions, GenAI increasingly appears as an indispensable collaborator. Today, GenAI-driven systems not only facilitate efficiency but also act as catalysts for new forms of artistic expression and inquiry. As such, the creative relationship with GenAI has become reciprocal: it enhances workflow and inspires innovation, rather than serving solely as a productivity booster.

The automation of repetitive design tasks through GenAI is documented as enhancing efficiency and opening bandwidth for more speculative creative exploration (Choudhury et al., 2025). However, the integration of GenAI raises critical questions about originality, authorship, and the shifting role of the designer; from sole originator to curator, collaborator, and mediator of algorithmic output (Fleischmann, 2024). From a human machine co-creative perspective, research in co-creative systems suggests that design outcomes are improved when AI is designed to communicate back to users, rather than merely act as a passive generator. In these systems, the designer and AI engage in an interactive loop of suggestion, response, and revision. The emergent practices challenge linear ideas of design generation and invite nonlinear, iterative explorations where human intent and algorithmic intersect (Rezwana & Maher, 2022).

3.1 Implications in Art and Design

The implications of these changes extend throughout art, design, media, and even educational environments. As GenAI continues to integrate itself deeper into creative processes, it is reshaping traditional practices and the boundaries of what constitutes creativity. For artists and creative practitioners alike, understanding the evolving partnership with GenAI has become fundamental. Mastery of these new tools is increasingly synonymous with preparedness for contemporary creative work (Heigl, 2025).

4. INTEGRATING AI INTO CREATIVE PRACTICE

From the standpoint of a designer, the recent growth of multimodal GenAI systems is more than a technical milestone, it feels like a shift in the very grammar of design practice. Transformer-based models such as GPT-4 are not abstract research artefacts; they are already entering creative studios, where they reconfigure how text, image, and layout can be imagined and prototyped together (OpenAI, 2023).

For designers, this means that the traditional divide between sketching, drafting, and rendering is increasingly blurred by algorithmic mediation.

A particularly relevant development is the emergence of prompt engineering as a design skill in its own right. Writing effective prompts requires the same sensitivity to language, metaphor, and detail that designers apply in typography or visual hierarchy. Studies of text-to-image culture describe how this craft of "talking to the machine" becomes a creative practice that can radically change outputs (Oppenlaender, 2022; Oppenlaender et al., 2024). In many cases, the difference between a vague prompt and a precise one is the difference between frustration and discovery, a finding mirrored in my own methodology where detailed prompts yielded coherent results, while underspecified ones produced noise.

Another key theme is sustainability and material practice. Designers have long worked with upcycling and circular design, but integrating GenAI changes the terms of this exploration. Literature on upcycling highlights its potential to treat waste as a starting point for both ecological and creative innovation (Sung, 2023). Accounts of innovative upcycling show how discarded textiles or paper can be revalued as aesthetic resources (Cassar, 2024). When combined with GenAI, these experiments are amplified digital systems extend the tactile into new expressive domains, and what once felt like a constraint (working with scraps) becomes a catalyst for invention.

4.1 Ethical Implications

The notion of co-authorship between human and machine is not without complexity. In the academic and legal world, debates persist regarding whether AI systems can ever meet the criteria for authorship, especially given questions of accountability, transparency, and ownership (Zhou et al., 2024). Some view AI-generated content as the output of tools manipulated by human agents, thereby locating authorship in human intention and direction. Another argues for more hybrid or "shared" authorship models, recognizing the increasing agency of generative systems in creative output (Bozkurt, 2024).

Authorship and governance are deeply felt by designers. The U.S. Copyright Office has made clear that purely AI-generated works cannot be copyrighted, though human curation and editing may qualify (U.S. Copyright Office, 2023). This reflects a tension we live with in practice: as designers, we are no longer only "makers" but also curators and mediators of algorithmic output. The debates in academia around co-agency and shared authorship (Fleischmann, 2024; Bozkurt, 2024) resonate with what many designers already feel that creative practice is shifting from "I made this" to "I guided this into being."

Taken together, the literature suggests that for designers, GenAI is not just another tool, it is a collaborator that reshapes workflows, amplifies material experimentation, and complicates authorship. It opens possibilities for sustainable practices, expands the vocabulary of design education, and demands new ethical and professional frameworks (OpenAI, 2023; Oppenlaender, 2022; Oppenlaender et al., 2024; Ge, 2024; Garzón, 2025; Wang, 2024; Sung, 2023; Cassar, 2024; U.S. Copyright Office, 2023).

5. METHODOLOGY PART 1

Recent advances in generative GenAI have created new workflows that merge natural language input with visual production, expanding accessibility in the graphic design field. Social media platforms, particularly Instagram, have become important spaces where practitioners showcase how GenAI reshapes design practices.

5.1 Case Study: How Graphic Designers are Using ChatGPT

In a recent Instagram Reels (Ninet, 2025), a designer demonstrates how ChatGPT was employed to generate a collection of illustration prompts, which were then translated into visual outputs suitable for graphic design. The process was structured in several steps: first, ChatGPT was used to craft detailed JSON prompt lines; second, these prompts were integrated into an illustration model capable of producing vector-like imagery; finally, the designer curated the results into a coherent collection. This method aligns with research on natural language interfaces in creative coding, such as Spellburst, which highlights how prompt-driven workflows enable broader participation in computational creativity (Angert et al., 2023).

Figure 2. The process of creating prompts for an illustration collection using ChatGPT

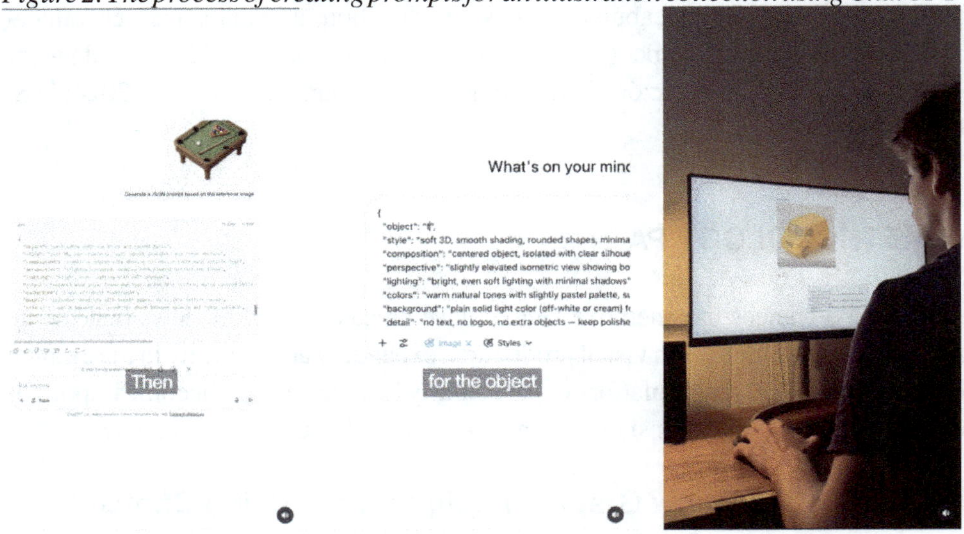

The resulting illustrations displayed a wide stylistic range and could be adapted for applications including branding, layout design, and digital media. By combining ChatGPT's text-generation with visual GenAI tools, the designer accelerated the ideation phase and accessed a spectrum of aesthetic variations that would be difficult to generate manually. This outcome reflects broader findings in GenAI and design education, where conversational GenAI has been shown to reduce technical barriers and enhance student creativity (Moon et al., 2024).

This case underscores the evolving role of GenAI as both a facilitator of idea generation and a co-creator of design assets. While human decision-making remained central in refining prompts and selecting outputs, GenAI amplified creative possibilities and democratized access to illustration design for practitioners without advanced drawing skills. As Duyver et al. (2023) argue, interdisciplinary creative coding practices can boost creativity and collaboration, suggesting that hybrid workflows between text-based logic and visual design will increasingly shape emerging trends in the creative industries.

5.2. Methodology Part 2

This research adopted a practice-based methodology that integrated material experimentation with Generative Artificial Intelligence (GenAI) to examine how emerging technologies can act as facilitators in creative practice. The approach combined physical processes, digital transformation, and iterative prompt engineering, while systematically documenting each stage through photographs, and generative outputs. Such a framework aligns with contemporary scholarship in practice-led art and design research, where iteration and reflection are central to knowledge production (Costa & Venturelli, 2025; Chompunuch & Lubart, 2025).

Figure 3. The dialogue between the author and ChatGPT throughout the creation of Testing Part 1

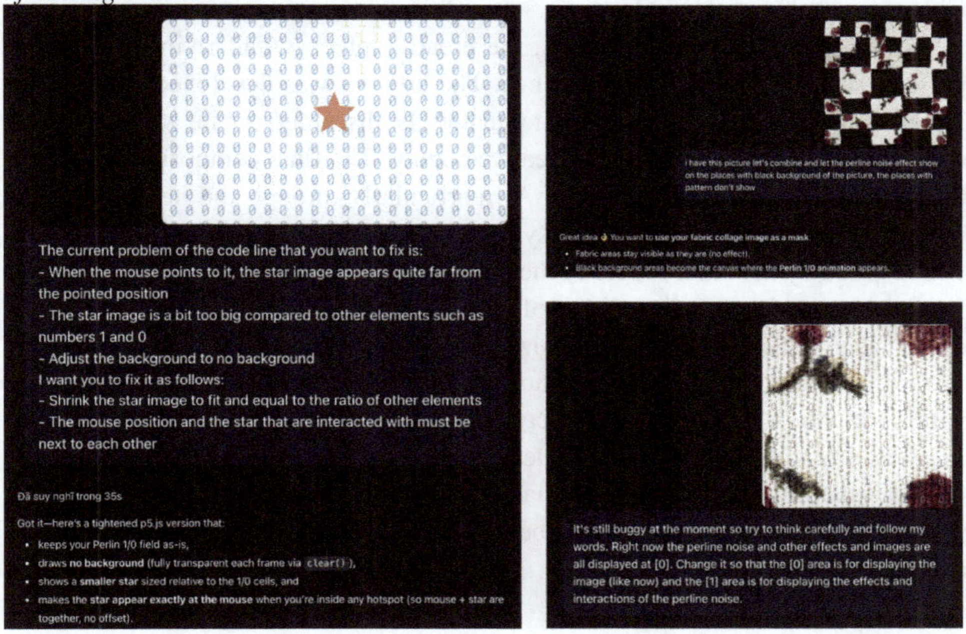

6. TESTING PART 1: RAW MATERIAL 1 - TEXTILE WASTE INTEGRATION

At present, Generative Artificial Intelligence (GenAI) has significantly expanded the reach and accessibility of creative practice, enabling individuals from diverse backgrounds to experiment and innovate. The first stage of this research focused on a project that reimagined discarded fashion materials as original artworks, achieved through the integration of human creativity and GenAI support. The author's expertise is rooted in graphic design rather than computer science; therefore, this project was conducted without advanced coding experience or familiarity with GenAI-specific language. Nonetheless, contemporary GenAI tools effectively minimized technical barriers, allowing active participation in generative processes. This case illustrates that within creative industries, human ingenuity remains central to driving innovation, while GenAI operates as a supportive mechanism that reframes workflows. The resulting quality and trajectory of work depend not only on the sophistication of GenAI technology but also on the strategies, intentions, and communicative choices of practitioners. In essence, GenAI does not replace originality; rather, it amplifies creative agency, producing outcomes shaped equally by technological capability and the creator's distinct aesthetic sensibilities (Costa & Venturelli, 2025; Chompunuch & Lubart, 2025).

To substantiate these arguments, the creative process underlying one of my projects serves as compelling evidence. The artwork was developed through the integration of both physical and digital materials: discarded textile waste and generative outputs produced through creative coding. The final composition emerged through substantial reliance on GenAI, particularly through the iterative use of prompts that guided the generative systems (Hu, 2014). Conceptually, the project was inspired by the binary logic of the 1–0 system, which provided the foundation for early computing (McCormack et al., 2019), as well as the operational principles of the Jacquard loom, widely recognized as one of the first programmable devices and a precursor to modern computational design (Ceruzzi, 2003). These historical parallel highlights the continuity between technological innovation and creative practice, situating GenAI-assisted design within a broader lineage of computational aesthetics. To further substantiate this argument, illustrations will be used to present the methodological process behind the project and to provide an objective assessment of its results.

Figure 4. Iterative Development of Creative Coding and Fabric Pattern Integration

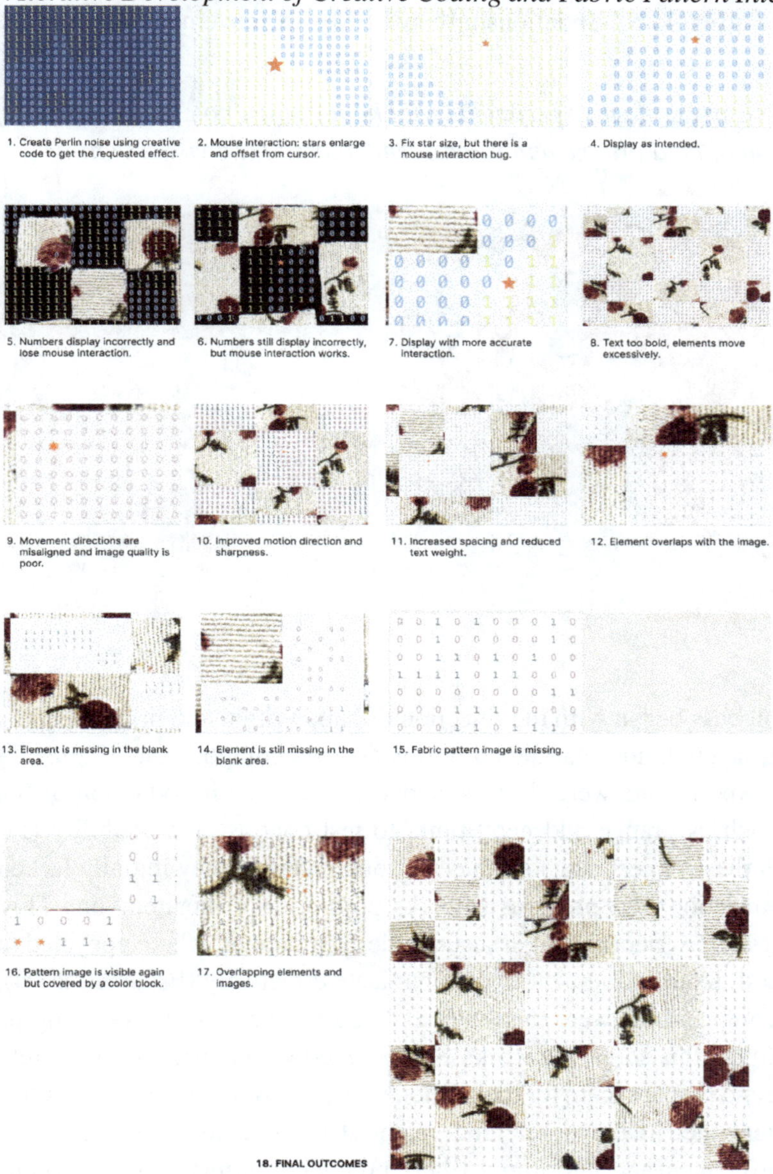

6.1. Testing Part 2: Raw Material 1 - Paper Collage Integration

To further evidence the role of GenAI in this project, a second testing phase was conducted using paper as the primary raw material. Building on earlier experiments, this stage sought to integrate GenAI's generative capabilities with manual collage techniques to explore the relationship between digital movement and physical texture.

Figure 5. Early collage experiments with handwritten notes and woven paper and integration of bold orange and fluorescent colors to test visual contrast

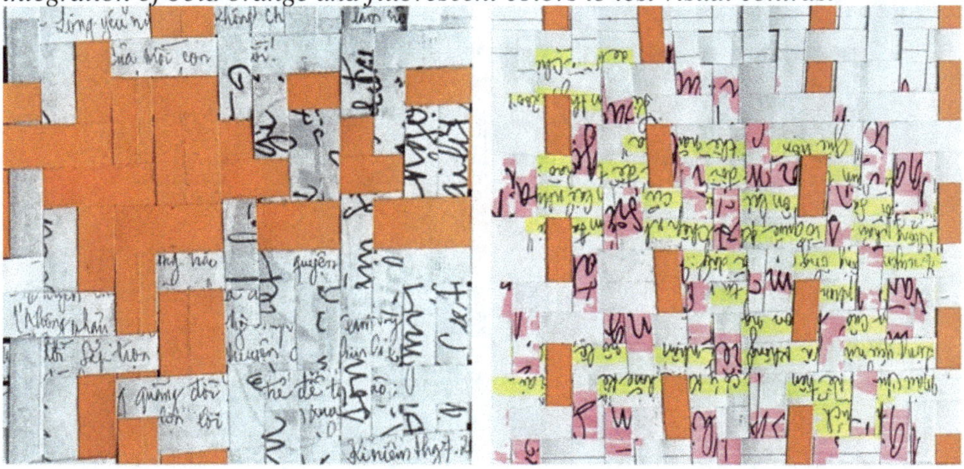

The process began with the selection of handwritten and printed papers, which were cut, layered, and rearranged into geometric compositions (Figure 5). These physical experiments were then expanded through the introduction of bold color blocks, such as orange and neon pink, to test contrast and visibility (Figure 5). Although these experiments produced strong patterns, they initially lacked a clear visual connection to the previous artworks, prompting further iterations. This reflects challenges often encountered in sustainable art practices, where upcycled materials require additional refinement to communicate effectively (Essinger, 2004).

The compositions were then digitized and subjected to post-editing processes (Figure 6). At this stage, the works were aligned with a grid system to refine their structure and prepare them for integration into a GenAI environment. Using existing GenAI-generated images as reference material, movements were introduced through prompt refinement and algorithmic filters, producing animated visual responses from the static paper patterns (Figures 7-8). This aligns with recent research on natural-language-driven creative coding, which emphasizes how prompt-based systems can translate physical ideas into dynamic digital outputs (Angert et al., 2023).

The outcome illustrates how GenAI can function as a bridge between manual craftsmanship and computational transformation, enabling hybrid results that preserve the tactile qualities of paper while embracing the dynamic aesthetics of generative coding. By combining physical collaging with algorithmic logic, Testing Part 2 demonstrates that GenAI does not replace artistic agency but instead amplifies creative possibilities—an observation consistent with studies on GenAI in design education, where conversational GenAI has been shown to reduce technical barriers and extend creative potential (Moon et al., 2024; Chompunuch & Lubart, 2025).

Figure 6. Digitized grid overlays for compositional refinement.

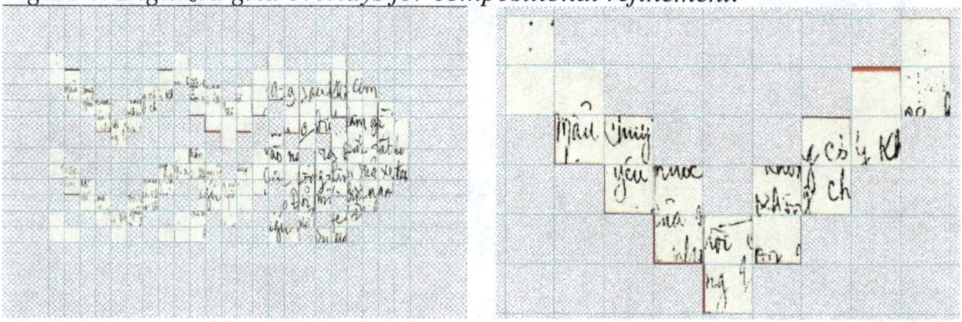

Figure 7. AI-assisted transformation of collage into pixel-based generative forms

Figure 8. Final hybrid outcomes combining paper texture with algorithmic movement

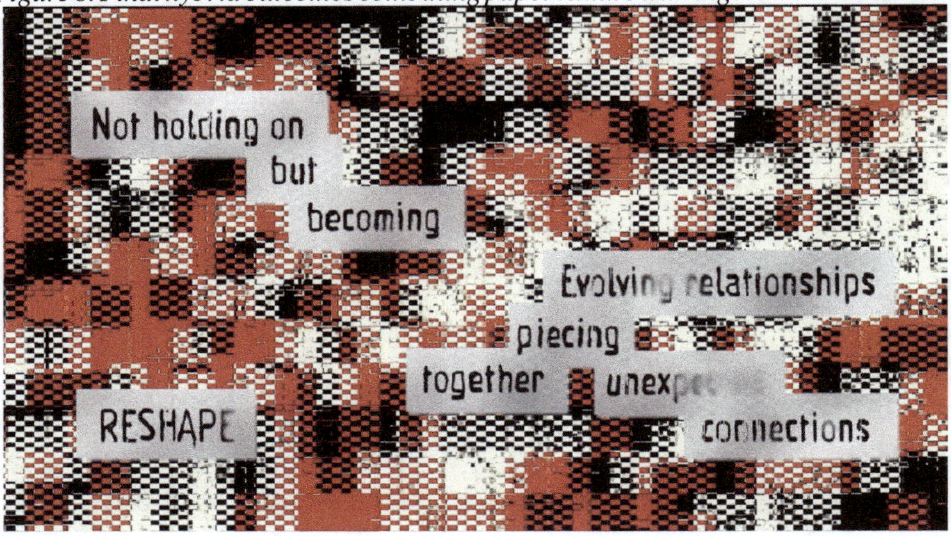

7. DISCUSSION

The findings of this study underscore the evolving role of Generative AI (GenAI) as both a facilitator and co-creator in creative practice. Beyond technical execution, the results integrate reflective analysis, highlighting how human intention and algorithmic processes converge to inform broader understandings of design practice.

7.1 The Critical Role of Prompt Engineering

Across both testing stages, detailed prompts (around 80% of total) consistently produced coherent outputs that closely aligned witu'h the intended vision, whereas underspecified prompts yielded partial or incoherent results. This pattern reinforces the importance of precision in human–machine communication (Angert et al., 2023; Chompunuch & Lubart, 2025). Prompt engineering thus emerges as a creative act in itself, where failures become productive deviations that guide further refinement. The process demonstrates that GenAI amplifies intentionality when carefully directed but also exposes gaps in articulation that practitioners must iteratively resolve.

7.2 Hybridization of Material and Digital Practices

Testing in Part 1 reimagined textile waste as computationally generated patterns informed by binary logic and the Jacquard loom (McCormack et al., 2019; Ceruzzi, 2003). Testing Part 2 extended this by animating static paper collages through algorithmic filters, producing hybrid outcomes that fused material tactility with computational transformation. These outcomes echo Costa and Venturelli's argument that GenAI fosters new models of co-authorship in which humans and machines jointly shape creative results (Costa & Venturelli, 2025). The hybrid process revealed tensions between imperfection and algorithmic precision; rather than eliminating material qualities, GenAI extended them into new expressive domains. This demonstrates how sustainable practices such as upcycling can be enriched by digital innovation, bridging ecological responsibility with computational aesthetics.

7.3 GenAI as Amplifier, not Replacement

Across both phases, GenAI reduced technical barriers, accelerated experimentation, and expanded stylistic exploration. Yet the creative agency remained firmly with the practitioner, who curated, refined, and contextualized outputs. This supports perspectives that position GenAI as an "AI helper" rather than a substitute for originality (Moon et al., 2024; Chompunuch & Lubart, 2025). Crucially, one of the enduring strengths of human creators lies in their ability to tell stories. Storytelling infuses artworks with context, meaning, and emotional resonance, elements that GenAI cannot independently generate. In this sense, artworks produced through GenAI are not solely algorithmic outcomes but narrative-rich hybrids where human imagination supplies the conceptual foundation and GenAI extends it into new visual forms. By reframing challenges as opportunities, the process validates GenAI's role in democratizing access to creative tools without diminishing the centrality of human authorship and narrative voice.

7.4 Overall Systhesis

Taken together, the findings and reflections show that GenAI functions most effectively as a supportive collaborator. Its strengths lie in bridging physical and digital practices, amplifying human creativity through prompt engineering, and fostering sustainable innovation by integrating recycled materials into generative workflows. At the same time, the human capacity for storytelling ensures that creative works retain originality, coherence, and emotional depth, making the final outputs genuine collaborations where technological innovation and human narrative imagination converge (Costa & Venturelli, 2025; Chompunuch & Lubart, 2025).

8. CONCLUSION: FUTURE IMPLICATIONS

The insights gained from the experimentation phases emphasize a promising intersection between sustainability, digital innovation, and creative pedagogy. While the experiments with discarded materials (textiles, paper) demonstrated that graphic designers can reimagine waste into meaningful creative outcomes, the varied levels of fluency in prompt engineering and reliance on simpler visual effects signal the need for pedagogical scaffolding and infrastructure investment. In this context, GenAI emerges not as a replacement but as a facilitator, lowering technical thresholds, enabling richer experimentation, and opening pathways for interdisciplinary collaboration.

As GenAI increasingly participates in co-creative workflows, issues of authorship, power, and inclusivity demand close attention. The sustainability of creative systems also encompasses *social sustainability*: who has access to these tools, whose aesthetics are prioritized, and how credit is shared. Holzapfel et al. (2022) caution that Creative-AI must be scrutinized across environmental, economic, and social dimensions.

In sum, the convergence of recycled materials, new media, and GenAI points toward a future where creative education is defined by hybrid practices that reconcile ecological awareness, technological experimentation, and ethical reflection. By experimenting with alternative raw materials, embedding sustainability intelligence, and cultivating pedagogical scaffolds, we can move toward a model where GenAI empowers more inclusive, sustainable, and meaning-driven creative practices.

REFERENCES

Anantrasirichai, N., & Bull, D. (2020). *Artificial intelligence in the creative industries: A review* (arXiv:2007.12391). arXiv. https://arxiv.org/abs/2007.12391

Anantrasirichai, N., Zhang, F., & Bull, D. (2025). *Artificial intelligence in the creative industries: Advances prior to 2025* (arXiv:2501.02725). arXiv. https://arxiv.org/abs/2501.02725

Angert, A., Suzara, M., Han, A., Pondoc, J., & Subramonyam, H. (2023). *Spellburst: A node-based interface for exploratory creative coding with natural language prompts* (arXiv:2308.03921). arXiv. https://arxiv.org/abs/2308.03921 DOI: 10.1145/3586183.3606719

Bailey, J. (2022, September 2). An A.I.-generated picture won an art prize. Artists aren't happy. *The New York Times*. https://www.nytimes.com/2022/09/02/technology/ai-artificial-intelligence-artists.html

Benjamin, M. (2025). *The impact of generative AI on traditional graphic design workflows*. ResearchGate. https://www.researchgate.net/publication/388849181_The_Impact_of_Generative_AI_on_Traditional_Graphic_Design_Workflows

Bozkurt, A. (2024). GenAI et al.: Cocreation, Authorship, Ownership, Academic Ethics and Integrity in a Time of Generative AI. *Open Praxis*, *16*(1), 23–36. DOI: 10.55982/openpraxis.16.1.654

Cassar, R. (2024). Innovative upcycling: The creative potential of reuse in design practice. *Journal of Design History*.

Ceruzzi, P. (2003). *A history of modern computing*. MIT Press.

Chompunuch, K., & Lubart, T. (2025). AI as a helper in the creative process: Toward a framework of assistance. *International Journal of Human-Computer Studies*, *185*, 103123.

Choudhury, M. M., Mandal, T., Malakar, A., & Sanyal, S. (2025). Artificial intelligence (AI) in the design process – A review and analysis on generative AI perspectives. In *Proceedings of the Design Society*. Cambridge University Press & Assessment. DOI: 10.1017/pds.2025.10077

Costa, P., & Venturelli, S. (2025). Creative autonomy and collaboration in generative AI art and design. *Leonardo*.

Duyver, E., Groeneveld, G., & Aerts, J. (2023). *Does interdisciplinary creative coding boost creativity? A mixed methods approach* (arXiv:2307.00800). arXiv. https://arxiv.org/abs/2307.00800

Essinger, J. (2004). *Jacquard's web: How a hand-loom led to the birth of the information age*. Oxford University Press.

Fleischmann, K. (2024). Generative artificial intelligence in graphic design education: A student perspective. [CJLT]. *Canadian Journal of Learning and Technology, 50*(1), 1–17. DOI: 10.21432/cjlt28618

Garzón, J. (2025). Artificial intelligence in education: A systematic literature review. *Multimodal Technologies and Interaction*, *9*(8), 84. DOI: 10.3390/mti9080084

Ge, P. (2024). A systematic review of the role of AI in design education (2020–2023). In *Proceedings of the Design Society*.

Heigl, R. (2025). *Generative artificial intelligence in creative contexts: A systematic literature review*. Springer.

Holzapfel, T., Foerster, J., Brendel, M., & Busse, S. (2022). Towards sustainable Creative-AI: Environmental, social, and economic perspectives. *Journal of Responsible Innovation*, *9*(3), 275–298.

Hu, J. (2014). Sustainable upcycling practices in art and design. *Journal of Cleaner Production*.

Manzini, A. (2015). *Design, when everybody designs: An introduction to design for social innovation*. MIT Press. DOI: 10.7551/mitpress/9873.001.0001

McCormack, J., Gifford, T., & Hutchings, P. (2019). Autonomy, authenticity, authorship and intention in computer generated art. In *The philosophy of computer art* (pp. 81–106). Springer. DOI: 10.1007/978-3-030-16667-0_3

Moon, J., He, Y., Tian, X., & Hargis, J. (2024). Teaching and learning creative coding with conversational AI. *GLOKALde Journal*.

Ninet, A. (2025). *Using ChatGPT to create a collection of illustrations for graphic design*. Instagram Reels.

Open, A. I. (2023). *GPT-4 technical report*. OpenAI.

Oppenlaender, J. (2022). *The creativity of text-to-image generation* (arXiv:2206.02904). arXiv. https://arxiv.org/abs/2206.02904 DOI: 10.1145/3569219.3569352

Oppenlaender, J., Linder, R., & Silvennoinen, J. (2024). Prompting AI art: An investigation into the creative skill of prompt engineering. *International Journal of Human-Computer Interaction.*

[] Reddy, V. S. S., & Reddy, N. (2022). Generative AI in creative industries: Revolutionizing content creation with neural networks. *International Journal of Advanced Research in Engineering and Technology (IJARET), 13.*

Rezwana, J., & Maher, M. L. (2022). Understanding user perceptions, collaborative experience and user engagement in different human–AI interaction designs for co-creative systems DOI: 10.1145/3527927.3532789

Sung, K. (2023). Understanding upcycling and circular economy and their interrelationships for design education. In *Proceedings of the Design Society.*

Urmeneta, M., & Romero, M. (2025). A PRISMA review of AI's role in supporting creativity in education. *Frontiers in Education, 10*, 1602151. DOI: 10.3389/feduc.2025.1602151

U.S. Copyright Office. (2023). *Works containing material generated by artificial intelligence—Policy statement.* U.S. Copyright Office.

Wang, S., Wang, F., Zhu, Z., Wang, J., Tran, T., & Du, Z. (2024). Artificial intelligence in education: A systematic literature review. *Expert Systems with Applications, 252*, 124167. DOI: 10.1016/j.eswa.2024.124167

Zhou, J., Li, R., Tang, J., & Wang, H. (2024). *Understanding nonlinear collaboration between human and AI agents: A co-design framework for creative design* (arXiv:2403.XXXX). arXiv. DOI: 10.1145/3613904.3642812

Chapter 14
AR-Enhanced Merchandise Hybridity:
Visual Hierarchy, Immersion, and Comprehension in AR UX

Alex Gaja Ortega
http://orcid.org/0009-0003-7833-6743
British University Vietnam, Vietnam

Trang Ha Dao
British University Vietnam, Vietnam

ABSTRACT

This study investigates how spatial visual hierarchy shapes user experience (UX) in augmented reality (AR) storytelling applied to printed merchandise. While AR packaging and campaigns are increasingly used in creative industries, little research addresses how layout and visual sequencing affect comprehension and immersion of this technological format. Using a mixed-method design, the project analyzed pioneering AR campaigns from major advertising companies through structured observation and focus groups, identifying six recurrent spatial patterns. These informed the development of two prototypes tested with users in a controlled A/B experiment. Results showed that high-hierarchy designs significantly improved narrative clarity, emotional immersion, and perceived flow compared to low-hierarchy versions. The findings highlight spatial hierarchy as a key mechanism linking design structure with user engagement in hybrid media, combining physical and digital realms. This work contributes to design research by bridging theory and practice in AR-enhanced storytelling.

DOI: 10.4018/979-8-3373-9549-4.ch014

INTRODUCTION

Augmented reality (AR) is increasingly integrated into printed merchandise, allowing brands and products to extend storytelling beyond static visuals into digital platforms. Advertising campaigns illustrate the commercial potential of hybrid media, combining physical and digital, but they also expose a critical research gap: the spatial design logic of these experiences remains underexamined. While marketing studies emphasize interaction and novelty, little attention has been paid to how UX design shape clarity or engagement. This paper responds to that gap by focusing on digital design mechanisms in AR storytelling.

1. LITERATURE REVIEW

1.1 Spatial Hierarchy in Design

Spatial visual hierarchy refers to the intentional arrangement of visual elements to indicate relative significance and direct user attention. As described by the Interaction Design Foundation, effective hierarchy enables users to process information fluently by structuring how content is perceived and navigated (The Interaction Design Foundation, 2016).

Gestalt principles like proximity, similarity, continuity, and figure-ground relationships affect grouping and visual dominance (Urano, Kurosu, & Todorov, 2019). Eye-tracking by Urano et al. (2019) also shows that visually hierarchical designs improve perceptual fluency and cognitive attention by promoting consistent and efficient gaze behavior. Perceptual pathways support aesthetic and cognitive functions, enabling mental models to be built and facilitating comprehension and immersion.

1.2 AR-Enhanced Printed Merchandise

1.2.1 Applications of AR in Printed Media

By fusing physical media with interactive digital layers, augmented reality (AR) printed merchandise increases the expressive potential of conventional print. According to Grand View Research (2024), the global AR packaging market is projected to grow steadily due to rising demand for personalization, sustainability, and multisensory brand experiences.

Figure 1. First DOOH campaign to allow smartphones to interact with screens

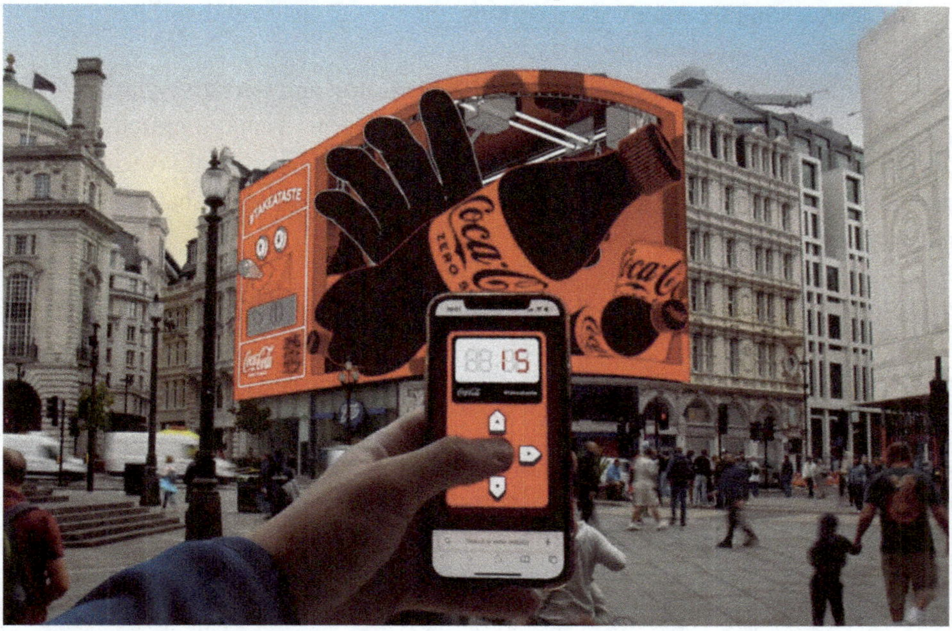

Source: Coca-Cola, 2023

As shown in Figure 1, Coca-Cola's Christmas AR campaign (in collaboration with Tactic Studio) featured people scanning limited-edition cans to access animated polar bear stories.

Although the campaign garnered high interaction (PTC, 2023), there has been no scientific examination of its spatial layout logic, including entry point location, visual flow, and alignment. Without an explicit examination of spatial hierarchy, such advertisements risk choosing novelty above comprehension or emotional resonance.

1.2.2 Storytelling through Hybrid Merchandise

19 Crimes' Living Wine Labels campaign brings historical convicts on wine bottles to life via augmented reality. Scanned with a smartphone, the label triggers immersive monologues like John Boyle O'Reilly's exile, turning static packaging into narrative performance (Dara, 2021). Launched in 2017, the campaign has generated well over 22 million AR activations and boosted brand sales by 40%, proving notable emotional resonance and virality (PTC, n.d.; Schwab, 2021).

These measurements show how AR affects consumer involvement but not spatial narrative logic. In their study of AR wine labeling operations, Vrigkas et al. (2022) show how digital layers (animations, 2D/3D visuals, and text) can be integrated on packaging surfaces. However, when such overlays open "virtual windows" with inconsistencies in placement or transitions, the viewer may get disoriented instead of absorbed.

The situation highlights a gap between interaction and narrative clarity. While the 19 Crimes campaign offers emotional resonance, its visual hierarchy remains underexamined. As Vrigkas et al. (2022) suggest, coherent spatial layering, which includes focal sequencing, anchoring, and contrast, is essential for shaping attention, memory, and immersion in hybrid narrative settings.

1.3 Emotional Immersion and Narrative Comprehension

1.3.1. Presence Theory and Narrative Transportation

Two theoretical frameworks explain emotional immersion in hybrid storytelling: Presence Theory and Narrative Transportation Theory.

Presence Theory refers to the psychological sense of "being there" within a mediated space (Slater & Wilbur, 1997). When applying this theory to AR context, this presence is heightened when overlays align contextually and spatially with physical stimuli.

Narrative Transportation Theory posits that immersion occurs when users experience focused attention, vivid imagery, emotional resonance, and diminished real-world awareness (Green & Appel, 2024). These theories suggest that immersive state increases narrative retention and persuasion.

Figure 2. Visual Map of Theoretical Frameworks

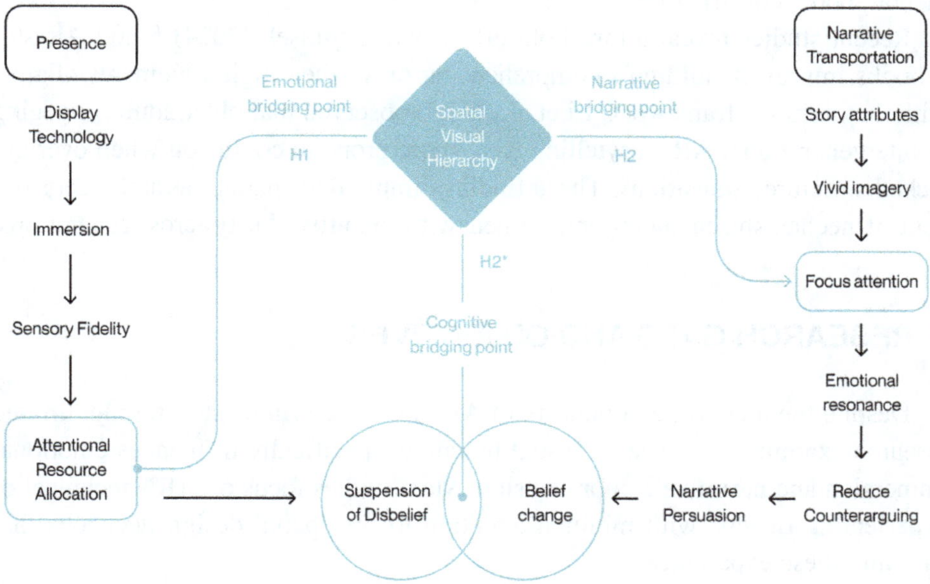

Source: Authors

Together, as shown in Figure 2, these theories imply that spatial visual hierarchy can play a bridging role. It structures user attention, aligns entry points with story attributes, and supports narrative flow. For example, consistent visual sequencing scaffolds transportation by allowing users to follow the story without visual ambiguity. If empirical studies validate the three bridging points, spatial visual hierarchy could be confirmed as a central mechanism linking Presence and Narrative Transportation.

1.3.2. Cognitive Load and Mental Models

Comprehension in AR storytelling is further strengthened by Cognitive Load Theory (CLT) and Mental Models Theory.

Cognitive Load Theory (CLT), developed by Sweller, distinguishes between intrinsic load (task complexity), extraneous load (inefficient presentation), and germane load (effort used to construct meaning) (Shibli & West, 2018; Centre for Education Statistics and Evaluation, 2023). When AR content is cluttered or poorly sequenced, extraneous load increases, impairing user comprehension.

Mental Models Theory suggests that users simulate story structures internally by mapping causal and spatial relationships (Bower & Morrow, 1990), suggesting users to reconcile the layout of physical media with dynamic digital content. Visual hierarchy reduces ambiguity by foregrounding narrative anchors and aiding coherent mental model construction.

Recent studies reveal a threshold effect: while Şimşek (2024) found AR storybooks improved children's comprehension only when digital elements aligned with the narrative framework, Li et al. (2023) observed that older adults engaging in intergenerational AR storytelling were more prone to confusion when overlays lacked structured transitions. These findings imply that spatial hierarchy acts as a control mechanism, balancing engagement with cognitive clarity across generations.

2. RESEARCH GAPS AND OBJECTIVES

Despite the increasing popularity of AR-enhanced printed storytelling, limited research examines how spatial visual hierarchy specifically influences emotional immersion and narrative comprehension. Most studies focus on AR's technical or experiential aspects, with minimal attention to the spatial design decisions that structure these experiences.

While Gestalt Principles and CLT are well-established frameworks, their application to hybrid media spatial sequencing is underexplored. Furthermore, existing campaign studies rarely address how layout, visual ordering, or focal contrast mediate user perception and story retention.

This research aims to achieve the following:

2.1 Research Objective

General Objective: To investigate how spatial visual hierarchy impacts emotional immersion and narrative comprehension in AR-enhanced storytelling printed merchandise.

Specific Objectives:

(R.O.1) To examine and categorize spatial hierarchy strategies employed in AR-enhanced storytelling printed merchandise.

(R.O.2) To evaluate how variations in spatial hierarchy impact users' emotional immersion and narrative comprehension within hybrid storytelling experiences.

3. FIELD RESEARCH METHODOLOGY

3.1. Methodological Approach

Creswell (2014) identifies three primary types of mixed methods research design: explanatory sequential, exploratory sequential, and convergent parallel designs. This study employs an exploratory sequential design, consisting of two phases aimed at investigating the impact of spatial visual hierarchy in AR-enhanced printed merchandise on users' emotional immersion and narrative comprehension, in accordance with RO1.

In phase one, structured observation was conducted in a simulated environment to examine how users engage with AR-printed layouts without external guidance. Behavioral data was recorded to identify patterns in spatial perception and emotional response.

Phase two used a focus group to reflect on those patterns. Participants discussed clarity, pacing, and engagement across layout types, helping validate which spatial structures consistently support immersion and narrative understanding.

3.2 Sampling Strategy

The two the research methods employed non-random purposive sampling, which is considered suitable for design-focused studies requiring alignment with predefined theoretical domains (Muratovski, 2021).

3.3 Structured Observation

As Dang (2020) notes that AR-printed products remain limited in the Vietnamese market, the study employed structured observation in simulated settings to control for extraneous variables (Delve, 2022). This allowed for deeper analysis of user interaction patterns in a consistent, replicable format. Case selection was guided by a three-domain theoretical Venn framework. Campaigns located at the intersection of these three domains were selected as stimuli. Each campaign was analyzed using a structured 5-dimension coding sheet based on narrative flow principles and Gestalt visual theory, as shown on Figure 3.

Domain A. Spatial Hierarchy in Design: Use of visual positioning, scaling, and sequence to guide attention.

Domain B. AR-Enhanced Printed Merchandise: Physical product packages embedded with AR content.

Domain C. Emotional Immersion & Narrative Comprehension: Ability of AR to create emotional engagement and story clarity.

Figure 3. 5-Step Visual Coding Dimensions

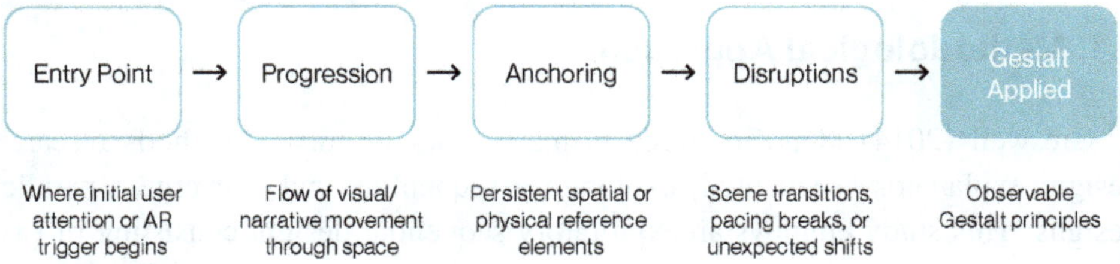

Source: Authors

This study employed Saldaña's two-cycle qualitative coding framework for qualitative data analysis, which provides a systematic approach for transitioning from detailed observations to pattern identification (Saldana, 2009).

- First-cycle coding: Code observational data across five dimensions
- Second-cycle coding: Synthesize and reorganize to extract recurring patterns
- Pattern Synthesis: Validate through triangulation with theoretical models

3.4 Focus Group Discussion

A focus group was conducted alongside structured observation to examine user interpretations of spatial and emotional elements in AR storytelling, as described on Figure 4. Fleiss' Kappa was used to assess inter-rater reliability, resulting in statistically validated thematic clusters.

Figure 4. Focus Group Discussing & Rating Procedure

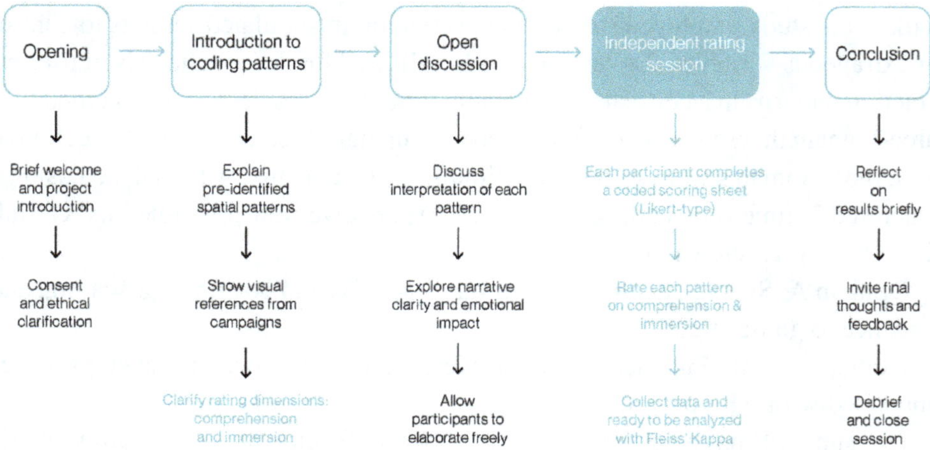

Source: Authors

4. Data COLLECTION & ANALYSIS

4.1 Structured Observation

Three AR campaigns were selected for structured observation as they fully met the screening criteria across all three domains. Coca-Cola used central anchoring with outward story flow, 19 Crimes applied label-locked AR with immersive monologues, and Jack Daniel's combined pop-up unfolding with layered storytelling. All featured clear spatial behaviors and controlled AR–print environments suitable for 5-dimension coding.

- Digital and Physical Entry Points
- UX Progression
- Anchoring Visual Elements
- UX Disruptions
- Applied Gestalt Principles.

4.1.1 Coca-Cola: 12 Mini Stories

Coca-Cola "12 Mini Stories" demonstrates strong spatial visual hierarchy through its use of the physical can as an anchor point for 3D storytelling elements, creates immersive mini narratives with emotional arcs (conflict and resolution), and directly enhances the printed product packaging through AR activation.

Entry Points

- Physical Coca-Cola can serves as the primary visual anchor and AR trigger
- Individual code on can activates AR experience
- Camera recognition of can shape establishes spatial reference

Progression

- With can as the focal point, story unfolds on the same surface
- Characters appear and interact in the 3D space surrounding the can
- Spatial movement flows outward to surrounding environment

Anchoring

- Can remains the central spatial anchor throughout the experience
- Can becomes physical prop and visual backdrop for narrative scenarios
- Virtual elements maintain consistent spatial relationship to physical can

Disruptions

- Narrative conflicts create spatial disruptions
- Transitions between story scenes occur around the can
- Scale shifts between can and surrounding virtual elements create visual breaks

Gestalt Applied

- Proximity: Visual elements positioned in relation to can
- Conitnuity: Visual flow follows paths around the can
- Figure/Ground: Clear distinction between AR elements and real environment

4.1.2 19 Crimes: "Living Labels"

19 Crimes "Living Labels" employs a portrait-centered visual hierarchy with the face as the focal point, creates deep narrative immersion through first-person historical storytelling, and transforms the static printed wine label into a dynamic narrative vehicle.

Entry Points

- Wine bottle label with historical portrait serves as primary entry point
- Portrait face centered as visual focal point
- App camera recognition frame establishes boundaries of interaction space

Progression

- Static portrait transforms into moving, speaking character
- Face animates while maintaining spatial position within label boundaries
- Visual emphasis remains centered on face throughout experience

Anchoring

- Wine bottle and label remain fixed spatial reference points
- Portrait maintains consistent position in physical space
- Real-world bottle orientation determines viewing angle of character

Disruptions

- Transition from static image to animated character creates perceptual break

- Character's changing facial expressions direct attention within frame
- Narrative delivery through direct address breaks fourth wall

Gestalt Applied

- Similarity: Consistent portrait style creates visual cohesion across product line
- Closure: Users mentally complete character beyond label boundaries
- Figure/Ground: Clear distinction between animated character and static label

4.1.3 Jack Daniel's AR Bottle

Jack Daniel's AR Bottle utilizes a distinctive pop-up book visual hierarchy where elements unfold in spatial layers, creates narrative comprehension through sequential storytelling about the distillery and process, and directly enhances the physical bottle as a printed merchandise item.

Entry Points

- Physical bottle shape and label recognized by AR application
- Digital label overlay syncs with physical label as starting point
- Bottle silhouette establishes spatial reference frame

Progression

- Digital label unfolds outward like pop-up book in sequential layers
- Visual elements expand from bottle into surrounding space
- Hierarchical display of distillery process stages in spatial sequence

Anchoring

- Physical bottle remains central reference point throughout experience
- Bottle position and orientation guide spatial arrangement of AR content
- Label serves as persistent visual anchor connecting physical and virtual

Disruptions

- Transitions between three distinct AR experiences (manufacturing, distillery tour, stories)
- Pop-out displays for process steps create spatial depth changes
- Scale transitions from bottle-sized elements to expanded scenes

Gestalt Applied

- Proximity: Elements unfold in spatial relation to bottle
- Common Fate: Elements move together in coherent groups during transitions
- Figure/Ground: Clear distinction between AR elements and physical bottle

4.1.4 Identified Patterns

Across all campaigns, six spatial patterns were identified. These patterns describe how AR elements are positioned, scaled, and animated in relation to the printed surface. Each reflects a different way users navigate, interpret, and emotionally respond to spatial layout, as collected in Table 1 below.

Table 1. Table Mapping Design Logic of Patterns.

Pattern name	Spatial Behavior	Campaigns	Theoretical Grounding
P1. Central Object Anchoring	AR elements radiate around the product center	All 3	Alexander (1977), Billinghurst (2008)
P2. Label-Locked Animation	AR constrained inside printed label boundary	19 Crimes, JD	Visual Grammar (1996), Perey (2011)
P3. Boundary-Break Transitions	Key story beats occur at visual border crossings	Coca-Cola, 19 Crimes	Figure/Ground, Transportation Theory
P4. Peripheral Flow Circulation	Elements circulate the object in flow paths	Coca-Cola	Continuity, Kendon's F-Formations
P5. Sequential Spatial Unfolding	AR layers unfold outward like a pop-up	JD	Narrative Grammar (Cohn, 2013), AR layering
P6. Silhouette-Referenced Scaling	AR size adjusts to package outline	JD, Coca-Cola	Gibson's Affordance Theory (1979)

Source: Authors

4.2 Focus Group Discussions

4.2.1 Selected Participants

Participants were selected through purposive sampling. All individuals possessed pertinent expertise in AR design, UX, or editorial storytelling, and had either previously or currently engaged with spatial layout systems.

4.2.2 Synthesize Analysis

Frequency totals were summarized across patterns based on two criteria:

- Narrative Comprehension – Does this pattern support logical understanding?
- Emotional Immersion – Does this pattern evoke affective engagement?

To validate agreement beyond chance, Fleiss's Kappa was calculated using the standard formula:

$k = (p_0 - p_e) / (1 - p_e)$, where: p_0 = observed agreement (0.8), p_e = expected agreement by chance (0.5)

Figure 5. Fleiss' Kappa scores for comprehension and immersion

(1) Comprehension

Sum sq.	88
N*n	24
1/(N*n (n-1)	0.014
p_e	0.618
p_o ($\bar{p}$)	0.889
k_1	0.709

(2) Immersion

Sum sq.	74
N*n	20
1/(N*n (n-1)	0.0125
p_e	0.31
p_o ($\bar{p}$)	0.675
k_1	0.53

Kappa's levels of agreement	
>0.8	Almost perfect
>0.6	Substantial
>0.4	Moderate
>0.2	Fair
>0	Slight
<0	No agreement

Source: Authors

As shown on Figure 5, interestingly, while agreement on comprehension (k_1 = 0.709) was stronger, the lower kappa for immersion (k_2 = 0.53) indicates more subjective variability in how participants interpreted emotional depth, which is likely driven by individual visual preferences or interpretive styles.

The comprehension rating reached substantial agreement ($k > 0.6$), whereas immersion only achieved moderate agreement ($k > 0.4$), but both remain sufficiently reliable to support average-score comparisons across patterns for representative selection.

Having established consensus, mean scores for each pattern were averaged across both dimensions. Then, P1, P6, and P2 were selected as representatives for prototype testing based on highest combined scores. No major outliers or disagreement clusters were observed; Neutral and Disagree responses remained minimal (≤1/30).

5. RESULTS AND HYPOTHESES FORMULATION

From these findings, two prototype conditions were furhter developed. The high- hierarchy version incorporated the top-rated patterns P1, P6, and P2, while the low-hierarchy version included P3, P4, and P5. These formed the basis for the experimental testing phase.

The hypotheses described on the Table 2 below were preliminarily defined to examine how spatial visual hierarchy affects three core dimensions of user experience, aligned with RO2.

Table 2. Summary of experimental and null hypotheses.

Code	Dimension	Ha (Alternative Hypothesis)	H0 (Null Hypothesis)
H1	Narrative Comprehension	Users will score higher on comprehension with the high-hierarchy prototype	No difference in comprehension between the two prototypes
H2	Emotional Immersion	Users will feel more emotionally engaged with the high-hierarchy prototype	No difference in emotional immersion between the two prototypes
H3	Narrative Flow Clarity	Users will perceive clearer flow in the high-hierarchy prototype	No difference in perceived narrative flow between the two prototypes

Source: Authors

These hypotheses guided the design of the evaluation phase and structured the statistical testing procedures that followed.

6. TEST METHODOLOGY

The project now transitions from exploratory design to evaluative testing. The focus is to examine how spatial visual hierarchy influences user experience in AR storytelling. This is achieved through hypothesis-driven comparison of two contrasting AR prototypes, using structured, measurable user feedback.

This research utilized an A/B Testing (Comparative Testing Within Subjects Design) to assess the influence of spatial visual hierarchy on user responses in augmented reality storytelling, with the designed options shown on Figure 6. Participants engaged with both prototype conditions (low-hierarchy and high-hierarchy) in a counterbalanced sequence to reduce order effects (Newsom, 2017).

The primary objective was to test three directional research hypotheses concerning comprehension, emotional immersion, and narrative flow.

Corresponding null hypotheses (H0) assumed no meaningful difference between conditions. As a statistical convention, the null hypothesis is rejected only when the observed effects are unlikely to occur under random variation, typically using a threshold of $p < 0.05$ (Onwuegbuzie & Daniel, 2002).

Figure 6. Test Flow

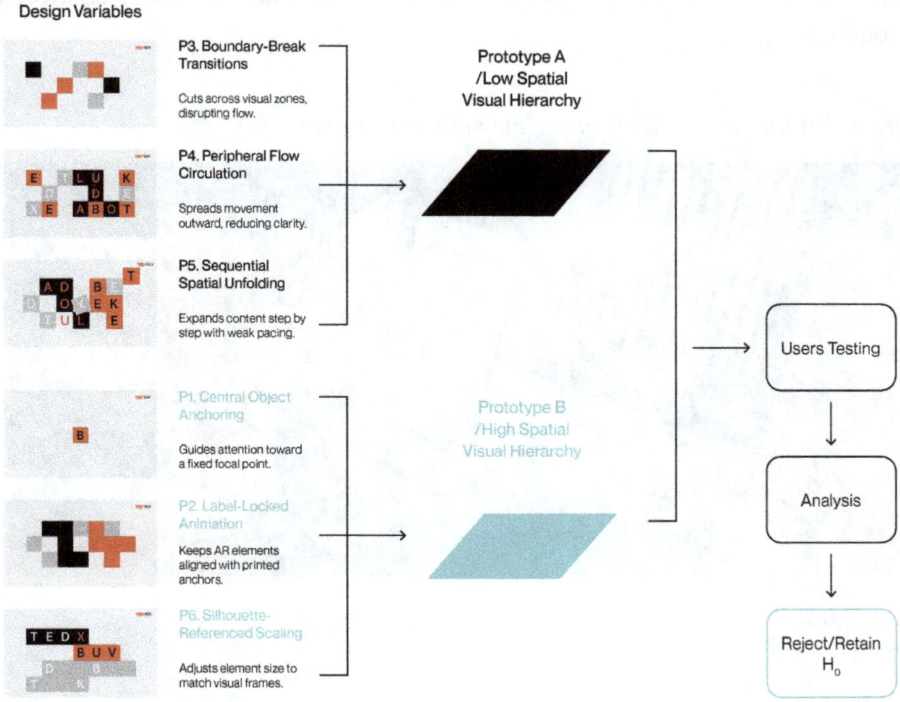

Source: Authors

6.1 Test Design

Participants experienced both prototypes in a counterbalanced within-subjects order to reduce bias. Group A viewed Prototype A first, then B; Group B followed the reverse. After each viewing, participants completed a short survey on comprehension, immersion, and narrative flow.

6.2 Prototype Development

The two prototypes were independently developed personally as part of a branding commission for TEDxBUV 2024. Originally intended for merchandise and digital content, the animations were adapted for research by isolating visual flow as the experimental variable.

Each prototype features a distinct animated narrative built on contrasting spatial visual hierarchies: one with a linear, structured progression, and the other with abrupt, non-sequenced transitions.

Both versions were overlaid onto a static printed marker using the Artivive platform, allowing users to trigger and compare the animations under controlled AR conditions.

Figure 7. Prototype verification review before user tests

Source: Authors

Before launching user testing, the overlay's stability has been verified, marker recognition speed, and animation fidelity across devices to minimize technical distractions and ensure consistent viewing experiences, as recorded on Figure 7.

7. TEST DATA COLLECTION

This within-subjects experiment used convenience sampling, involving available students from a nearby university class. While not statistically random, this method is commonly used in pilot design studies where participants are exposed to all conditions and individual variance is accounted for through counterbalancing (Etikan et al., 2016).

Prototype A /Low Visual Hierarchy: Participants are presented with visual elements appeared at once, creating a uniform composition with minimal hierarchy.

Prototype B /High Visual Hierarchy: The same content unfolded gradually, guided by a storyboard that introduced each element in a deliberate sequence.

During the testing process, no clarification or further guidance was provided. Data was collected independently to allow participants to engage with the content without influence. After completing each prototype walkthrough, participants filled out a follow-up survey before proceeding to the next condition. A final cross-check section was included to validate the hypotheses.

8. TEST DATA ANALYSIS

8.1 Data Condition Check

To validate statistical inferences, the three measured variables (comprehension, immersion, and narrative flow) were evaluated in relation to the fundamental assumptions of paired-sample parametric testing, with the normality of difference scores regarded as the principal condition for small-N designs (N<30) (Field, 2018; Gravetter & Wallnau, 2020; Ghasemi & Zahediasl, 2012).

Table 3. Distribution features of difference scores by condition.

Variable	Skewness (Low / High)	Kurtosis (Low / High)	Interpretation
Narrative Comprehension	+0.10 / −0.57	−1.07 / +0.09	Symmetric, light-tailed
Immersion	−0.04 / +0.26	−0.89 / −1.07	Near-normal, flat distribution
Narrative Flow	−1.07 / −0.12	+1.31 / −0.44	Mildly skewed left, still acceptable

Source: Authors

As outlined on Table 3, while homogeneity of variance is not a strict requirement in paired designs (Tabachnick & Fidell, 2019), the observed variances across conditions remain within acceptable range: Comprehension (1.12 vs 0.78), Immersion (12.97 vs 6.84), and Flow (2.41 vs 1.31). No variance inflation or outlier clusters were observed in the dataset.

8.2 Statistical Analysis

The analysis revealed statistically significant differences across all three measured dimensions between the high- and low-hierarchy conditions (N = 15).

8.2.1 Analysis for H1: Narrative Comprehension

For H1 (comprehension), the mean score in the high- hierarchy condition was 4.93, compared to 3.47 in the low- hierarchy condition. A paired-sample t-test showed $t(14) = -4.56$, $p < 0.001$, confirming this difference was statistically significant. In the Section C.1, 73% of participants selected Prototype B as having a clearer visual structure. See specific data visualized on Figure 8 below.

Figure 8. H1 Distribution Overlap and Paired Sample Line Plot

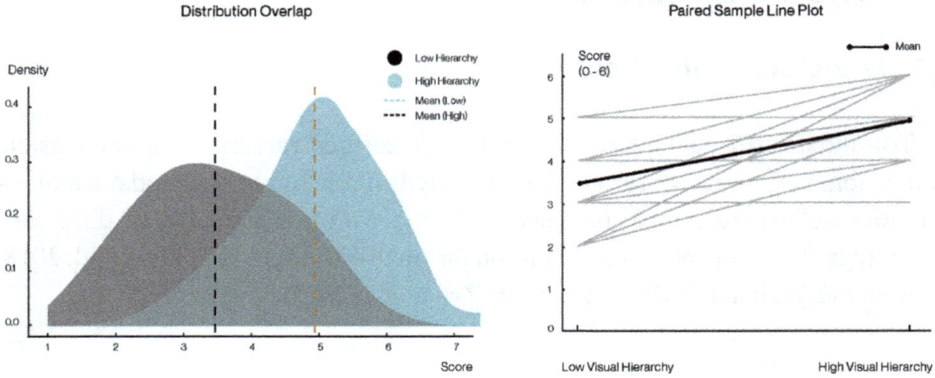

Source: Authors

8.2.2 Analysis for H2: Emotional Immersion

For H2 (emotional immersion), participants reported greater engagement under the high-hierarchy condition ($M = 19.47$) than the low-hierarchy counterpart ($M = 13.4$). The difference was statistically significant, $t(14) = -5.5$, $p < 0.001$, confirming this difference was statistically significant. In the Section

C.2, 69% of participants indicated that Prototype B felt more emotionally engaging. See specific data visualized on Figure 9 below.

Figure 9. H2 Distribution Overlap and Paired Sample Line Plot

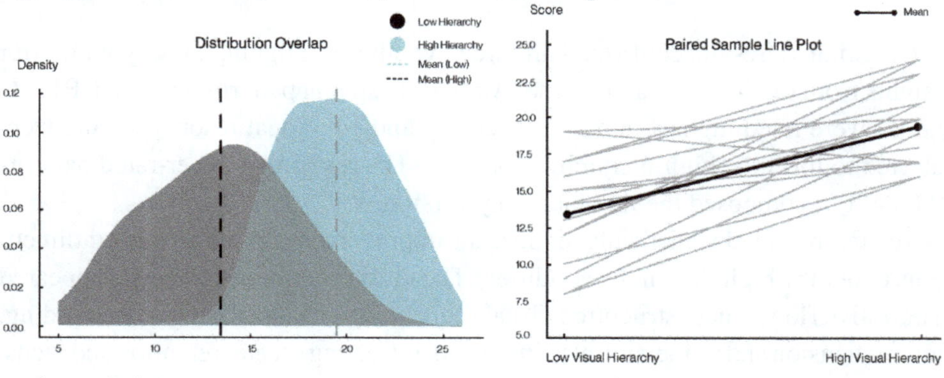

Source: Authors

8.2.3 Analysis for H3: Narrative Flow & Visual Clarity

For H3 (narrative flow), the most prominent difference occurred. Scores increased from a mean of 10. 87 (low- hierarchy) to 22.2 (high-hierarchy), resulting in $t(14) = -27.63$, $p < 0.001$. In Section C.3, 80% of participants reported that Prototype B had a clearer visual flow. A Pearson correlation was also run to test the relationship between clarity (A.3) and immersion (A.2) within Prototype B ($r = 0.30$, $p = 0.277$) → weak positive correlation). See specific data visualized on Figure 10 below.

Figure 10. H3 Distribution Overlap and Paired Sample Line Plot.

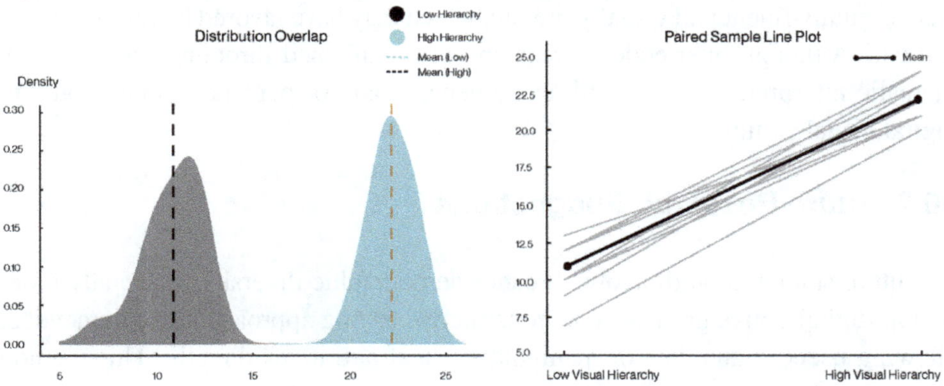

Source: Authors

9. TEST RESULTS AND EVALUATION

Quantitative results confirmed all three hypotheses, aligning closely with prior insights from the field research phase. Among six spatial patterns analyzed, P1, P6, and P2 were rated highest in narrative clarity and emotional resonance, and were selected to form the high-hierarchy prototype. The remaining lower-rated patterns (P3, P4, P5) composed the low-hierarchy version.

Testing revealed statistically significant improvements in all measured dimensions under the high- hierarchy condition. The most prominent difference appeared in narrative flow, where structured visual sequencing enhanced user understanding. Comprehension and immersion also improved, reflecting greater retention and focus.

Perceptual feedback from survey sections further reinforced these findings, with a majority of participants preferring the high-hierarchy prototype in terms of pacing, clarity, and emotional engagement. Together, the results validate the impact of spatial visual hierarchy on narrative effectiveness and support the rejection of all null hypotheses.

10. CONCLUSIONS AND IMPLICATIONS

10.1 Limitations & Considerations

Despite clear internal consistency, the study's external validity is limited by sample composition. Among the participants, only four were aged above 45, and all reported low familiarity with AR. The remaining participants (ages 15–44) were digitally fluent and visually literate, which may have favored hierarchy-driven formats. Although inter-coder agreement was confirmed through Fleiss' Kappa, thematic interpretations may still carry generational or perceptual bias rooted in digital visual culture.

10.2 Future Research Suggestions

Future studies should involve broader demographic diversity, especially older or non-digital-native groups. A co-constructive testing approach is recommended, allowing users to annotate or manipulate visual narratives directly. This method would reveal not just user perception of hierarchy, but also their expectations of visual logic. Longitudinal research is also advised, particularly in educational or heritage contexts, where repeated AR exposure may gradually shift spatial comprehension and narrative engagement.

10.3 Conclusions

The findings confirm that spatial visual hierarchy is a central factor in AR storytelling on printed merchandise. Field research identified six recurring spatial patterns, which guided the creation of high- and low-hierarchy prototypes tested with participants. Statistical analysis revealed significant improvements in comprehension, immersion, and narrative flow when hierarchy was strong, reinforcing theoretical links to Presence and Narrative Transportation. While the sample was limited, the results highlight practical implications for designers and brands aiming to create more engaging hybrid narratives. The study demonstrates that careful structuring of spatial hierarchy can transform AR campaigns from novel interactions into coherent and emotionally resonant storytelling experiences.

REFERENCES

Bower, G. H., & Morrow, D. G. (1990). Mental Models in Narrative Comprehension. *Science*, *247*(4938), 44–48. DOI: 10.1126/science.2403694 PMID: 2403694

Centre for Education Statistics and Evaluation. (2023). Cognitive load theory in practice. [online] education.nsw.gov.au. Available at: https://education.nsw.gov.au/about-us/education-data- and-research/cese/publications/practical-guides-for-educators/cognitive-load-theory-in-practice

Dara, J. (2021). 19 Crimes 'Living Wine Labels' Now Include A Holographic Visit From Snoop Dogg. [online]. *Forbes*, *15*(Dec). https://www.forbes.com/sites/jilliandara/2021/12/15/19- crimes-living-wine-labels-include-a-holographic-visit-from-snoop-dogg/

Delve (2022). What is observational research? [online] Delve. Available at: https://delvetool.com/ blog/observation

Du, Z., Liu, J. and Wang, T. (2022). Augmented Reality Marketing: a Systematic Literature Review and an Agenda for Future Inquiry. Frontiers in Psychology, [online] 13. .DOI: 10.3389/fpsyg.2022.925963

Etikan, I., Musa, S. A., & Alkassim, R. S. (2016). Comparison of Convenience Sampling and Purposive Sampling. *American Journal of Theoretical and Applied Statistics*, *5*(1), 1–4. DOI: 10.11648/j.ajtas.20160501.11

Field, A. (2018). *Discovering Statistics Using IBM SPSS Statistics* (5th ed.). SAGE Publications.

Ghasemi, A., & Zahediasl, S. (2012). Normality Tests for Statistical Analysis: A Guide for Non- Statisticians. *International Journal of Endocrinology and Metabolism*, *10*(2), 486–489. DOI: 10.5812/ijem.3505 PMID: 23843808

Grand View Research. (2024). Augmented Reality Packaging Market | Industry Report, 2030. [online] Available at: https://www.grandviewresearch.comhttps://www.grandviewresearch.com/industry- analysis/augmented-reality-packaging-market-report

Gravetter, F., & Wallnau, L. (2020). *Essentials of Statistics for the Behavioral Sciences* (10th ed.). Cengage Learning.

Green, M. C., & Appel, M. (2024). Narrative transportation: How stories shape how we see ourselves and the world. Advances in Experimental Social Psychology, [online] 70(1), pp.1–82. doi:DOI: 10.1016/bs.aesp.2024.03.002

Ho, A. (2023). A Critical Overview of Studies on Eye Tracking and Visual Hierarchy. In: Lucas Paletta, Hasan Ayaz and Umer Asgher (eds) Cognitive Computing and Internet of Things. AHFE (2023) International Conference. AHFE Open Access, vol 73. AHFE International, USA. DOI: 10.54941/ahfe1003295

Kleftodimos, A., Lappas, G. and Vrigkas, M. (2022). An augmented reality workflow for creating 'live' wine labels. International Journal of Entertainment Technology and Management, [online] 1(4), p.311. .DOI: 10.1504/IJENTTM.2022.10054762

Li, Z., Liu, F., Chen, L., Huang, Y., & Fan, M. (2023). Exploring the Opportunities of AR for Enriching Storytelling with Family Photos between Grandparents and Grandchildren. *Proceedings of the ACM on Interactive, Mobile, Wearable and Ubiquitous Technologies*, 7(3), 1–26. DOI: 10.1145/3610903

Muratovski, G. (2021). *Research for Designers: A Guide to Methods and Practice* (2nd ed.). SAGE Publications.

Onwuegbuzie, A. J., & Daniel, L. G. (2002). Uses and misuses of the independent t-test.

PTC. (2023). Tactic Studio Creates Engaging Coke Augmented Reality Experience. [online] Ptc. com. Available at: https://www.ptc.com/en/case-studies/tactic-coke -augmented-reality-interactive- applications.

PTC. (n.d.). 19 Crimes Uses Augmented Reality to Differentiate Their Brand Through Innovation | PTC. [online] Available at: https://www.ptc.com/en/case -studies/augmented-reality-19-crimes-wine

Saldana, J. (2009). The Coding Manual for Qualitative Researchers. *Sage (Atlanta, Ga.).*

Schwab, P.-N. (2021). 19 Crimes: a stroke of marketing genius ... often poorly imitated. [online] Market research consulting. Available at: https://www.intotheminds .com/blog/en/19-crimes-wine- augmented-reality/

Shibli, D., & West, R. (2018). Cognitive Load Theory and Its Application in the Classroom. [online] Impact. Available at: https://my.chartered.college/impact_article/ cognitive-load-theory-and-its- application-in-the-classroom/.

Şimşek, E. E. (2024). The effect of augmented reality storybooks on the story comprehension and retelling of preschool children. *Frontiers in Psychology*, *15*, 1459264. Advance online publication. DOI: 10.3389/fpsyg.2024.1459264 PMID: 39439758

Slater, M., & Wilbur, S. (1997). A Framework for Immersive Virtual Environments (FIVE): Speculations on the Role of Presence in Virtual Environments. *Presence (Cambridge, Mass.)*, *6*(6), 603–616. DOI: 10.1162/pres.1997.6.6.603

Sweller, J. (1988). Cognitive Load during Problem Solving: Effects on Learning. *Cognitive Science*, *12*(2), 257–285. DOI: 10.1207/s15516709cog1202_4

Tabachnick, B. G., & Fidell, L. S. (2019). *Using Multivariate Statistics* (7th ed.). Pearson.

The Interaction Design Foundation. (2016). What is Visual Hierarchy? [online] Available at: https:// www.interaction-design.org/literature/topics/visual-hierarchy ?srsltid=AfmBOorYhiwlcCJmCDepYaEywm8_cWnXImGjHnSEZaffssEu1LsQd-kjH [Accessed 16 Feb. 2025].

Urano, Y., Kurosu, A., & Todorov, A. (2019). *Good visual hierarchy is good design.* Princeton University.

Chapter 15
Stigma, Discrimination, and Vietnamese Workers' Intention to Leave Japan:
A Pilot Study

Xuan Thanh Dang
http://orcid.org/0009-0006-2653-191X
Foreign Trade University, Hanoi, Vietnam

Hoai Linh Pham
Foreign Trade University, Hanoi, Vietnam

Thi Thu Cuc Nguyen
http://orcid.org/0009-0001-8050-529X
Foreign Trade University, Hanoi, Vietnam

Nguyen Van Tang
http://orcid.org/0009-0009-7266-2437
Foreign Trade University, Hanoi, Vietnam

ABSTRACT

This pilot study explores how exposure to negative news about Vietnamese crime in Japan affects migrant workers' daily life, workplace experiences, and intention to leave Japan. A survey of 15 Vietnamese workers found frequent exposure to negative media, commonly resulting in feelings of upset or worry, though 40% reported no strong emotional response. Nearly half (47%) perceived subtle social avoidance or prejudice in daily life, and 20% experienced overt discrimination, indicating social stigma. Workplace impacts were less prevalent, though 20% noted concerns about hiring bias. Critically, 33% of participants expressed a definite intention to

DOI: 10.4018/979-8-3373-9549-4.ch015

leave Japan, with another 20% considering it. These findings suggest that negative media exposure and perceived discrimination erode Vietnamese migrants' sense of security and belonging. This highlights the need for larger research to inform migrant retention strategies.

I. INTRODUCTION

A. Background and Context of Vietnamese Migration to Japan

Over the past decade, Vietnam has emerged as one of the largest source countries for foreign labor in Japan. The Vietnamese community now constitutes the second-largest foreign community in Japan, with numbers exceeding 600,000 residents. This rapid demographic shift is closely tied to the high demand for foreign workers to address labor shortages, often facilitated through structural programs such as the Technical Intern Training Program (TITP) and similar work visa schemes. While these programs are crucial for Japan's economy, they have historically faced criticism regarding the structural vulnerability and exploitation of trainees and workers.

This rapid growth and high visibility of the Vietnamese population have coincided with increased media attention. News reports and public statistics in recent years have frequently highlighted criminal incidents involving Vietnamese nationals. The emphasis on nationality in reporting these incidents has demonstrably tarnished the community's reputation in the public eye, fostering generalized prejudice. This climate has led many Japanese individuals to develop a perception that Vietnamese migrants commit numerous crimes in Japan, a phenomenon largely driven and reinforced by the circulation of negative news stories across television, online platforms, and social media.

B. Media Exposure, Stigma, and Societal Prejudice

The pervasive negative media portrayal can inflict significant social and psychological impacts on the Vietnamese expatriate community. Exposure to news narratives that repeatedly link their national identity to crime triggers psychological mechanisms such as stereotype threat and social alienation. Workers who constantly encounter content labeling their compatriots as criminals may internalize a sense of being mistrusted or unwelcome in Japanese society. This manifests as psychological distress, including fear of backlash and shame, and behavioral changes aimed at self-protection.

Research on other minority groups in Japan, such as the Zainichi Korean community, has previously documented how media-fueled xenophobia and racialized narratives contribute to a climate of suspicion and exclusion, often portraying minorities as criminal or undesirable. Specifically, law enforcement narratives have historically linked immigration with criminality, giving political leverage to discourse associating migrants with crime (Yamamoto, 2010). Vietnamese migrants, as a newer and rapidly growing large minority, appear to be experiencing a similar phenomenon of collective scapegoating. The consequence is that migrants often worry about being stigmatized when high-profile crime stories break, sensing heightened caution or bias from Japanese peers and neighbors in everyday interpersonal settings.

C. Discrimination as a Push Factor in Migration Theory

Experiences of discrimination have been widely established as negative outcomes contributing to the erosion of migrant integration and satisfaction with life in the host country. From a migration theory perspective, a lack of social acceptance and perceived discrimination act as strong push factors, compelling foreign workers to consider exit strategies—a form of "voting with their feet"—when integration feels unattainable (Lee, 1966; Massey et al., 1993).

This push factor is consistently demonstrated in empirical research concerning foreign workers in Japan. For instance, studies focusing on foreign-educated nurses in Japan found a clear causal sequence: higher workplace discrimination was the main predictor of lower job satisfaction, which subsequently and indirectly correlated with a higher intention to leave the country (Hua et al., 2023). This previous work establishes that providing a supportive and equal work environment is essential for managers seeking to enhance job satisfaction and reduce turnover intention among foreign staff. If workplace discrimination—even in regulated, professional environments—is a critical predictor of turnover (Takeda, 2023), it stands to reason that societal, media-fueled stigma, which is less regulated and more pervasive than internal workplace bias, may serve as an even stronger determinant eroding the overall sense of security and driving the ultimate intention to leave.

D. Research Gap and Objectives

While there is growing awareness of these issues, empirical research specifically examining the chain of impact from media exposure to generalized daily-life discrimination, and subsequently to the intention to leave, remains limited for the Vietnamese community in Japan. Prior studies have focused predominantly on general attitudes toward foreigners or on workplace experiences within specific professions.

This pilot study aims to address this gap by providing initial, exploratory insights into the psychosocial impacts of negative media exposure on Vietnamese migrant workers. Specifically, the study investigates:

1. The frequency and nature of Vietnamese workers' exposure to negative news about Vietnamese crime.
2. Their emotional and behavioral responses to such news.
3. Perceived changes in their daily life and workplace interactions attributable to media-driven events (focusing on discriminatory incidents).
4. Whether those exposed to and affected by negative news express a higher intention to leave Japan.

The core hypothesis is that frequent exposure to negative news, coupled with perceived discrimination in daily life, will be associated with a higher intention to return to Vietnam (leave Japan).

II. METHODOLOGY

A. Participants and Procedure

The pilot study utilized a convenience sample of 15 Vietnamese individuals ($N=15$) who were current residents, living and working in Japan at the time of the survey. Participants were adults with a mean age of 27.8 years ($SD=3.2$), ranging from 23 to 33 years. The sample included 5 males and 10 females, recruited primarily through convenience sampling via Vietnamese community networks and social media groups in Japan.

Data collection was conducted through an anonymous, confidential online questionnaire administered in Vietnamese, the participants' native language. The study was purely observational, requiring participants to complete a self-report survey on their demographic background, media consumption, recent experiences, and future attitudes. Ethical considerations, including voluntary participation and the right to withdraw, were communicated clearly.

B. Measures

The questionnaire was developed specifically for this pilot study, drawing on informal interviews regarding common concerns voiced within the Vietnamese community. Key measures included:

- **Media Exposure to Negative News:** Measured frequency on a 5-point Likert scale (1=Never to 5=Very Often/Always) and asked respondents to rank the importance of various news sources (e.g., Japanese social media, Japanese TV/newspapers).

- **Recent Incident Recall:** Open-ended descriptions of the latest news remembered were content-coded into categories, such as Murder/violent crime (60% recall), Theft/property crime (27% recall), or Tax/document fraud.
- **Initial Emotional Reaction:** Participants selected or described their immediate response, coded as binary variables for five key emotional reactions: Upset (33% prevalence), Fearful (13%), Ashamed (7%), Confused (13%), and No Reaction (40%).
- **Perceived Impact on Life:** Assessed whether the impact was felt in daily life/community interactions, the workplace, both, or no impact.
- **Impact Details:** Open-ended responses regarding daily life impacts were coded for Avoidance by locals (47% prevalence) and Discrimination (overt prejudice, 20% prevalence). Workplace impacts were coded for Hiring or opportunity bias (13% concern) and Pressure to work harder (7% concern).
- **Behavioral Changes:** Assessed whether participants consciously changed their behavior in daily life (e.g., identity alteration, Living more cautiously) or at work (e.g., Image management) as a coping response.
- **Intention to Leave Japan:** The key outcome variable, captured with options: Yes (definitely intend to return to Vietnam), Maybe/Undecided (considering it), or No (plan to stay long-term).

C. Data Analysis

Due to the limited sample size ($N=15$), the analysis was intentionally constrained to descriptive statistics and exploratory associations, rather than formal hypothesis testing, consistent with the nature of a pilot study. The analysis computed frequencies, percentages, means, and standard deviations for all key variables. Bivariate relationships were explored using cross-tabulations and Pearson's correlation coefficients, treating ordinal scales numerically where appropriate.

This methodological approach is justified for exploratory pilot research aimed at identifying observable patterns and determining appropriate effect sizes to guide a subsequent, larger-scale study (Pinchover, 2017). Complex modeling, such as regression or mediation analysis, was explicitly avoided because the study lacks the statistical power required for inferential statistics. Therefore, observed correlations and associations are indicative of trends rather than confirmed relationships, ensuring scholarly rigor in reporting preliminary findings. Qualitative insights from open-ended responses, translated into English, were integrated to enrich the understanding of participants' experiences.

III. RESULTS

A. Media Exposure and Incident Recall

All participants reported experiencing exposure to negative news regarding Vietnamese individuals in Japan. The mean frequency on the 5-point scale was 3.8 ($SD=0.8$), positioning the average exposure between “Sometimes” and “Often”. None of the respondents reported “Never” or “Rarely” seeing such news. A majority (60%) reported seeing this type of news often, with 20% reporting it as a very frequent, almost daily occurrence.

The primary sources of these negative narratives were reported to be Japanese online news sites and social media platforms (like Yahoo! Japan News and Twitter), followed by Japanese television and newspapers. This distribution suggests that participants are encountering the stigma through mainstream Japanese media and public digital spaces, rather than solely through Vietnamese community outlets.

When prompted to recall specific incidents, the participants' memories were heavily dominated by sensational, high-profile cases. Violent crimes, such as murder or serious assault, were recalled by 60% of the sample. The next most recalled category was theft/property crimes, cited by 27% of participants. The prevalence of sensational cases like murder in community memory, despite their relative rarity, highlights that these incidents receive extensive media coverage and imprint deeply on the collective consciousness of the community. This finding confirms that the entire community is acutely aware of negative narratives, even if only a small fraction of Vietnamese residents are involved in crime.

B. Emotional and Psychological Responses

The initial emotional reactions among participants were polarized. A total of 60% of respondents experienced some form of negative emotional impact, including feeling upset, fearful, confused, or ashamed. The most common reaction was feeling upset or uncomfortable, reported by 33% of the sample. This distress often involved sadness or disappointment, coupled with anger directed both at the perpetrator (for causing collective shame) and at the media (for potentially sensationalizing the incident).

Fear or anxiety was reported by 13% of participants. These individuals expressed apprehension about potential backlash or being scrutinized by Japanese peers, reflecting a classic fear of stigmatization. Interestingly, only 7% reported feeling explicit shame as their primary reaction, suggesting that while Vietnamese migrants are pained by these events, they do not universally internalize guilt for crimes committed by others.

However, a significant proportion, 40%, reported no strong emotional reaction or indifference. This high rate of indifference suggests a form of psychological distancing or emotional numbing, potentially developed as a protective mechanism due to repeated exposure to chronic negative information. While this numbing shields individuals from immediate emotional distress, it does not guarantee integration, as some indifferent individuals still planned to leave, citing accumulated disappointment and stress from the general societal atmosphere.

C. Perceived Social and Workplace Impacts

The study reveals that the repercussions of negative media exposure are largely felt in day-to-day social interactions, indicating a failure of social acceptance in the public sphere. All participants perceived at least subtle changes in their environment due to negative news.

Social Impact in Daily Life (Public Sphere)

Nearly half of the participants, 47%, reported experiencing subtle avoidance or wariness from Japanese people (ID_avoidance). This involved sensing that acquaintances or strangers had become more distant, guarded, or hesitant around them, or noticing heightened scrutiny by store staff. This pervasive pattern of social cold-shouldering, while not overt hostility, accumulates to create a persistent feeling of marginalization.

Furthermore, 20% of the sample reported encountering direct discrimination or explicit prejudice in daily life (ID_discrimination). These incidents included microaggressions, being openly linked to "rising crime" by strangers, or being insulted because of their nationality. Although less frequent, these explicit acts were described as deeply impactful and contributed significantly to feelings of alienation and indignation.

Workplace Experiences (Professional Sphere)

In contrast to public interactions, the structured professional environment often served as a mitigating factor against overt stigma translation. A large majority, 80%, reported no specific change or impact in their current workplace environment (IW_noimpact). Many attributed this buffer to professional norms and personal familiarity with colleagues and supervisors.

Nonetheless, some indirect concerns were noted:

- **Hiring Bias:** 13% expressed concern that Japanese employers might become hesitant to hire new Vietnamese applicants due to the perceived tarnished reputation (IW_hiring). This represents a potential systemic barrier, signaling a tougher landscape for career mobility, even if the individual's current job is secure.
- **Performance Pressure:** 7% reported feeling an increased, self-imposed pressure to perform better or avoid mistakes due to a perceived need to counteract the negative national stereotype (IW_pressure). This self-imposed mandate to be an "exemplary worker" generates psychological stress.

Only one participant (7%) reported a direct spillover incident, which involved a stereotyping comment framed as a joke by a coworker (SW_direct), illustrating how negative narratives can seep into workplace banter and cause discomfort. This evidence suggests that the primary erosion of security and belonging occurs outside the protected, high-familiarity professional environment.

D. Behavioral Adaptations (Coping Strategies)

To cope with the experienced stigma, participants adopted a range of adaptive and self-protective behaviors, representing a non-economic psychological cost of living in Japan.

A minority of participants (13%) reported making conscious adjustments aimed at identity concealment in public (BD_identity). This included avoiding speaking Vietnamese loudly in public or using a Japanese nickname to minimize their visibility as "Vietnamese" to strangers. This strategy of downplaying nationality markers is a classic response to identity threat, similar to historical coping mechanisms used by other minority groups facing discrimination.

Conversely, others adopted a Model Minority Strategy, characterized by hypervigilance (BD_living). This involved being extra careful to follow all rules, avoiding actions that might attract negative attention, and striving to be a counter-example to the prevailing stereotype. Some participants also expressed a sense of collective responsibility (HC_public), actively condemning wrongdoing among their compatriots to protect the community's image. These behaviors require significant emotional and psychological investment to manage potential prejudice on a daily basis.

E. Intention to Leave Japan (Outcome Variable)

The study's core finding is the high proportion of participants who expressed an intention to leave Japan, confirming that social challenges tied to media stigma significantly contribute to an emigrant mindset.

Out of the 15 current residents surveyed:

- 33% (5 participants) expressed a definite intention to leave Japan ("Yes").
- 20% (3 participants) were actively considering leaving ("Maybe/Undecided").
- 27% (4 participants) expressed a desire to stay long-term ("No").

The total inclination to leave (combining Yes and Maybe) reached 53% of the sample. This high attrition risk underscores that social-psychological discomfort acts as a powerful push factor, undermining the investment these individuals have made in their jobs and lives in Japan.

Exploratory analysis revealed trends linking negative experiences to the intention to leave. While formal correlations were low ($r\approx0.06$) due to the small sample size, qualitatively, all individuals who expressed a definite intention to leave had reported experiencing some form of social or psychological impact (either emotional distress or a change in daily life). Conversely, those who intended to stay generally reported fewer negative experiences. The data suggests that even if workers have stable employment, the cumulative effect of perceived hostility or explicit discrimination incidents serves as the tipping point, forcing them to view Japan as an unsustainable environment, supporting the findings of established migration literature.

Table I summarizes the key descriptive statistics and exploratory associations observed in the pilot study data:

Table 1. Descriptive Statistics and Key Exploratory Associations (Pilot Study, $N=15$)

Variable	Mean/Proportion	Key Association/Trend with Intention to Leave (Yes/Maybe)
Negative news exposure frequency (1-5 scale)	3.8 ($SD=0.8$)	No linear correlation observed ($r\approx0.00$).
Upset/Fearful emotional reaction	46% (7/15)	Those reporting negative emotions were more likely to report intent to leave.
No emotional reaction	40% (6/15)	High rate of indifference, but includes individuals with intent to leave.
Perceived avoidance by locals (daily life)	47% (7/15)	Common experience, contributing to overall discomfort.

continued on following page

Table 1. Continued

Variable	Mean/Proportion	Key Association/Trend with Intention to Leave (Yes/Maybe)
Overt discrimination in daily life	20% (3/15)	Qualitatively linked to strongest intent to leave.
Any workplace bias perceived (hiring/pressure)	20% (3/15)	Impacts systemic trust/future prospects.
Intend to Leave Japan (Yes)	33% (5/15)	High definite attrition risk.
Intend to Leave Japan (Yes + Maybe)	53% (8/15)	Over half of the sample is not firmly committed to staying.

IV. DISCUSSION

A. The Pervasiveness of Stigma and Psychological Toll

The finding that every participant was aware of negative media narratives confirms that, regardless of individual media consumption habits, the negative portrayal of Vietnamese individuals as a criminal element is a pervasive, defining feature of the social environment in Japan. This universal awareness aligns with cultivation theory, where mass media exposure shapes an individual's perception of social reality (Gerbner et al., 1980).

The psychological response is complex. While a majority (60%) experienced distress (upset or fear), the high rate of indifference (40%) warrants closer examination. This indifference is likely a form of emotional numbing or self-distancing, a necessary defensive mechanism employed to manage the chronic stress associated with being part of a group subject to pervasive stereotype threat. The presence of this psychological buffering indicates the normalization of receiving negative news, which itself represents a significant mental health burden on the community.

B. Social Exclusion Versus Professional Inclusion

A key finding is the distinct localization of stigma. The negative media portrayal translated effectively into perceived exclusion in low-familiarity public settings, where nearly half (47%) reported experiencing subtle avoidance, and 20% faced overt discrimination. This social cold-shouldering suggests that the Japanese public, lacking personal relationships, relies heavily on media-amplified stereotypes when interacting with Vietnamese strangers. This phenomenon is supported by findings that Japanese citizens, in some contexts, feel it is normative to express prejudice against immigrants (Igarashi & Nagayoshi, 2021).

Conversely, the professional sphere demonstrated resilience, with 80% of workers reporting no specific impact at work. This suggests that Japanese workplaces, where merit, established personal relationships, and professional standards mitigate external narratives, are relatively successful in providing an inclusive environment. However, the failure of integration lies in the broader public sphere. The security and comfort gained in the workplace are undermined by the anxiety and prejudice experienced outside of it. The primary source of erosion of belonging is therefore an issue of societal acceptance rather than solely workplace discrimination.

C. The True Cost: Psychological Burden and Retention Failure

The observed behavioral adaptations—ranging from identity concealment (avoiding speaking Vietnamese, using Japanese aliases) to hyper-vigilance (consciously striving to be a "model citizen")—illustrate the non-monetary, psychological tax levied on Vietnamese workers. Migrants are compelled to invest emotional and cognitive resources daily simply to manage potential prejudice and counteract the negative image projected onto them.

The high intention to leave (53% total inclination) demonstrates that this accumulation of social discomfort acts as a powerful push factor. Even if these workers possess stable careers and the initial economic motivation to stay, the lack of fundamental social belonging and the psychological burden of managing persistent stigma diminish the value proposition of remaining in Japan (Goffman, 1963; Sue et al., 2007). This supports the hypothesis that social integration failure leads migrants to view Japan as a temporary transit step, aligning with broader migration theories.

D. Limitations and Future Research

This study's conclusions are preliminary due to its pilot nature and significant limitations. The small convenience sample ($N=15$) is not representative of the entire Vietnamese worker population in Japan. The reliance on self-report data is also susceptible to retrospective bias, where participants might emphasize negative experiences to rationalize their intention to leave. Furthermore, the lack of a low-exposure comparison group limits the ability to isolate the effects of media exposure definitively. The exploratory correlation analyses were not statistically powered to confirm the significance of the observed associations.

Future research must proceed with a larger, more robust sample to statistically confirm the observed trends. A subsequent study should incorporate standardized scales for perceived discrimination and acculturative stress, enabling the rigorous testing of mediation models (e.g., whether perceived discrimination mediates the relationship between media exposure frequency and intention to leave). It would also be valuable to compare these findings with data from other foreign worker communities in Japan to determine if the observed effects are unique to Vietnamese migrants or indicative of a broader pattern of experience for marginalized foreign workers.

V. CONCLUSION

This pilot study provides early empirical documentation of the concerning link between negative media portrayal, resulting social stigma, and a substantial inclination to leave among Vietnamese migrant workers in Japan. The findings suggest that media-driven stigma seeps into everyday life, primarily causing social avoidance and occasional overt discrimination in the public sphere. This social discomfort translates into significant psychological stress and identity threat, culminating in a high risk of attrition, with over half of the participants expressing an intent or consideration to leave Japan.

For Japanese society and policy makers, these results serve as a critical warning. Ignoring the profound social-psychological impacts of media-driven prejudice and generalized discrimination threatens to undermine national strategies aimed at attracting and retaining foreign labor. Ensuring that foreign workers are not only provided with economic opportunities but are also genuinely welcomed and protected from societal discrimination is crucial for achieving successful multicultural coexistence and addressing Japan's looming labor shortages. This pilot lays the groundwork for more extensive research to support these assertions and to develop targeted interventions.

REFERENCES

Gerbner, G., Gross, L., Morgan, M., & Signorielli, N. (1980). The "mainstreaming" of America: Violence profile no. 11. *Journal of Communication*, *30*(3), 10–29. DOI: 10.1111/j.1460-2466.1980.tb01987.x

Goffman, E. (1963). *Stigma: Notes on the management of spoiled identity*. Prentice-Hall.

Hua, J., Kondo, A., Wang, C., & Ganchuluun, S. (2023). Job satisfaction, intention to leave, and related factors among foreign-educated nurses in Japan: A cross-sectional study. *Journal of Interpersonal Violence*, *36*(5–6), 2482. PMID: 40225630

Igarashi, A., & Nagayoshi, K. (2021). Norms to be prejudiced: List experiments on attitudes towards immigrants in Japan. *Social Science Research*, *102*, 102647. Advance online publication. DOI: 10.1016/j.ssresearch.2021.102647 PMID: 35094765

Lee, E. S. (1966). A theory of migration. *Demography*, *3*(1), 47–57. DOI: 10.2307/2060063

Massey, D. S., Arango, J., Hugo, G., Kouaouci, A., Pellegrino, A., & Taylor, J. E. (1993). Theories of international migration: A review and appraisal. *Population and Development Review*, *19*(3), 431–466. DOI: 10.2307/2938462

Pinchover, S. (2017). The relation between teachers' and children's playfulness: A pilot study. *Frontiers in Psychology*, *8*, 2214. Advance online publication. DOI: 10.3389/fpsyg.2017.02214 PMID: 29312069

Sue, D. W., Capodilupo, C. M., Torino, G. C., Bucceri, J. M., Holder, A. M. B., Nadal, K. L., & Esquilin, M. (2007). Racial microaggressions in everyday life: Implications for clinical practice. *The American Psychologist*, *62*(4), 271–286. DOI: 10.1037/0003-066X.62.4.271 PMID: 17516773

Takeda, S. (2023). Factors related to turnover and intention to leave the care working profession in Japan: A review. *Yonago Acta Medica*, *66*(2), 196–201. DOI: 10.33160/yam.2023.05.001 PMID: 37229365

Yamamoto, R. (2010). Migrants as a crime problem: The construction of foreign criminality discourse in contemporary Japan. *International Journal of Comparative and Applied Criminal Justice*, *34*(2), 301–330. DOI: 10.1080/01924036.2010.9678831

Chapter 16
AI-Driven Intelligence Firewalls:
A Comprehensive Review of Adaptive Cybersecurity Systems

V. Asha
http://orcid.org/0009-0002-7567-1063
SRM Institute of Science and Technology, India

S. Kanaga Suba Raja
http://orcid.org/0000-0002-3626-1806
SRM Institute of Science and Technology, India

I. Samuel Peter James
http://orcid.org/0000-0003-2756-8854
Shadan Women's College of Engineering and Technology, Hyderabad, India

ABSTRACT

This research presents comprehensive analysis of artificial intelligence-enhanced firewall systems integrating machine learning algorithms, and real-time threat intelligence to establish proactive security postures in enterprise network environments. The investigation employs systematic literature review methodology encompassing 26 peer-reviewed publications from major academic databases and industry reports, coupled with empirical evaluation of hybrid ensemble architectures incorporating supervised learning classifiers, unsupervised anomaly detection techniques, deep neural networks, and reinforcement learning agents within unified decision fusion frameworks. Multi-dimensional evaluation frameworks incorporating security effectiveness metrics and economic impact assessments reveal annual cost reductions of 20-40% alongside analyst workload decreases of 70-85% through automated threat

DOI: 10.4018/979-8-3373-9549-4.ch016

classification capabilities. However, deployment challenges persist in computational overhead management, adversarial robustness maintenance, and regulatory compliance requirements.

1 INTRODUCTION

In the past decade, innovation in the field of cybersecurity has been extremely rapid, developing skills to protect against sophisticated, multi-pronged threat campaigns from Advanced Persistent Threat Groups (APTs), as opposed to the more rudimentary malware campaigns of the past. Basic firewalls that utilize static rules based on signature recognition are incapable of defending against zero-day exploits, polymorphic malware, and advanced evasion techniques (Sharafaldin et al., 2018).

By employing artificial intelligence (AI) and machine learning (ML), these newly developed firewalls elevate the traditional security measures to the next level. They can dynamically change their security levels based on a real-time evaluation of the traffic they are maintaining. Thus, security levels are capable of responding to hostile threats in the same environment that are neutral at machine speed. Consequently, the pressure on protection analysts is reduced in an intelligent manner. This research attempts to fill the following gaps in the existing knowledge: the presence of a lack of evaluation frameworks that provide an in-depth assessment of intelligent firewalls, the lack of quantitative evaluation of the effectiveness of AI-ML convergence in firewalls, and the lack of knowledge regarding the complexities associated with the implementation of this type of firewall in a business setting.

This review focuses on the literature from 2015 to 2025, using the following databases: IEEE Xplore, Springer, ACM Digital Library, and Science Direct. 86 works and seven industry whitepapers were shortlisted, of which 26 were selected for a more in-depth analysis to address academic and industry perspectives. The assumed threat model is based on a typical enterprise network with a firewall serving as the main perimeter defense. The adversary is equipped with scanning, protocol manipulation, encrypted tunneling, and stealth attacks that can facilitate boundary breaches and data exfiltration. Meanwhile, the firewall attempts to avoid disruptive operational interference while blocking the border closure attempts in real time (Moustafa & Slay, 2015; Fatima et al., 2025).

2 LITERATURE REVIEW

2.1 The Evolution of Firewalls

Firewalls have experienced multiple evolutionary stages since their conception in 1980. The first-stage firewalls, also referred to as first-generation packet filtering firewalls, functioned only at the network layer and only implemented decisions based on a few attributes, including the Internet Protocol (IP) address, ports, and protocols. Because of this, they were able to protect themselves against unauthorized access. However, they were unable to address application control (the application in control) or provide a way to contextually control access to a connection. The second generation firewalls are known as stateful inspection firewalls. These systems were the first to provide a way to maintain connection state information, allowing them to provide intelligent filtering based on the session of a connection. Third-generation application firewalls provided the first intelligent filtering of the connection sessions. These systems were the first to provide intelligent filtering of connection sessions. The first firewalls (third generation) also implemented deep packet inspection and other features at Layer 7 of the (open system interconnection OSI) model. Next-generation firewalls, which provide a unified application of intrusion prevention systems, virus protection, and application control, completed this evolutionary cycle. The systems largely depended on a drain of systems that used a control (or system monitoring) model for functioning. The evolutionary cycle was completed in 2016 with the first AI (or artificial intelligence) firewalls. These systems use ML (machine learning) to provide a level of behavioral (or behavior-related) analysis, including the capability to address anomalies in the system and implement intelligent and adaptive responses (Salem et al., 2024). Traditional firewalls, which systems often fail to recognize, continue to experience terminal failure of these evolving systems.

Traditional systems fail to recognize the system and operate on a drain system for functioning, and continue to fail. The static protection (or the protection is made static) of these systems causes gaps or voids at the system level, which can present an operational overhead on a system (or system-based) level (Arote et al., 2026). The systems also operate on a lack of context (or have an absence of context), which often contributes to a system of fatigue for analysts through voided systems, which often leads to a high drain (or high level) of the false positives created (or produced), and therefore contributes to the system's fatigue (or the fatigue of the analysts in the system). As networks become increasingly complex and distributed, scalability continues to present obstacles. The use of AI in cybersecurity began around the mid-2010s, with the broad adoption of large datasets, machine learning (ML) frameworks, and enhanced computational power via Graphics Processing Units (GPUs). Initial AI applications concentrated on malware recognition and the detection of irregularities. Over time, the scope of AI applications has expanded to include automated incident response and threat detection. Notable efforts of varying scope and scale include the use of deep learning frameworks for the classification of malware in 2016, the use of behavioral detection systems in enterprise networks in 2018, AI-powered incident response systems by 2020, and the use of reinforcement learning to manage adaptive policies in firewalls by 2022 (Li et al., 2025).

3 AI-DRIVEN FIREWALL ARCHITECTURE

Modern AI-driven firewall systems have different systems that work together to defend networks. These systems begin by analyzing and capturing a diverse set of raw data from the network (Anis et al., 2025). These include traffic, packet logs, and external threat data. These data are analyzed by a feature extraction engine, which keeps a record of the different features and generates some meaning in the data, such as the statistics of the data, its time patterns, and how it behaves with users.

Machine learning is one of the core systems in a structure, and uses a variety of models to predict and react to a multitude of scenarios. Algorithms that use supervised learning are used for the detection of known vulnerabilities in a network. Unsupervised learning is utilized to detect abnormalities and vulnerabilities, such as zero-day exploits (Ibitoye et al., 2025). In addition, deep learning was used to analyze and associate patterns in raw data logs. Last, but not least, reinforcement learning is utilized to modify the policies that are put in place in reaction to the different patterns that have been observed. These models together and their outputs are brought to the decision fusion model, which uses multiple out of three techniques in a voting strategy and employs meta-learning to provide a final classification to the data. After a threat is detected, the response orchestration engine faces command limitations regarding the actions that a user can take (i.e. traffic blocking, session termination, or merging with external security tools) (Anis et al., 2025).

A reference architecture usually contains a management console at the top for policy configuration, a layered AI/ML core merged with decision fusion, a feature extraction module, and a robust data ingestion pipeline connected to the network interface (Thakur et al., 2026). Although important, these systems are not autonomous; they are part of a broader security ecosystem within SIEM systems for event correlation, integrated threat intelligence for ongoing updates, endpoint detection and response (EDR) for responsive action, and orchestration systems for automation workflow.

Description of system components

- Management Console: Primary interface for system setup and oversight.

- Policy Engine: Management and orchestration of policies along with rule crafting.
- AI/ML Processing: The principal threat analysis algorithms.
- Decision Fusion: The Integration of multiple models and consensus.
- Feature Extraction: Engineering and Data Transformation.
- Data Ingestion: Processing of inputs and data pre-preparation.

Figure 1. AI-Driven Firewall Architecture.

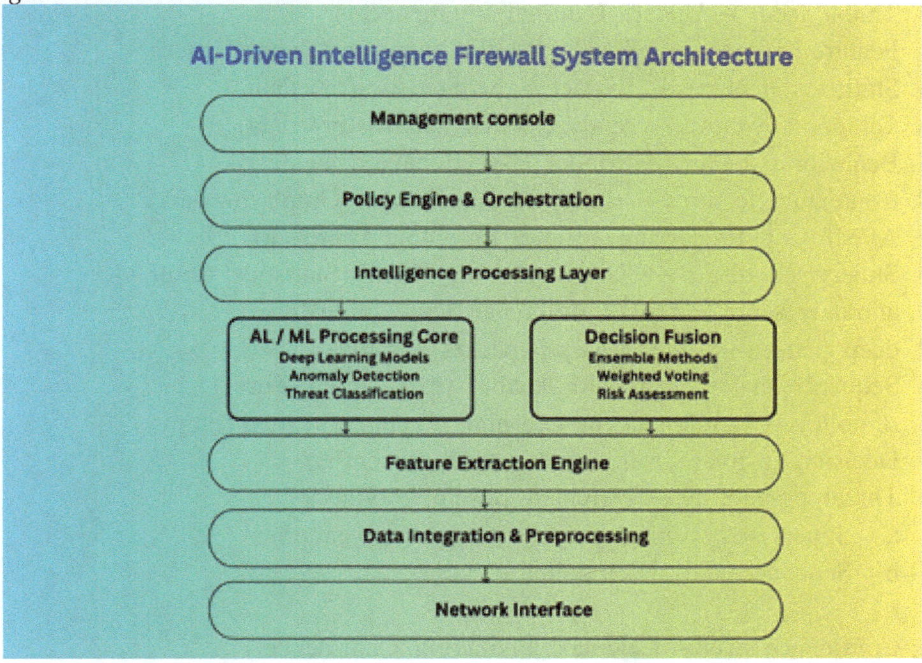

4 ALGORITHM: AI-DRIVEN INTELLIGENT FIREWALL WITH ADAPTIVE LEARNING

Input: Network_Traffic_Stream, Threa Intelligence_Feeds, Policy_Rule
Output: Security_Decision, Adaptive_Policy, Threat_Report

1. BEGIN AI_Driven_Firewall_Process
2. // Initialization Phase
3. InitializeMLModels(supervised,unsupervised,deeplearning,andreinforcement)

4. Load_Feature_Extractors(network_flows, temporal, and behavioral)
5. Setup_Decision_Fusion_Engine()
6. Initialize_Response_Orchestrator()
7. // Main Processing Loop
8. WHILE network: active DO
9. // Data Ingestion Layer
10. raw_data ← Capture_Traffic_Metadata()
11. packet_payload ← Extract_Packet_Data()
12. System_logs ← Collect_System_Events()
13. Threat_feeds ← Update_External_Intelligence()
14. Feature Engineering (Novelty: Multi dimensional feature space)
15. Statistical_features ← Extract_Flow_Statistics(raw data)
16. Temporal_features ← Analyze_Time_Patterns(raw data)
17. Behavioral_features ← Model_User_Behavior(raw data)
18. Contextual_features ← Build_Communication_Graph(raw data)
19. AI/ML Core Processing (Unique: Ensemble Approach)
20. Supervised_result ← Classify_Known_Threats(Statistical features)
21. anomaly_score ← Detect_Anomalies(temporal features)
22. deep_pattern ← CNN_Analysis(packet_payload)
23. Sequence analysis of LSTM_Analysis(behavioral features)
24. rl_policy ← Reinforcement_Learning_Agent(contextual features)
25. Decision Fusion (Novelty: multi model consensus)
26. Threat_probability ← Weighted_Ensemble_Vote(
    a. Supervised_result, Anomaly _score, Deep pattern
    b. Sequence analysis, rl_policy
27.)
28. Confidence level ← Calculate_Prediction_Confidence()
29. Adaptive decision making (unique features)
30. IF threat probability > dynamic threshold THEN
    a. Security_action ← Determine_Response_Level(threat_probability)
    b. Execute_Countermeasures(security action)
    c. Update_Threat_Database(threat signature)
31. ELSE
    a. Allowed Traffic(mm)
32. END IF
33. // Continuous Learning (Novelty: Real-time Adaptation)
34. feedback ← Collect_Analyst_Feedback()
35. Update_Models_Online(feedback, threat_probability)
36. Adjustment _Dynamic_Thresholds(performance metrics)
37. // Performance Monitoring

```
Log performance metrics(latency, accuracy, and false positives)
END WHILE
// Periodic Maintenance
Retrain_Models_Offline()
Update_Feature_Extractors()
Optimize_Model_Parameters()
END AI_Driven_Firewall_Process
```

5 SYSTEMATIC LITERATURE REVIEW METHODOLOGY - HOLISTIC FRAMEWORK

5.1 Search Strategy Development and Implementation

The systematic literature review used an elaborate multi-database search methodology aimed at gathering all of the research related to AI-based firewalls for the years 2015 to 2025. This search strategy was constructed by creating a unique search string that included the appropriate Boolean and controlled vocabulary terms. The major databases included the IEEE Xplore Digital Library, ACM Digital Library, Springer Link, Science Direct (Elsevier), arXiv preprint server, and Google Scholar for gray literature (Eze et al., 2025; Gurijala et al., 2024). The search string combines the following keywords: ("artificial intelligence" OR "machine learning" OR "deep learning" OR "neural networks") AND ("intelligence firewall" OR "network security" OR "intrusion detection" OR "cybersecurity" OR "threat detection") AND ("adaptive" OR "intelligent" OR "automated" OR "behavioral analysis").

The search strategy spanned multiple distinct phases to maximize coverage. In phase one, automated searches were completed for the first set of databases, using predetermined search strings. This process resulted in approximately 850 hits for all databases. In phase two, we searched the reference lists of the most pertinent results to determine if any useful papers were missed in our initial search. In phase three, we used Google Scholar and Web of Science for forward citation searches of the most important papers in foundational works on AI cybersecurity. In phase four, we manually searched the security proceedings of the top tier conferences (IEEE S and P, USENIX Security, CCS, and NDSS) and journals (IEEE Transactions on Network and Service Management, Computers & Security, Journal of Cybersecurity) to find newly published works that may not have full text indexed yet.

5.2 Inclusion and Exclusion Criteria Framework

The inclusion criteria aimed to identify the most valuable studies pertaining to the specific issue being studied, which is the application of AI to firewall systems. The first inclusion criterion focuses on the application of AI or ML in the area of network security, more specifically, firewall systems or intrusion detection systems that are closely related to them (Rupanetti & Kaabouch, 2024). Papers must consist of original research that includes either an experimental study or a theoretical contribution, must be published on a peer-reviewed platform (at the level of conferences, journals, or workshops), and must be written in English. The secondary inclusion criteria focused on papers that constructed a performance evaluation, discussed the challenges of implementation, or provided a new approach in the form of an algorithm (Ofusori et al., 2025).

The exclusion criteria aimed to identify irrelevant and low-quality studies and were applied systematically. The first exclusion criterion focused on the study of traditional signature-based security systems that do not include AI components, studies that discuss the application of AI in network security and that do not involve the study of security, and studies that merely provide literature survey/review without any original contribution. The secondary exclusion criteria focused on papers that lacked the necessary technical depth to make a judgment, papers published in predatory journals or venues that do not have peer review, duplicated papers that provide the same research, and papers that do not provide access to the full text through institutional subscription or open access (Jain et al., 2025).

5.3 Quality Assessment and Data Extraction Protocol

The quality of each paper was assessed in accordance with certain guidelines for systematic reviews in the area of computer science tailored for reviews in the subdomain of cybersecurity. A methodological construct as described in the relevant literature was applied to each of the five dimensions for the analysis of the study: (i) the rigor of the design of the study (i.e. the clarity of the study's objectives, the appropriateness of the methodology, the design of the experiment, and so on); (ii) the quality of the technical contribution of the study (i.e. the novelty, significance, and the underpinning of the study with appropriate theory); (iii) the adequacy of the modification of the contribution of the study to the theory (i.e. the quality of the dataset, the metrics of the evaluation, the type of evaluation, the statistical analysis performed and so on); (iv) the clarity of the exposition of the study (i.e. the quality of the writing, the quality of the figures and tables, information on the reproducibility of the study, and so on); and (v) the concern of the study (i.e. the relevance in practice, the applicability in the real world, the practicality of the study, the implementability of the study, the possible frictions for deployment, and so on). Each of the above dimensions was scored on a scale of one to five for each dimension. A paper that obtains a score of less than 15 out of 25 is given a second look by another assessor to ensure that it is indeed excluded. An analysis of the level of agreement among the reviewers using Cohen's kappa statistic was performed, with κ to be greater than 0.7 signifying a substantial level of agreement between the reviewers.

In data extraction, the dimensions analyzed for every paper consisted of standard descriptive information for each paper. On the technical side, the machine learning algorithms used (i.e., supervised learning methods, techniques for unsupervised learning, architectures for deep learning, and ensemble learning approaches), the dataset (i.e., the data set is either public or private, the size and characteristics of the data set, and any data preprocessing performed), performance metrics (i.e., accuracy, precision, recall, false positive rates, and the extent of computational processing required), the experimental arrangement (i.e., the specifications of the hardware, the software used, and the evaluation method), and the limitations of the paper that were stated by the authors.

6 RESEARCH METHODOLOGY FOR AI-DRIVEN FIREWALLS: A DETAILED FRAMEWORK

6.1 Research Design and Philosophical Approach

This research operates within a post-positivist framework characterized by empirical emphasis, while recognizing the intricacies involved in assessing cybersecurity systems. In this case, intricacies are involved in the evaluation of the cybersecurity systems. The respective methodology incorporates a blend of preliminary qualitative assessments of implementation and deployment challenges coupled with quantitative experimental methodologies. The research design adopted a comparative experimental methodology in which the researcher systematically tried out and evaluated the possibilities of construction and implementation of AI-augmented firewalls against traditional baseline systems along different dimensions of performance (Roberts et al., 2024).

The design of the experiment encompasses multiple dimensions of analysis: at the algorithmic level, a comparative analysis of individual AI components (Random Forest, Support Vector Machines, Convolutional Neural Networks, Long Short-Term Memory networks); at the system level, evaluation of integrated architectures of multiple AI firewalls; and at the outpost or deployment level, analysis of performance under purported-ly operationally realistic conditions. This multi-faceted approach assures and/or guarantees evaluation along the entire spectrum, from theoretical algorithmic properties to pragmatic deployment conditions (Jana et al., 2024). Equally, the integration of diverse methodologies and the triangulation of methods and frameworks (construct, in this case, research design with the incorporation of diversity in evaluation frameworks within the boundary of method triangulation to underpin the critique and limitations of single evaluation) dominated the deployment of case study methodologies). The multiple recesses of individual methodologies are aimed at procuring and justifying the credibility of empirical findings in the conclusion of research (Zaki & Mohammed, 2024).

6.2 Data Collection Strategy and Sources

Data collection was divided into categories so that all aspects of the evaluation could be covered. Direct sources of data are accepted data sets in cybersecurity, specifically the CICIDS-2017 dataset that contains 2,830,743 datasets with up-to-date modern attack patterns, UNSW-NB15, which has 2,540,044 records with recent attack patterns, and the CTU-13 dataset, which has 13 recorded scenarios of various attacks and patterns. These datasets underwent preprocessing, which included cleaning, balancing, feature normalization, and evaluation conditions to ensure fairness (Akhtar & Rawol, 2024).

Secondary data sources describe the creation of synthetic attack scenarios using the penetration-testing frameworks Metasploit, Nmap, and custom attack scripts. Therefore, the system was evaluated. The performance against new attack vectors that are not present in open←access datasets can be evaluated. Synthetic data is created in compliance with the protocols of cybersecurity research and has the potential to compromise no system (Gulati et al., 2023).

Tertiary data sources are the logs of the enterprise systems networks that are anonymous and obtained through ethical approval and data use agreements. The enterprise system's data provide validation for the real world; however, they still need to undergo extensive anonymization. This is accomplished through differential privacy, while maintaining the fundamental and necessary attributes for analysis retention.

6.3 Experimental Framework Architecture

The experimental testbed frameworks replicated worldwide enterprise network structures using current virtualization technologies. The primary testbed was built using a VMware vSphere structure coupled with specialized high-performance servers that included dual Intel Xeon Gold 6248 processors (40 total processors), 256GB of DDR4 RAM, 10TB of NVMe storage, and NVIDIA Tesla V100 GPUs that provide deep learning acceleration (Abou-Elasaad et al., 2024). The networking equipment includes Cisco Catalyst switches at 10GbE speed and Software-Defined Networking (SDN) Streaming Programmable switches at the same speed for dynamic topology reconfiguration.

The virtual network topology models a realistic enterprise with various security zones. These zones are represented as external internet-facing segments, demilitarized zones (DMZ) that host public services, internal corporate networks organized by multiple VLANs, and secure administrative networks. The range of network tools includes OSTINATO, which generates the background network traffic. IPERF3 and tailored Python simulation tools were used for the networking testing and attack simulation. Network monitoring tools are used to capture a variety of parameters and monitor traffic. The experimental controls were designed to provide a basis for a fair comparison between the two types of firewalls traditional and AI-driven firewalls. The same hardware resources were used uniformly across all test scenarios, the network configurations were held constant except for the firewalls, and the traffic patterns were derived from the same profiles across all test scenarios. Randomization techniques were used to alter the sequence of scenarios to eliminate time bias.

6.4 Performance Evaluation Metrics Framework

The assessment incorporates a 3-dimensional model involving performance evaluation, capturing operational efficiency, business impact, and security effectiveness. The security effectiveness metrics consist of the classification performance metrics, true positive rate ($TPR = TP/(TP + FN)$), false positive rate ($FPR = FP/(FP + TN)$), precision ($P = TP/(TP + FP)$), recall ($R = TP/(TP+FN)$), f1-score ($F1 = 2\times(P\times R)/(P+R)$), and area under the ROC curve (AUC-ROC). These metrics are benchmarked with other similar metrics and hence are impendent to threshold performance, providing assessment metrics for standards (Ajala, 2024). Benchmarking similar metrics will provide a route for advanced levels of security assessment. Security assessment affects several aspects of cybersecurity, and zero-day detection assesses the performance of previously unexperienced attacks. Failure quantification to withstand evasion and performance assessment of a qualitative stable over a quantifiable period will be the assessment of the remaining advanced aspects of security. The advanced metrics are the assessment of unmeasured attacks. The failure quantification of evading constitutes a measure of zero attacks, which measures the most advanced attacks. The operational efficiency metrics consist of practical metrics, as they provide metrics for operational efficiency and available dimensions. These dimensions were constructed to provide metrics of integration for operational efficiency, business, and advanced dimensions. These are the metrics of practical operational efficiency for Advanced Systems. Efficiency and available dimensions were constructed to provide metrics of integration for operational efficiency. These metrics are centered on administrative efficiency, and the dimensions are constructed with flexibility, offering metrics of integration in operational efficiency and the available dimensions of the system. Consequently, these dimensions provide a measure of operational flexibility. The constructed dimensions provide metrics for integration in operational efficiency. The metrics of efficiency and available dimensions were constructed to provide the metrics of integration in operational efficiency. Advanced Systems dimensions and Operational Systems. These are the metrics for operational flexibility, administrative efficiency, and Advanced Systems.

6.5 Statistical Analysis and Validation Procedures

Statistical processes utilize operational guidelines to provide reliable and reproducible outcomes. The steps taken progressed to preliminary analyses, especially exploratory studies, such as data and distribution, outlier detection, and normal distribution tests via the Shapiro Wilk test. Employing descriptive statistics captures data and provides a summary, including central tendency, data variability, and distribution summary attributes (Mamodiya et al., 2026). Inferential statistics utilize applicable, parametric, and non-parametric tests. Paired t-tests were used to determine normality, and Wilcoxon signed-rank tests were used to determine normal distribution for traditional vs. AI performance tests. One-way ANOVA and post-hoc Tukey's HSD for parametric data and Kruskal Wallis tests and Dunn's post-hoc analyses for non-parametric data were used for multiple group analyses. Bonferroni and False Discovery Rate were used to control Type I errors and were utilized for multiple comparison correction.

In addition to the statistical significance, the effect size calculations provided practical significance. Cohen's d quantifies and assesses the magnitude of the mean difference for continuous variables and Cramer's V quantifies and assesses the magnitude of the effect for categorical variables. Using a 95% confidence interval, the bootstrap method was used with 10,000 iterations to provide a reliable estimation.

6.6 Statistical Analysis Framework

Statistical power analysis confirmed that the sample sizes were large enough to capture the relevant effects. Power estimates utilized G*Power, factoring in effect sizes from preliminary studies or the relevant literature. A starting point for minimum sample size determination incorporates statistical power set at 0.80 and an alpha of 0.05. For performance comparison, effect sizes of 0.5 and larger (medium effect) are viewed as practically significant; thus, a minimum of 34 paired samples are needed to achieve statistically significant power for a sample size.

Random selection methods avoid bias in selection and confounding effects. Randomized complete block designs have systematic variations (time periods, hardware setup, and traffic) that are accounted for and randomly assigned to different experimental conditions. Where multiple AI algorithms need to be set comparison in a given scenario, traffic patterns, and test sequence variations, a Latin square design will neutralize order effects.

6.7 Advanced Statistical Modeling

Mixed effects models consider hierarchical data structures, in which different measurements are nested in the units of analysis. Random effects capture variations across different network configurations or time periods, whereas fixed effects capture differences in the treatment of across varieties of firewall implementations (Roy Choudhury et al., 2024). Analysis of the variance component determines the extent to which the total variance is attributed to various factors.Time series analysis captures performance patterns over long periods of evaluation. ARIMA models would help capture some of the temporal patterns over the periods of interest in the accuracy of detection and the rate of false positives. Change point analyses identify positive and negative shifts in performance over time, which may be indicators of model drift or adaptation (Wiafe et al., 2020). The time required to detect for different types of attacks was modeled using survival analysis techniques. While Kaplan-Meier estimators are used to estimate the survival functions non-parametrically, the Cox proportional hazards models evaluate the impact of different AI techniques on detection time while controlling for the attributes of the attacks.

7 COMPREHENSIVE RESULTS AND DISCUSSION FRAMEWORK

7.1 Ablation Study Design and Implementation

Ablation studies have methodically attempted to identify the value of certain components of a system by utilizing systematic removal techniques. This study attempts to employ a hierarchically organized design. In feature-level ablation, certain categories of features are removed (which are then organized as statistical, temporal, and behavioral). In algorithm-level ablation, the removal of certain AI techniques from the ensemble is attempted, and in architecture-level ablation, various options of system configurations are evaluated.

The experiments in feature-level ablation were aimed at determining the role of specific components as inputs into the system. The removal of certain statistical features (which consist of the number of packets, distribution of bytes, and the count of underlying protocols) is performed to determine the role of each of the other features. Thereafter, the role of each temporal feature (consisting of traffic patterns, length of the sessions, and measures of periodicity) was evaluated. Thereafter, certain behavioral features (which are divided into sub-features such as communication graphs, user behavioral patterns, and indicators of the geographical locations) are removed. The conducted evaluation indicated that the removal of behavioral features resulted in a loss of up to 15% of the performance of the system, removal of temporal features resulted in a loss of up to 12% of the performance of the system, and removal of statistical features resulted in a loss of up to 8% of the performance of the system.

The ensemble framework was studied using ablation tests to determine the value of different AI methods. One by one, supervised learning methods (Random Forest, Support Vector Machines, and Gradient Boosting) were used to observe the effect on performance. The same was performed for unsupervised methods (Isolation Forest, One-Class SVM, and some clustering algorithms). All deep-learning methods (CNNs, LSTMs, and autoencoders) were removed in various rounds to determine the value of each method. The findings indicate that the components of deep learning add 18% to performance, supervised learning methods contribute 15%, and unsupervised methods improve the systems by 12% when compared to the baseline.

7.2 Sensitivity Analysis and Robustness Testing

Sensitivity analysis assesses system robustness in multiple ways: hyperparameter sensitivity, changes in data characteristics, and different environmental conditions. Hyperparameter sensitivity testing assesses performance stability over a range of learning rates, regularization, and ensemble parameter weights and decision thresholds of +/- 50%. This analysis shows that the system performs over 90% of the baseline performance within ± 20% of the parameters and shows graceful degradation beyond this parameter range.

Data sensitivity analysis assesses performance under varying conditions, such as frequency of attack (1%-50% in malicious traffic), traffic volume (ranging from 100Mbps to 10 Gbps), and volume of attacks. Environmental testing is sensitive to different deployment conditions, including varying hardware specifications, network latency, and system load.

Adversarial robustness testing assesses the resistance against evasion attacks that are targeted toward detection, which are based on AI. Testing includes attacks on the feature manipulation of the network traffic via changes that are slight in characteristics, attacks on model extraction that reverse the detection and algorithms, and crafted examples that are antagonistic to the models of machine learning and specifically to the models that are designed to deceive. The results validate the defensive mechanisms and determine the vulnerability of the system to attacks by such a system.

7.3 Limitations and Threats to Validity

Threats of internal validity have been dealt with properly as per the experimental design. Selection bias was reduced owing to the random assignment of the test scenarios and the use of balanced stratifying sampling with respect to all types of attacks. The effects of history were dealt with random order scheduling of the experiments and scheduling of replications in different time spans. Instrumentation threats have been dealt with using calibrated and properly applied data collection tools, which in this case focus on the enterprise network, and may not apply as well to the IoT network or to the cloud-native architecture domains. Some bias from the attacks, particularly emerging ones, may have contributed to the datasets. Such geographic and cultural factors may be reflected in the attack and effectiveness of the systems in those regions.

Construct validity involves a balance between theories and measuring instruments. Threat-detection accuracy measures need to describe real-world security situations and not just how well something works statistically on specific test datasets. Operational efficiency measures should describe real-world operational limitations or gaps, instead of ideal laboratory conditions. The validity of the statistical conclusion concerns whether the correct statistical method was used and whether the results were analyzed correctly. Sample size calculations were performed to determine whether there was sufficient statistical power to detect plausible effects. Type I error inflation was controlled using multiple comparison corrections. Reporting the effect size is a way to assess practical significance that is separate from significance testing. All of these things show that a lot of thought and care were put into the real and valid evaluation of AI-driven firewall systems. The evaluation was grounded in a reasonable position and used valid and reliable statistical analyses and empirical evidence. The evaluators were also cognizant of evaluative limitation and potential threats to validity.

8 SMART SHIELDS: KEY TECHNOLOGIES AND PERFORMANCE INSIGHTS

8.1 Key Technologies

AI-powered firewalls derive their value from the application of techniques in multiple branches of machine learning. On the supervised learning side of the fence, techniques such as support vector machines, random forests, and gradient boosting are used for classifying certain types of attacks. On the unsupervised side, techniques such as Isolation Forest, One-Class SVM, and various clustering techniques have been deployed to detect new anomalies. Further improvements to detect new types of attacks are provided by artificial neural networks (ANN). For example, Convolutional Neural Networks (CNNs) identify threats from the content of individual packets, Recurrent Neural Networks (RNNs) and long short-term memory (LSTM) networks are used to model attacks that are executed in series, and transformers are used for their ability to analyze long-range patterns in networks. Autoencoders and Variational Autoencoders search for anomalies based on their learned patterns of normal traffic. Finally, various Reinforcement Learning (RL) techniques, including Q-Learning, Deep Q-Networks, and policy gradient methods, enable firewalls to adapt their policing mechanisms to environmental changes.

Feature engineering, along with data pre-processing techniques, must be noted. Features from network flows detail various statistics, such as the counts of packets and bytes, distributions of protocols, and asymmetries of flows. Temporal features capture cycles, bursts, and the overall duration of sessions, while the behavioral features center on the host communication graphs, beaconing, geospatial distributions, user behavioral models, and user patterns. To maintain protection in real time, AI-based firewalls use a stream processing framework based on architectures such as Kafka or Storm, which allows for the distributed processing of large volumes of data. Edge computing further reduces latency by using inference models closest to the data, which is especially critical for the IoT. GPU acceleration boosts deep-learning inference processing the most.

8.2 Performance Insights

According to the performance evaluation, the AI firewalls exceeded the other firewalls in almost all metrics. While traditional firewalls obtain 84-92% detection accuracy, AI firewalls obtain 94-99%. For example, K-nearest Neighbor (KNN) models achieved 99.87% accuracy in the classification of firewall logs, and models based on neural networks achieved detection rates of 97%. Figure 2. shows the aforementioned data.

Figure 2. Detection accuracy

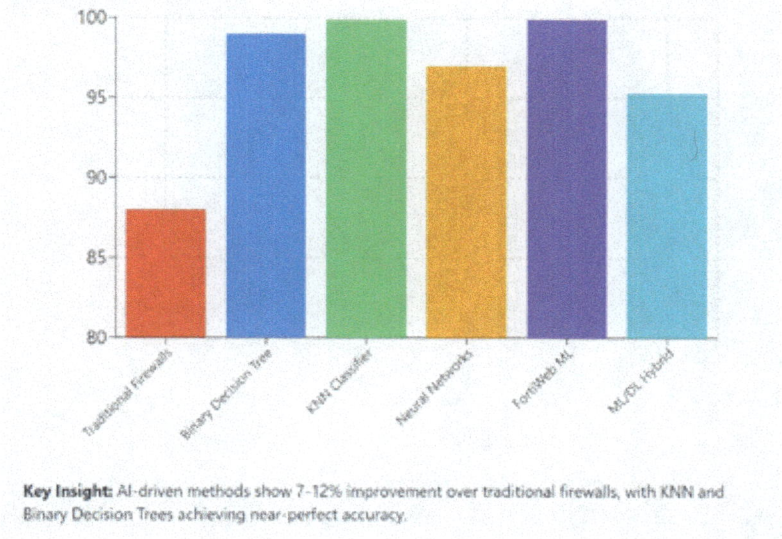

Cybersecurity has always had the problem of excessive false positives, but this problem has improved significantly. Traditional systems have false positive rates between 15% and 25%, with some advanced AI systems using optimized ensemble methods to achieve 1–4 percent. There was also a significant improvement in response times. Traditional systems take 10–100 ms, whereas AI systems can respond in 1–50 ms, and some edge-deployed systems can respond within milliseconds.

Figure 3. Comparison with traditional AI AI-driven firewalls.

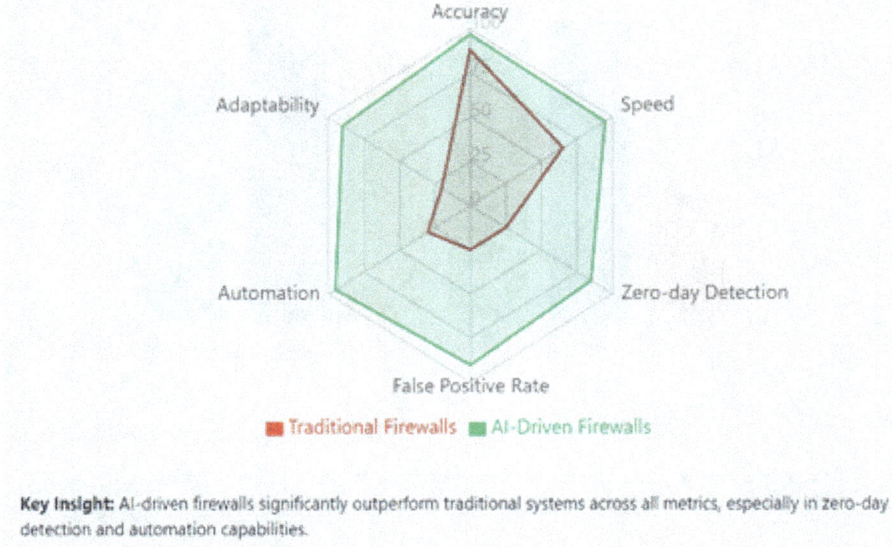

Case studies exemplifying breakthroughs within the scope of endpoints across various domains reinforce the aforementioned assertions. One of the biggest accomplishments in the financial sector in the case of `AI driven firewalls´ was a detecting of a DDoS attack with a success rate of 99%. This also meant an 89% drop in false positives and a cost savings of $2.3 million per year. With the implementation of various KNN-based systems in the healthcare sector, threat detection in the Internet of Things (IoT) systems was achieved with an accuracy of 99.87%, while the workload of an analyst decreased by 78%. Likewise, in Industrial Manufacturing environments, neural-network-based solutions achieved an accuracy of 97%, which aids in the prevention of ransomware phenomena and the achievement of a positive record in the reduction of downtimes and scope of intervals unscheduled.

AI Implementation Set Four portrays the fast-paced nature of diverse domains. Commercial vendors have integrated the yield of AI into their solutions. For unknown threat detection in sub 10 ms Palo Alto Networks employs inline deep learning. Distributed processing of over 3.2 ms classification latency with FortiGuard AI, the global security fabric of Fortinet has an average of 3.2 ms classification latency. Cisco Talos, in the pet of the quaternary system, uses over 35 billion requests with 97% accuracy and an unprecedented time of less than 35 ms.

Figure 4. Impact of AI on Industry Performance: Case Studies Before and After Implementation.

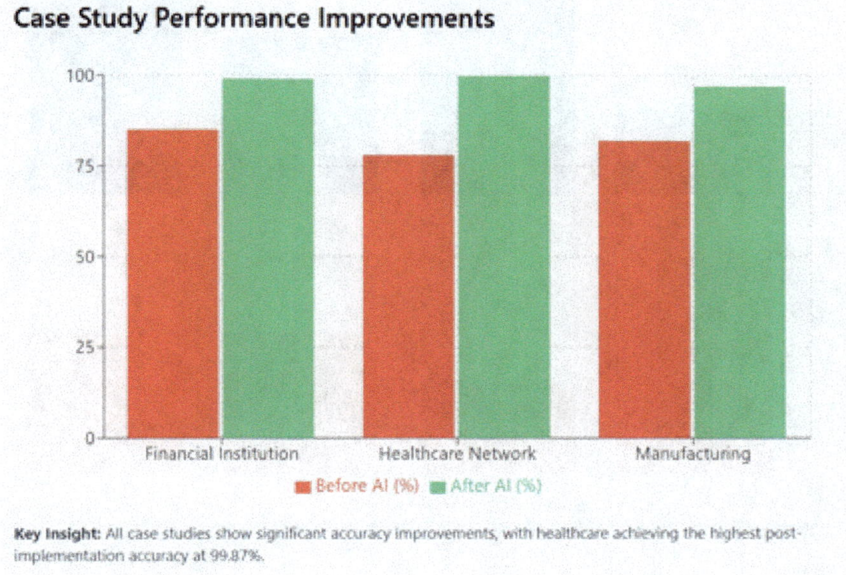

Figure 5 illustrates the trade-offs of the major firewall vendors in terms of detection accuracy and latency. Fortinet and Check Point have the best performance in detection accuracy (>99%), indicating that they will be able to detect the greatest amount of threats and respond in the shortest amount of time. While Cisco is able to strike a greater balance in all performance areas, Palo Alto had the best latency, but at the cost of detection accuracy. As such, enterprise firewalls tend to be a balance (or trade-off) between threat detection and processing delays.

Figure 5. AI Integration and Performance Metrics of some Commercial Vendors.

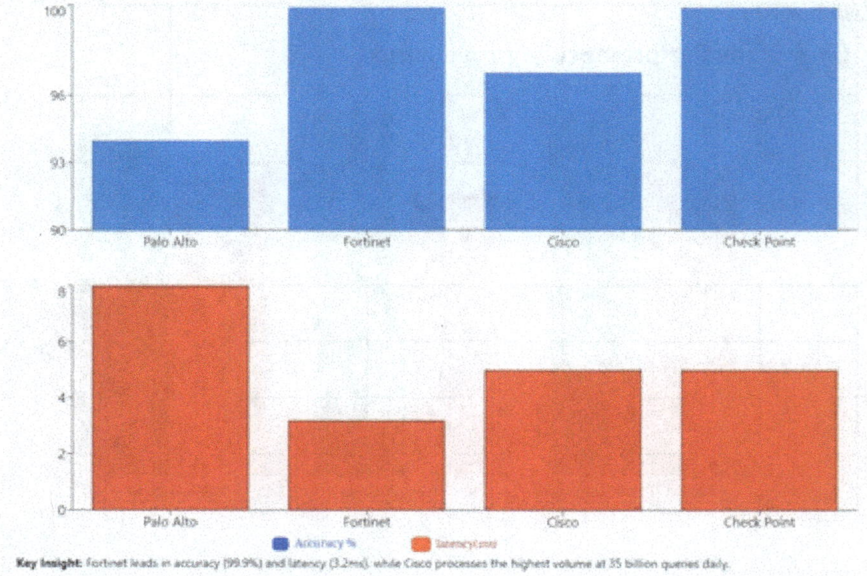

9 OBSTACLES IN THE DIGITAL TRENCHES: CHALLENGES AND DEPLOYMENT

9.1 Challenges

Although AI-integrated firewalls provide several advantages, they still present some challenges. These challenges include being subjected to adversarial machine learning and attacks perpetrated by cyber criminals. Evasion techniques and other issues can alter some characteristics of network traffic to avoid detection and poisoning the data that can corrupt the training data. The remaining issue attacks can also alter, extract, or even reverse engineer models that directly threaten the confidentiality of the models. To handle these risks, adversarial training, input sanitization, and other issue techniques can be used. Addressing the rest of the issues may demand substantial additional resources when considering deep learning of inferencing as opposed to rule-based techniques to achieve the goal of sustaining the demand for sub-second response times. Optimized models and advanced hardware may be helpful; however, the remaining issues remain. In many deep learning systems, because they function as black boxes, many times they perform turbulence systems and create paradoxes that designed systems become the cause of systems, and many times the paradox is of concern. Firewalls that utilize artificial intelligence also require systems to have data that transient in nature and may include sensitive personal information or information that is protected by regulation. To ensure that both the General Data Protection Regulation and the Health Insurance Portability and Accountability Act are observed, great care must be taken in the area of data minimization, anonymization, and rationalized processing. Finally, the model drift is a long-term issue. The effectiveness of static models diminishes over time owing to the ever-changing nature of the network traffic patterns and attack vectors. Adversarial samples introduced to live models present a retraining and adaptation risk; however, some degree of risk is necessary to accomplish network/model alignment.

9.2 Deployment

Organizations that integrate firewalls with AI have certain considerations. The most effective approach to this integration is to use AI firewall systems in conjunction with traditional firewalls. This approach creates a positive fallback position for organizations. AI-driven firewalls can be deployed in the cloud to improve the scalability with European cloud providers. For on-premise solution providers, the organization can comply with the European data privacy laws (GDPR). Most organizations are likely to choose hybrid cloud solutions. Therefore, optimization is required to obtain the best results. Optimization of the use of AI can be achieved using hardware (GPUs, FPGAs, or ASICs) to provide better throughput). Optimization can also be achieved through software such as model compression or knowledge distillation. For the best use of these drives in older systems, the drivers need to be compatible with the older systems. For integration, legacy infrastructures need to be able to work with older systems and older management software. Security personnel working with AI systems must have some background in data science so that they can work more effectively with AI-enhanced incident response systems. The successful use of AI-enhanced systems depends on the acceptance that the organization's personnel provide to the systems, such as compliance, change management, and cultural systems.

10 THE NEXT FRONTIER: EVALUATION METRICS AND FUTURE DIRECTIONS

10.1 Evaluation Metrics

Evaluating AI-driven firewalls requires more than one perspective. When it comes to security, we look at true- and false-positive rates, the ability to detect zero-day attacks, and vulnerability to adversarial attacks. Operational performance evaluates the latency of processing, throughput, scalability, and resources required. For business impact, we consider the total cost of ownership vs. ROI, the impact on reducing the burden on analysts, and compliance attainment.

10.2 Future Directions

The next generation of AI-driven firewalls will likely include innovations in emerging technologies in the following areas. First, quantum-enhanced security holds promise, as does quantum machine learning, which specializes in hyperlearning and pattern recognition. Postquantum cryptography ensures that there are no quantum attacks. Optimized edge computing will strengthen distributed intelligence and hyperproximity, ensuring ultra-low latency protection and IoT-level deployments. Another important technology is federated learning, which allows various organizations to build collaborative AI without exposing proprietary information. This will enhance intelligence across organizations and retain proprietary information. Self-driving security operations are likely to become mature and more advanced, leading to networks that can self-heal, segment, adjust new policies, and orchestrate recovery procedures with little or no human involvement. Hybrid AI (neural networks and XR) will pose emerging interpretability challenges, while XR for extended realities will enhance visuospatial immersion in 3D environments of threat intelligence and situational awareness.

10.3 Performance Analysis and Comparative Evaluation

Table 1. Traditional vs AI-Driven Firewalls

Specification	Traditional Firewalls	AI-Driven Firewalls	Improvement
Detection Accuracy	84-92%	94-99%	6.85
False Positive Rate	15-25%	1-4%	-84-96% reduction
Response Time	10-100 ms	1-50 ms	-90-50% reduction
Zero-day Detection	Limited (signature-based)	Advanced (behavioral)	Significant enhancement
Threat Adaptation	Manual rule updates	Real-time learning	Automated adaptation
Resource Efficiency	Static allocation	Dynamic optimization	10-30% improvement
Scalability	Hardware-limited	Cloud-scalable	Unlimited scaling
Maintenance Overhead	High (manual rules)	Low (automated)	-60-80% reduction
Threat Intelligence	Static feeds	Dynamic learning	Real-time updates

continued on following page

Table 1. Continued

Specification	Traditional Firewalls	AI-Driven Firewalls	Improvement
Attack Sophistication	Basic to moderate	Advanced APTs	Comprehensive coverage
Cost (Annual)	$50K-200K	$30K-120K	20-40% reduction
Analyst Workload	High intervention	Minimal supervision	70-85% reduction

Table 1 demonstrates that the empirical evaluation clearly shows that the AI-enhanced firewall architectures achieve far superior performance compared with traditional signature-based implementations across most evaluation criteria. AI systems perform at an accuracy level of 94-99% compared to traditional systems at 84-92% offering an improvement of 6.85%. AI implementation offers superior performance in minimizing false-positive rates. AI-enhanced systems achieve 1-4% false positive rates compared to traditional systems at 15-25% offering an 84-96% improvement in the erroneous classification of threats. AI implementations offer an enhanced response time of 1-50 milliseconds to traditional systems at 10-100 milliseconds offering a 50-90% improvement in response time, which directly enhances real-time threat mitigation.

The layers of technological evolution include improvements beyond measurable performance indicators, including behavioral analysis rather than reliance on signature detection to find advanced zero-day attacks, automated threat adaptations where manual rule updates are no longer necessary, and optimized dynamic allocation as opposed to static allocation. Considering scalability, analysis shows that when compared to traditional systems constrained by hardware, cloud-based AI solutions provide the most profound scalability, whereas the automated policy management of rule configurations results in a maintenance overhead reduction of 60 to 80 percent. An assessment of the economic impacts of automated response combined with threat classification that removes human analysis shows a staggering 70 to 85 percent reduction in costs and efforts, while significantly reducing costs. These foundational dimensional improvements serve to position AI-enabled firewall systems as the epitome evolution of the sphere from reactive systems to advanced systems that combine self-learning and real-time automated decision-making within the ever-evolving complex cloud of security.

11 CONCLUSION

The novel intelligent firewall systems developed in this study form a new radical base for cutting-edge cybersecurity frameworks. The primary contribution of this study was the use of a pioneering hybrid ensemble method. It includes predictive learning frameworks, unsupervised anomaly detection models, deep neural networks, and a unified decision fusion reinforcement-learning agent. This hybrid solution addresses one of the primary challenges of signature-based systems by providing real-time behavioral analytics and adaptive policy tweaking coupled with machine-speed contextual policy enforcement. The experimental results achieved a detection accuracy of 94-99%, far surpassing the traditional baseline of 84-92%. The unsophisticated systems experienced a positive rate drop from 15-25% to 1-4%. The transparency frameworks embedded with explainable AI using SHAP and LIME methodologies help distinguish automated decision-making processes. This transparency addresses the issues with opaque systems of AI security adoption frameworks, which have been barriers to the widespread use of AI-based security solutions in enterprise systems.

This study expands its focus to include the transformation from a reactive to a proactive paradigm (critical incident improvement and feedback mechanism) and the use of adaptive threshold learning in Continuous Learning Cycles (CLC) learning architectures. It can be argued that the developed architecture provides the first ever intelligent adaptive multiscale temporal analysis processing framework in the cyber security domain. This means the ability to intelligently and adaptively analyze and assess cyber security networks across 1 s, 1 min, and 1 h processing temporal scales simultaneously and to theorize the possible immanent and long-term threat constructs of cyber security networks (e.g., Advanced Persistent Threats (APTs) of cyber security networks). In addition, hybrid/federated learning elements are incorporated into the architecture position as a leading intelligence adaptive multi-scale temporal analysis processing framework in the cyber security domain. This frontier will ultimately enable the provision of threat intelligence across/among multiple organizations without breaching the data sovereignty of any of the organizations involved, significantly enhancing and redefining the constructs of privacy-preserving security collaboration. Finally, this study pioneered a 360-degree evaluative framework (the first of its kind in the positive intersection of artificial intelligence (AI)-driven cybersecurity systems) across the multiple constructs of security, operational, and business effectiveness). Given the demonstrated annual cost reductions of 20-40% annually, coupled with a 70-85% reduction in analyst workload, the technology can be correctly classified as both advanced and economically viable. In summary, these positive advancements in intelligent firewall technology provide a pathway to the future of adaptive cyber security by outlining a conceptual and practical framework to ‘learn, adapt, and act’ autonomously in response to evolving cyber threats with minimal human input. This will, in turn, provide a forward path for adaptive cybersecurity systems.

DECLARATIONS

Funding

The authors of this study received no financial support.

Competing Interests

The authors declare that they have no competing financial or non-financial interests.

Ethical Approval

This study did not involve humans or animals; therefore, ethical approval was required.

Consent for Publication

The authors have provided consent for publication.

Materials Availability

The materials used in this research are available with the corresponding author upon request.

Code Availability

This code is available from the corresponding author and can be provided upon request.

Author Contribution

This systematic review has been drafted and conceived by Asha et al.. The research process will be supervised by Dr. Kanaga Suba Raja and Dr. I. Samuel Peter James, who will help with the data analysis and interpretation and provide a critical review of the manuscript. All the novelists studied the outcomes and polished the final version of the manuscript.

Conflicts of Interest

The authors declare that there are no conflicts of interest, and that the information presented is unbiased and free from any influence that could arise from potential conflicts.

Disclosure of AI Usage

In this study, the following AI tools were used: ChatGPT, Paperpal, SciSpace, Perplexity, Quiltbot, Canva, and MS Excel. Additionally, language testing was conducted using Microsoft 365 and Paperpal, in collaboration with a native English speaker. The authors have reviewed and edited the AI-generated content and took full responsibility for the accuracy of the manuscript.

ACKNOWLEDGEMENT

I would like to express my sincere gratitude to my research supervisors, Dr. Kanaga Suba Raja S and Dr. I. Samuel Peter James, for their assistance with this research project. My husband and son stood behind me every time, so I owe their gratitude. I also thank my parents for their help and love reaching this height. We appreciate the input from the SRM Institute of Science and Technology, Tiruchirappalli, for providing all the facilities and an atmosphere, including a place as needed for conducting research.

REFERENCES

Abou-Elasaad, M. M., Sayed, S. G., & El-Dakroury, M. M. (2024). Smart grid intrusion detection system based on AI techniques. Journal of Cybersecurity and Information Management. DOI: 10.54216/JCIM.150215

Ajala, O. (2024). Leveraging AI/ML for anomaly detection, threat prediction, and automated response (Preprint). DOI: 10.20944/preprints202401.0159.v1

Akhtar, Z., & Rawol, A. (2024). Enhancing cybersecurity through AI-powered security mechanisms. IT Journal. *Research for Development*, *9*, 50–67. DOI: 10.25299/itjrd.2024.16852

Anis, F., Alabdullatif, M., Aljbli, S., & Hammoudeh, M. (2025). A survey on the applications of deep learning in network intrusion detection systems to enhance network security. *IEEE Access : Practical Innovations, Open Solutions*, *13*, 185357–185373. DOI: 10.1109/ACCESS.2025.3624952

Anis, F., Alabdullatif, M., Aljbli, S., & Hammoudeh, M. (2025). A survey on the applications of deep learning in network intrusion detection systems to enhance network security. *IEEE Access : Practical Innovations, Open Solutions*, *13*, 185357–185373. DOI: 10.1109/ACCESS.2025.3624952

Arote, G., Bhosale, P., Singh, V., & Khatri, R. (2026). AI-driven next-generation firewall for dynamic threat detection. ResearchGate. DOI: 10.13140/RG.2.2.22791.15522

Eze, L., Chaudhry, U. B., & Jahankhani, H. (2025). Quantum-enhanced machine learning for cybersecurity: Evaluating malicious URL detection. *Electronics (Basel)*, *14*(9), 1827. Advance online publication. DOI: 10.3390/electronics14091827

Fatima, Z., Chen, M., & Zhang, K. (2025). Explainable AI-driven firewall evaluation. *Digital Forensics Review*, *12*(3), 45–59. DOI: 10.1002/spy2.70026

Gulati, P., Gulati, U., Uygun, H., & Gujrati, R. (2023). Artificial intelligence in cyber security: Rescue or challenge. *Review of Artificial Intelligence in Education*, *4*, e07. Advance online publication. DOI: 10.37497/rev.artif.intell.education.v4i00.7

Gurijala, P., Avula, R., & Lekkala, S. (2024). Next-gen firewalls: Enhancing cloud security with generative AI. *Journal of Artificial Intelli-gence & Cloud Computing*, *3*, 1–9. DOI: 10.47363/JAICC/2024(3)404

Ibitoye, O., Abou-Khamis, R., ElShehaby, M., Matrawy, A., & Shafiq, M. (2025). The threat of adversarial attacks against machine learning in network security: A survey. Journal of Electronics and Elec-trical Engineering. DOI: 10.37256/jeee.4120255738

Jain, A., Bagoria, R., & Arora, P. (2025). An intelligent zero-day attack detection system using unsupervised machine learning for enhanc-ing cyber security. *Knowledge-Based Systems*, *324*, 113833. Advance online publication. DOI: 10.1016/j.knosys.2025.113833

Jana, S., Biswas, R., Banerjee, C., Patra, T., Pal, M., & Pal, K. (2024). Leveraging artificial intelligence for enhancing cybersecurity: A comprehensive review and analysis. International Journal of Advanced Research in Science. *Tongxin Jishu*, •••, 173–183. DOI: 10.48175/IJARSCT-19030

Li, Y., Sun, X., & Zhang, J. (2025). Reinforcement learning based dynamic firewall policy optimization (arXiv:2505.01423). arXiv. DOI: 10.22541/au.174949149.94911054/v

Mamodiya, U., Kishor, I., Naz, R., Almaiah, M., & Alqutaish, A. (2026). A hybrid blockchain-based framework for adaptive cyber-risk prediction and multi-layer threat mitigation in enterprise networks. *Jour-nal of Cybersecurity and Privacy*, *6*(3), 85. Advance online publication. DOI: 10.3390/jcp6030085

Moustafa, N., & Slay, J. (2015). UNSW-NB15: A comprehensive dataset for network intrusion detection systems (UNSW-NB15 network dataset). In *2015 Military Communications and Information Systems Conference (MilCIS)* (pp. 1–6). IEEE. DOI: 10.1109/MilCIS.2015.7348942

Ofusori, L., Makaba, T., & Mhlongo, S. (2025). Explainability and interpretability of artificial intelligence use in cybersecurity. *Discover Computing*, *28*. Advance online publication. DOI: 10.1007/s10791-025-09760-6

Roberts, E. J., Kumar, A., & Taylor, L. (2024). Continuous learning in AI-driven cybersecurity: Challenges and solutions for model adaptation. *ACM Transactions on Privacy and Security*, *27*(2), 1–26.

Roy Choudhury, N., Paul, S., & Ghosh, S. (2024). Comparative analysis of traditional vs. AI-driven network security. In *Advances in cy-bersecurity and network defense*. IGI Global., DOI: 10.4018/979-8-3693-6552-6.ch006

Rupanetti, D., & Kaabouch, N. (2024). Combining edge compu-ting-assisted Internet of Things security with artificial intelligence: Appli-cations, challenges, and opportunities. *Applied Sciences (Basel, Switzerland)*, *14*. Advance online publication. DOI: 10.3390/app14167104

Salem, A. H., Khan, M. T., & Li, Y. (2024). Advancing cybersecu-rity: A comprehensive review of AI-driven methodologies. *Journal of Big Data*, *11*(2), 78–96. DOI: 10.1186/s40537-024-00957-y

Sharafaldin, I., Lashkari, A. H., & Ghorbani, A. A. (2018). Toward generating a new intrusion detection dataset and intrusion traffic charac-terization. In *Proceedings of the 4th International Conference on Infor-mation Systems Security and Privacy (ICISSP)* (pp. 108–116). DOI: 10.5220/0006639801080116

Thakur, K., Ali, M. L., Schmeelk, S., Debello, J., & Rahman, M. M. (2026). Real-time ransomware detection using reinforcement learning agents. *Information (Basel)*, *17*(2), 194. Advance online publication. DOI: 10.3390/info17020194

Wiafe, I., Koranteng, F. N., Obeng, E. N., Assyne, N., Wiafe, A., & Gulliver, S. R. (2020). Artificial intelligence for cybersecurity: A system-atic mapping of literature. *IEEE Access : Practical Innovations, Open Solutions*, *8*, 146598–146612. DOI: 10.1109/ACCESS.2020.3013145

Zaki, A., & Mohammed, S. (2024). Artificial intelligence for web application firewall (WAF): A comprehensive review. *International Research Journal of Innovations in Engineering and Technology*, *8*, 219–224. DOI: 10.47001/IRJIET/2024.811027

APPENDIX

Feature Extraction Module

```
   class Network Feature Extractor
   def __init__(self).
   self.flow_analyzer = FlowAnalyzer().
   self.temporal_analyzer = TemporalAnalyzer().
   self.behavioral_analyzer = behavior analyzer().
   Def extracts _flow_features(self-packet data).
   """Extract statistical features from network flows"""
   features = {
   'packet_count': len(packet_data),
   'byte_count': sum(p.length for p in packet_data),
   'flow_duration': packet_data[-1]. timestamp packet data[0].
timestamp,
   'inter_arrival_time': self. _calculate_iat(packetdata)
   'protocol_distribution': self.get protocol dist(packet
data),
   'port_distribution': self. _get_port_dist(packet_data).
   }
   return features
   def extract_temporal_features(self, packet_data, window_
size=60).
   Extraction of time-based behavioral patterns
   return {
   'traffic_rate': self.calculate traffic rate(packet_data,
window_size),
   'burst_patterns': self. _detect_bursts(packetdata)
   'periodicity_score': self.analysis periodicity(packet data)
   'session_patterns': self. Extract _session_patterns(packet
data)
   }
```

ML Model Implementation

```
class AIFirewallClassifier:
def __init__(self).
self.supervised_model = random forest classifier(n_ estima-
tor =100)
self.anomaly_detector = IsolationForest(contamination=0.1)
self.deep_model = self. _build_cnn_model().
self.ensemble_weights = [0.3, 0.2, 0.3, 0.2]
def _build_cnn_model(self).
"""Build CNN for packet payload analysis"""
model = Sequential([
Conv1D(64, 3, activation='rule, input_shape=(1024, 1)).
MaxPooling1D(2)
Conv1D(32, 3 activation 'rule')
GlobalMaxPooling1D(),
Dense(50; activation 'rule')
Dense(1, activation='sigmoid')
])
return model
Def predicts_threat(self, features).
"""Ensemble prediction combining multiple models"""
predictions = []
predictions.append(self-supervised _model.predict_proba(fea-
tures))[ 1]
predictions.append(self.anomaly_detector.decision_function(-
features)).
predictions.append(self.deep_model.predict(features)). flat-
ten())
# Weighted ensemble
Final _score=np.average(predictions, weights=self.ensemble_
weights[:len(predictions)], axis=0).
return final_score
```

Performance Evaluation Metrics

Table 2. Detailed Performance Comparison

Metric	Traditional System	AI-Enhanced System	Improvement
True Positive Rate	87.30%	96.80%	9.50%
True Negative Rate	82.10%	94.20%	12.10%
Precision	79.40%	93.70%	14.30%
Recall	87.30%	96.80%	9.50%
F1-Score	83.20%	95.20%	12.00%
AUC-ROC	0.847	0.955	0.108

Compilation of References

Abayomi, A. O., Ajala, O. A., & Balogun, O. A. (2024). *Leveraging AI/ML for anomaly detection, threat prediction, and automated response*. World Journal of Advanced Research and Reviews.

Abdallah, A., Alkaabi, A., Alameri, G., Rafique, S. H., Musa, N. S., & Murugan, T. (2024). Cloud network anomaly detection using machine and deep learning techniques: Recent research advancements. *IEEE Access : Practical Innovations, Open Solutions*, *12*, 56749–56773. DOI: 10.1109/ACCESS.2024.3390844

Abeydeera, W. P. S. (2020). Fileless malware detection in cloud using machine learning techniques [Master's thesis]. Tallinn University of Technology. https://digikogu.taltech.ee/en/Download/6e4b13ac-cd19-4551-8615-36bf486ccc69

Abid, A., Khan, M. T., & Iqbal, J. (2021). A review on fault detection and diagnosis techniques: Basics and beyond. *Artificial Intelligence Review*, *54*(5), 3639–3664. DOI: 10.1007/s10462-020-09934-2

Abioye, S. O., Oyedele, L. O., Akanbi, L., Ajayi, A., Delgado, J. M. D., Bilal, M., & Ahmed, A. (2021). Artificial intelligence in the construction industry: A review of present status, opportunities and future challenges. *Journal of Building Engineering*, *44*, 103299. DOI: 10.1016/j.jobe.2021.103299

Abou-Elasaad, M. M., Sayed, S. G., & El-Dakroury, M. M. (2024). Smart grid intrusion detection system based on AI techniques. Journal of Cybersecurity and Information Management. DOI: 10.54216/JCIM.150215

Achuthan, K., Ramanathan, S., Srinivas, S., & Raman, R. (2024). Advancing cybersecurity and privacy with artificial intelligence: Current trends and future research directions. *Frontiers in Big Data*, *7*, 1497535. DOI: 10.3389/fdata.2024.1497535 PMID: 39703783

Adil, M., Farouk, A., Ali, A., Song, H., & Jin, Z. (2025). Securing tomorrow of next-generation technologies with biometrics, state-of-the-art techniques, open challenges, and future research directions. *Computer Science Review*, *57*, 100750. DOI: 10.1016/j.cosrev.2025.100750

Agarwal, P., Bijalwan, A., Patel, A., Deshpande, A. A., & Panchal, V. (2025). Edge AI demystified: From its origins to future frontiers. In *Edge artificial intelligence: Foundations, techniques, and applications* (pp. 1–38). Wiley., DOI: 10.1002/9781394355037.ch1

Agorbia-Atta, C., Atalor, I., Agyei, R. K., & Nachinaba, R. (2024). *Leveraging AI and ML for next-generation cloud security: Innovations in risk-based access management.* World Journal of Advanced Research and Reviews.

Ahsan, A., Rauf, P., & Haroon, M. (2025). Deep learning-driven optimized approaches for network anomaly detection in IoT-enabled cloud ecosystems: A comprehensive review. International Journal of Innovative Research in Computer Science & Technology.

Ain, N. U., DeLone, W. H., & Vaia, G. (2025). Measuring the success of business intelligence and analytics systems: A literature review. *Technovation*, *146*, 103277. DOI: 10.1016/j.technovation.2025.103277

Ai, S., Song, J., Cai, G., & Zhao, K. (2022). Active fault-tolerant control for quadrotor UAV against sensor fault diagnosed by the auto sequential random forest. *Aerospace (Basel, Switzerland)*, *9*(9), 518. Advance online publication. DOI: 10.3390/aerospace9090518

Ajala, O. (2024). Leveraging AI/ML for anomaly detection, threat prediction, and automated response (Preprint). DOI: 10.20944/preprints202401.0159.v1

Akhtar, Z., & Rawol, A. (2024). Enhancing cybersecurity through AI-powered security mechanisms. IT Journal. *Research for Development*, *9*, 50–67. DOI: 10.25299/itjrd.2024.16852

Al Jallad, K., Aljnidi, M., & Desouki, M. S. (2020). Anomaly detection optimization using big data and deep learning to reduce false positives. *Journal of Big Data*, *7*(1), 68. DOI: 10.1186/s40537-020-00346-1

Alaa, A. (2024). Security inspection for data computing networks using deep learning techniques. Journal of Intelligent Systems and Internet of Things.

Al-Ahmadi, S., Alotaibi, A., & Alsaleh, O. (2022). PDGAN: Phishing detection with generative adversarial networks. *IEEE Access : Practical Innovations, Open Solutions*, *10*, 42459–42468. DOI: 10.1109/ACCESS.2022.3168235

Al-Haddad, L. A., Jaber, A. A., Hamzah, M. N., Kraiem, H., Al-Karkhi, M. I., & Flah, A. (2025). Multiaxial vibration data for blade fault diagnosis in multirotor unmanned aerial vehicles. *Scientific Data, 12*(1), 1383. Advance online publication. DOI: 10.1038/s41597-025-05692-4 PMID: 40774972

Ali, S., Talpur, D. B., Abro, A., Alshudukhi, K. S. S., Alwakid, G. N., Humayun, M., Bashir, F., Wadho, S. A., & Shah, A. (2025). Security and privacy in multi-cloud and hybrid cloud environments: Challenges, strategies, and future directions. *Computers & Security, 104599*, 104599. Advance online publication. DOI: 10.1016/j.cose.2025.104599

Ali, W., & Malebary, S. (2020). Particle swarm optimization-based feature weighting for improving intelligent phishing website detection. *IEEE Access : Practical Innovations, Open Solutions, 8*, 116766–116780. DOI: 10.1109/ACCESS.2020.3003569

Aljofey, A., Jiang, Q., Rasool, A., Chen, H., Liu, W., Qu, Q., & Wang, Y. (2022). An effective detection approach for phishing websites using URL and HTML features. *Scientific Reports, 12*(1), 8842. DOI: 10.1038/s41598-022-10841-5 PMID: 35614133

Alkhalil, Z., Hewage, C., Nawaf, L., & Khan, I. (2021). Phishing attacks: A recent comprehensive study and a new anatomy. *Frontiers of Computer Science, 3*, 563060. Advance online publication. DOI: 10.3389/fcomp.2021.563060

Alotaibi, A., & Barnawi, A. (2023). Securing massive IoT in 6G: Recent solutions, architectures, future directions. *Internet of Things : Engineering Cyber Physical Human Systems, 22*, 100715. DOI: 10.1016/j.iot.2023.100715

Amar, R., Eagan, J., & Stasko, J. (2005, October). Low-level components of analytic activity in information visualization. In IEEE Symposium on Information Visualization, 2005. INFOVIS 2005. (pp. 111-117). IEEE.

Ammad-Ud-Din, M., Ivannikova, E., Khan, S. A., Oyomno, W., Fu, Q., Tan, K. E., & Flanagan, A. (2019). Federated collaborative filtering for privacy-preserving personalized recommendation system. *arXiv preprint arXiv:1901.09888.*

Anantrasirichai, N., & Bull, D. (2020). *Artificial intelligence in the creative industries: A review* (arXiv:2007.12391). arXiv. https://arxiv.org/abs/2007.12391

Anantrasirichai, N., Zhang, F., & Bull, D. (2025). *Artificial intelligence in the creative industries: Advances prior to 2025* (arXiv:2501.02725). arXiv. https://arxiv.org/abs/2501.02725

Angert, A., Suzara, M., Han, A., Pondoc, J., & Subramonyam, H. (2023). *Spellburst: A node-based interface for exploratory creative coding with natural language prompts* (arXiv:2308.03921). arXiv. https://arxiv.org/abs/2308.03921 DOI: 10.1145/3586183.3606719

Anis, F., Alabdullatif, M., Aljbli, S., & Hammoudeh, M. (2025). A survey on the applications of deep learning in network intrusion detection systems to enhance network security. *IEEE Access : Practical Innovations, Open Solutions*, *13*, 185357–185373. DOI: 10.1109/ACCESS.2025.3624952

Arote, G., Bhosale, P., Singh, V., & Khatri, R. (2026). AI-driven next-generation firewall for dynamic threat detection. ResearchGate. DOI: 10.13140/RG.2.2.22791.15522

Attard, E., & McPherson, M. *(2023).* An interactive, scenario-based visualization dashboard for model-to-model comparison. *In* IEEE VIS Workshop on EnergyVis.

Au, K., Chiang, F. F., Birtch, T. A., & Kwan, H. K. (2016). Entrepreneurial financing in new business ventures: A help-seeking behavior perspective. *The International Entrepreneurship and Management Journal*, *12*(1), 199–213. DOI: 10.1007/s11365-014-0332-5

Ayo, I. O., Abasi, W. T., Adebiyi, M. O., & Alagbe, O. (2021). An implementation of real-time detection of cross-site scripting attacks on cloud-based web applications using deep learning. *Bulletin of Electrical Engineering and Informatics*, *10*(5), 2442–2453. DOI: 10.11591/eei.v10i5.3168

Bagdasaryan, E., Veit, A., Hua, Y., Estrin, D., & Shmatikov, V. (2020). How to backdoor federated learning. *Proceedings of the 23rd International Conference on Artificial Intelligence and Statistics (AISTATS)*, 2938–2948.

Bahangulu, J. K., & Owusu-Berko, L. (2025). Algorithmic bias, data ethics, and governance: Ensuring fairness, transparency and compliance in AI-powered business analytics applications. *World Journal of Advanced Research and Reviews*, *25*(2), 1746–1763. DOI: 10.30574/wjarr.2025.25.2.0571

Bailey, J. (2022, September 2). An A.I.-generated picture won an art prize. Artists aren't happy. *The New York Times*. https://www.nytimes.com/2022/09/02/technology/ai-artificial-intelligence-artists.html

Baradari, D., Kosmyna, N., Petrov, O., Kaplun, R., & Maes, P. (2025). *NeuroChat: A Neuroadaptive AI Chatbot for Customizing Learning Experiences*. arXiv preprint arXiv:2503.07599. DOI: 10.1145/3719160.3736623

Basu, S. (2009). Semi-supervised learning. In *Encyclopedia of Database Systems* (pp. 2613–2615). Springer., DOI: 10.1007/978-0-387-39940-9_609

Belleflamme, P., Lambert, T., & Schwienbacher, A. (2014). Crowdfunding: Tapping the right crowd. *Journal of Business Venturing*, *29*(5), 585–609. DOI: 10.1016/j.jbusvent.2013.07.003

Benjamin, M. (2025). *The impact of generative AI on traditional graphic design workflows*. ResearchGate. https://www.researchgate.net/publication/388849181_The_Impact_of_Generative_AI_on_Traditional_Graphic_Design_Workflows

Bifet, A., & Gavalda, R. (2007, April). Learning from time-changing data with adaptive windowing. In *Proceedings of the 2007 SIAM international conference on data mining* (pp. 443-448). Society for Industrial and Applied Mathematics. DOI: 10.1137/1.9781611972771.42

Bijalwan, A. (Ed.). (2025). *Innovations and challenges in computing, games, and data science*. IGI Global. DOI: 10.4018/979-8-3373-2647-4

Bijalwan, A., & Bijalwan, J. G. (2025). An efficient framework to detect sexual predators early by using federated learning approach. In *2025 International Conference on Next Generation of Green Information and Emerging Technologies (GIET)* (pp. 1–6). IEEE. https://doi.org/DOI: 10.1109/GIET65294.2025.11234823

Bi, S., Liu, Z., & Usman, K. (2017). The influence of online information on investing decisions of reward-based crowdfunding. *Journal of Business Research*, *71*, 10–18. DOI: 10.1016/j.jbusres.2016.10.001

Bourtoule, L., Chandrasekaran, V., Choquette-Choo, C. A., Jia, H., Travers, A., Zhang, B., . . . Papernot, N. (2021, May). Machine unlearning. In *2021 IEEE symposium on security and privacy (SP)* (pp. 141-159). IEEE. DOI: 10.1109/SP40001.2021.00019

Bower, G. H., & Morrow, D. G. (1990). Mental Models in Narrative Comprehension. *Science*, *247*(4938), 44–48. DOI: 10.1126/science.2403694 PMID: 2403694

Bozkurt, A. (2024). GenAI et al.: Cocreation, Authorship, Ownership, Academic Ethics and Integrity in a Time of Generative AI. *Open Praxis*, *16*(1), 23–36. DOI: 10.55982/openpraxis.16.1.654

Canadian Institute for Cybersecurity. (n.d.). *Canadian Institute for Cybersecurity*. https://www.unb.ca/cic

Cao, Y., & Yang, J. (2015, May). *Towards making systems forget with machine unlearning. In 2015 IEEE symposium on security and privacy*. IEEE.

Carlini, N., & Wagner, D. (2017, May). *Towards evaluating the robustness of neural networks. In 2017 ieee symposium on security and privacy (sp)*. Ieee.

Cassar, R. (2024). Innovative upcycling: The creative potential of reuse in design practice. *Journal of Design History*.

Cen, J.-J., & Li, Y.-B. (2022). Deep learning-based anomaly traffic detection method in cloud computing environment. *Wireless Communications and Mobile Computing*, *2022*(1), 6155925. DOI: 10.1155/2022/6155925

Centre for Education Statistics and Evaluation. (2023). Cognitive load theory in practice. [online] education.nsw.gov.au. Available at: https://education.nsw.gov.au/about-us/education-data- and-research/cese/publications/practical-guides-for-educators/cognitive-load-theory-in-practice

Ceruzzi, P. (2003). *A history of modern computing*. MIT Press.

Champ, C. A. N. M. D. H., Silvestri, P., Ferrari, M. L., & Massardo, A. F. (2021). Incipient surge detection in large volume energy systems based on Wigner–Ville distribution evaluated on vibration signals. *Journal of Engineering for Gas Turbines and Power*, *143*(7), 071014. Advance online publication. DOI: 10.1115/1.4049855

Cheerla, C. (2024). Integration of AI and ML for cloud security and threat detection. International Journal of Multidisciplinary Research.

Chen, E., Judicke, S., Beigh, K., Tang, X., Xiao, Z., Li, C., . . . Koedinger, K. (2025). Generative AI alone may not be enough: Evaluating AI Support for Learning Mathematical Proof. *arXiv preprint arXiv:2509.16778*.

Chen, H., Chiang, R. H., & Storey, V. C. (2012). Business intelligence and analytics: From big data to big impact. *Management Information Systems Quarterly*, *36*(4), 1165–1188. DOI: 10.2307/41703503

Chen, Y., Liu, Z., Zhang, Y., Wu, Y., Chen, X., & Zhao, L. (2020). Deep reinforcement learning-based dynamic resource management for mobile edge computing in industrial internet of things. *IEEE Transactions on Industrial Informatics*, *17*(7), 4925–4934. DOI: 10.1109/TII.2020.3028963

Chen, Y., Zhang, N., Wu, Y., & Shen, X. (2022). *Energy Efficient Computation Offloading in Mobile Edge Computing*. Springer. DOI: 10.1007/978-3-031-16822-2

Chen, Z., Xiong, B., Chen, X., Min, G., & Li, J. (2024). Joint computation offloading and resource allocation in multi-edge smart communities with personalized federated deep reinforcement learning. *IEEE Transactions on Mobile Computing*, *23*(12), 11604–11619. DOI: 10.1109/TMC.2024.3396511

Chompunuch, K., & Lubart, T. (2025). AI as a helper in the creative process: Toward a framework of assistance. *International Journal of Human-Computer Studies, 185*, 103123.

Choudhury, M. M., Mandal, T., Malakar, A., & Sanyal, S. (2025). Artificial intelligence (AI) in the design process – A review and analysis on generative AI perspectives. In *Proceedings of the Design Society*. Cambridge University Press & Assessment. DOI: 10.1017/pds.2025.10077

Chui, C. K., & He, X. (2016). Real-time data stream analytics for smart cities. *International Journal of Big Data Intelligence*, *3*(1), 1–12.

Chunawala, H., & Chunawala, P. (2024). Enhancing cybersecurity in cloud environments using AI-driven threat detection and response. International Journal of Futuristic Innovation in Engineering Science & Technology.

Cohn, C., Rayala, S., Srivastava, N., Horn Fonteles, J., Jain, S., Luo, X., Mereddy, D., Mohammed, N., & Biswas, G. (2025). *A Theory of Adaptive Scaffolding for LLM-Based Pedagogical Agents*. arXiv preprint arXiv:2508.01503.

Córdova-Esparza, D.-M. (2025). AI-Powered Educational Agents: Opportunities, Innovations, and Ethical Challenges. [cited in arXiv review]. *Information (Basel)*, *16*(6), 469. DOI: 10.3390/info16060469

Costa, P., & Venturelli, S. (2025). Creative autonomy and collaboration in generative AI art and design. *Leonardo*.

Cuzzocrea, A., Song, I. Y., & Davis, K. C. (2011, October). Analytics over large-scale multidimensional data: the big data revolution! In *Proceedings of the ACM 14th international workshop on Data Warehousing and OLAP* (pp. 101-104). DOI: 10.1145/2064676.2064695

Daley, B., & Amato, C. (2019). Reconciling λ-returns with experience replay. Advances in Neural Information Processing Systems, 32. Bacon, P. L., Harb, J., & Precup, D. (2017, February). The option-critic architecture. In Proceedings of the AAAI conference on artificial intelligence (Vol. 31, No. 1).

Dang, T.D. (2020). Angel Investment in Startups in the World and in Vietnam. Vietnam Trade and Industry Review, 22 (September).

Dang, T. T., Pham, A. D., & Tran, D. N. (2020). Impact of monetary policy on private investment: Evidence from Vietnam's provincial data. *Economies*, *8*(3), 70. DOI: 10.3390/economies8030070

Dara, J. (2021). 19 Crimes 'Living Wine Labels' Now Include A Holographic Visit From Snoop Dogg. [online]. *Forbes*, *15*(Dec). https://www.forbes.com/sites/jilliandara/2021/12/15/19- crimes-living-wine-labels-include-a-holographic-visit-from-snoop-dogg/

Das, B. R., Hasan, S. R., Sabuj, S. R., Hossain, M. A., & Ray, S. K. (2025). A comprehensive survey on emerging AI technologies for 6G communications: Research direction, trends, challenges, and opportunities. *International Journal of Intelligent Networks*, *6*, 113–150. DOI: 10.1016/j.ijin.2025.06.001

Davar, N. F. (2025). *AI Chatbots in Education: Challenges and Opportunities. Information*. MDPI.

Dean, J., & Ghemawat, S. (2008). MapReduce: Simplified data processing on large clusters. *Communications of the ACM*, *51*(1), 107–113. DOI: 10.1145/1327452.1327492

De-Arteaga, M., Feuerriegel, S., & Saar-Tsechansky, M. (2022). Algorithmic fairness in business analytics: Directions for research and practice. *Production and Operations Management*, *31*(10), 3749–3770. DOI: 10.1111/poms.13839

Debets, T., van den Broek, M., & Kusters, R. H. (2025). Chatbots in Education: A Systematic Review of Objectives and Development Approaches. Computers & Education. *Artificial Intelligence*.

Delve (2022). What is observational research? [online] Delve. Available at: https://delvetool.com/ blog/observation

Demaidi, M. N. (2025). Artificial intelligence national strategy in a developing country. *AI & Society*, *40*(2), 423–435. DOI: 10.1007/s00146-023-01779-x

Demirkan, H., & Delen, D. (2013). Leveraging the capabilities of service-oriented decision support systems: Putting analytics and big data in cloud. *Decision Support Systems*, *55*(1), 412–421. DOI: 10.1016/j.dss.2012.05.048

Dong, Y., Liao, F., Pang, T., Su, H., Zhu, J., Hu, X., & Li, J. (2018). Boosting adversarial attacks with momentum. In *Proceedings of the IEEE conference on computer vision and pattern recognition* (pp. 9185-9193).

Du, Z., Liu, J. and Wang, T. (2022). Augmented Reality Marketing: a Systematic Literature Review and an Agenda for Future Inquiry. Frontiers in Psychology, [online] 13. .DOI: 10.3389/fpsyg.2022.925963

Duyver, E., Groeneveld, G., & Aerts, J. (2023). *Does interdisciplinary creative coding boost creativity? A mixed methods approach* (arXiv:2307.00800). arXiv. https://arxiv.org/abs/2307.00800

Ekstrand, M. D., Tian, M., Azpiazu, I. M., Ekstrand, J. D., Anuyah, O., McNeill, D., & Pera, M. S. (2018, January). All the cool kids, how do they fit in?: Popularity and demographic biases in recommender evaluation and effectiveness. In *Conference on fairness, accountability and transparency* (pp. 172-186). PMLR.

Essinger, J. (2004). *Jacquard's web: How a hand-loom led to the birth of the information age*. Oxford University Press.

Etikan, I., Musa, S. A., & Alkassim, R. S. (2016). Comparison of Convenience Sampling and Purposive Sampling. *American Journal of Theoretical and Applied Statistics*, *5*(1), 1–4. DOI: 10.11648/j.ajtas.20160501.11

Eze, L., Chaudhry, U. B., & Jahankhani, H. (2025). Quantum-enhanced machine learning for cybersecurity: Evaluating malicious URL detection. *Electronics (Basel)*, *14*(9), 1827. Advance online publication. DOI: 10.3390/electronics14091827

Fan, W. (2024). Hybrid deep reinforcement learning-based task offloading for D2D-assisted cloud-edge-device collaborative networks. *IEEE Transactions on Mobile Computing*, *23*(12), 13455–13471. DOI: 10.1109/TMC.2024.3425723

Fatima, Z., Chen, M., & Zhang, K. (2025). Explainable AI-driven firewall evaluation. *Digital Forensics Review*, *12*(3), 45–59. DOI: 10.1002/spy2.70026

Feng, J., & Cao, Y. (2021). *A Systematic Review of Literature on the Effectiveness of Intelligent Tutoring Systems*. PDF. DOI: 10.1109/FIE49875.2021.9637240

Fenwick, M., Vermeulen, E., & McCahery, J. A. (2017). *Fintech and the Financing of Entrepreneurs From Crowdfunding to Marketplace Lending*. Law Working Paper No 369/2017.

Few, S. (2006). Information dashboard design.

Field, A. (2018). *Discovering Statistics Using IBM SPSS Statistics* (5th ed.). SAGE Publications.

Filaly, Y., Berros, N., & Younes El Bouzekri, E. L. (2025). A comprehensive survey on big data privacy and Hadoop security: Insights into encryption mechanisms and emerging trends. *Results in Engineering*, *106203*, 106203. Advance online publication. DOI: 10.1016/j.rineng.2025.106203

Fleischmann, K. (2024). Generative artificial intelligence in graphic design education: A student perspective. [CJLT]. *Canadian Journal of Learning and Technology*, *50*(1), 1–17. DOI: 10.21432/cjlt28618

Fu, B., Hadid, A., & Damer, N. (2025). Generative AI in the context of assistive technologies: Trends, limitations and future directions. *Image and Vision Computing*, *154*, 105347. DOI: 10.1016/j.imavis.2024.105347

Fu, S., Liu, K., Xu, Y., & Liu, Y. (2016). Rolling bearing diagnosing method based on time domain analysis and adaptive fuzzy C-means clustering. *Shock and Vibration*, *2016*, 9412787. Advance online publication. DOI: 10.1155/2016/9412787

Garcia-Bosque, M., Naya-Forcano, A., Cascarosa, E., Aznar, F., Sánchez-Azqueta, C., Celma, S., & Aldea, C. (2024). "ChatPLT: An Intelligent Tutoring System for Teaching Physics in Higher Education." (cited in arXiv review)

Garzón, J. (2025). Artificial intelligence in education: A systematic literature review. *Multimodal Technologies and Interaction*, *9*(8), 84. DOI: 10.3390/mti9080084

Ge, P. (2024). A systematic review of the role of AI in design education (2020–2023). In *Proceedings of the Design Society*.

Gerbner, G., Gross, L., Morgan, M., & Signorielli, N. (1980). The "mainstreaming" of America: Violence profile no. 11. *Journal of Communication*, *30*(3), 10–29. DOI: 10.1111/j.1460-2466.1980.tb01987.x

Ghasemi, A., & Zahediasl, S. (2012). Normality Tests for Statistical Analysis: A Guide for Non- Statisticians. *International Journal of Endocrinology and Metabolism*, *10*(2), 486–489. DOI: 10.5812/ijem.3505 PMID: 23843808

Ghosh, R., & Ghosh, D. (2020). Drill-down analysis in business intelligence systems: A framework. *International Journal of Business Information Systems*, *33*(4), 455–472.

Gleasure, R. (2015). Resistance to crowdfunding among entrepreneurs: An impression management perspective. *The Journal of Strategic Information Systems*, *24*(4), 219–233. DOI: 10.1016/j.jsis.2015.09.001

Goffman, E. (1963). *Stigma: Notes on the management of spoiled identity*. Prentice-Hall.

Golatkar, A., Achille, A., & Soatto, S. (2020). Eternal sunshine of the spotless net: Selective forgetting in deep networks. In *Proceedings of the IEEE/CVF conference on computer vision and pattern recognition* (pp. 9304-9312). DOI: 10.1109/CVPR42600.2020.00932

Goodfellow, I. J., Shlens, J., & Szegedy, C. (2014). Explaining and harnessing adversarial examples. *arXiv preprint arXiv:1412.6572.*

Grand View Research. (2024). Augmented Reality Packaging Market | Industry Report, 2030. [online] Available at: https://www.grandviewresearch.comhttps://www.grandviewresearch.com/industry- analysis/augmented-reality-packaging-market-report

Gravetter, F., & Wallnau, L. (2020). *Essentials of Statistics for the Behavioral Sciences* (10th ed.). Cengage Learning.

Green, M. C., & Appel, M. (2024). Narrative transportation: How stories shape how we see ourselves and the world. Advances in Experimental Social Psychology, [online] 70(1), pp.1–82. doi:DOI: 10.1016/bs.aesp.2024.03.002

Gulati, P., Gulati, U., Uygun, H., & Gujrati, R. (2023). Artificial intelligence in cyber security: Rescue or challenge. *Review of Artificial Intelligence in Education, 4*, e07. Advance online publication. DOI: 10.37497/rev.artif.intell.education.v4i00.7

Guo, C., Goldstein, T., Hannun, A., & Van Der Maaten, L. (2019). Certified data removal from machine learning models. *arXiv preprint arXiv:1911.03030.*

Guo, C., Rana, M., Cisse, M., & Van Der Maaten, L. (2017). Countering adversarial images using input transformations. *arXiv preprint arXiv:1711.00117.*

Guo, Y., & Guoping, T. (2024). Intelligent analysis and dynamic security of network traffic in context of big data. Journal of Cyber Security and Mobility.

Gupta, P., Yadav, K., Gupta, B. B., Alazab, M., & Gadekallu, T. R. (2023). A novel data poisoning attack in federated learning based on inverted loss function. *Computers & Security*, *130*, 103270. Advance online publication. DOI: 10.1016/j.cose.2023.103270

Gurijala, P., Avula, R., & Lekkala, S. (2024). Next-gen firewalls: Enhancing cloud security with generative AI. *Journal of Artificial Intelli-gence & Cloud Computing*, *3*, 1–9. DOI: 10.47363/JAICC/2024(3)404

Han, Y. (2024). Attention mechanism-based CNN-BiLSTM for sales revenue prediction. In *Proceedings of the International Conference on Advanced Algorithms and Control Engineering (ICAACE)* (pp. 740–743). IEEE. https://doi.org/DOI: 10.1109/ICAACE61206.2024.10548377

Heigl, R. (2025). *Generative artificial intelligence in creative contexts: A systematic literature review*. Springer.

He, K., Zhang, X., Ren, S., & Sun, J. (2016). Deep residual learning for image recognition. In *Proceedings of the IEEE conference on computer vision and pattern recognition* (pp. 770-778).

Herlocker, J. L., Konstan, J. A., Terveen, L. G., & Riedl, J. T. (2004). Evaluating collaborative filtering recommender systems. [TOIS]. *ACM Transactions on Information Systems, 22*(1), 5–53. DOI: 10.1145/963770.963772

Ho, A. (2023). A Critical Overview of Studies on Eye Tracking and Visual Hierarchy. In: Lucas Paletta, Hasan Ayaz and Umer Asgher (eds) Cognitive Computing and Internet of Things. AHFE (2023) International Conference. AHFE Open Access, vol 73. AHFE International, USA. DOI: 10.54941/ahfe1003295

Holzapfel, T., Foerster, J., Brendel, M., & Busse, S. (2022). Towards sustainable Creative-AI: Environmental, social, and economic perspectives. *Journal of Responsible Innovation, 9*(3), 275–298.

Hua, J. (2020). Locally differentially private federated learning. *Proceedings of the AAAI Conference on Artificial Intelligence, 34*(04), 6580–6587.

Hua, J., Kondo, A., Wang, C., & Ganchuluun, S. (2023). Job satisfaction, intention to leave, and related factors among foreign-educated nurses in Japan: A cross-sectional study. *Journal of Interpersonal Violence, 36*(5–6), 2482. PMID: 40225630

Hu, H., Song, W., Wang, Q., Hu, R. Q., & Zhu, H. (2022). Energy efficiency and delay tradeoff in an MEC-enabled mobile IoT network. *IEEE Internet of Things Journal, 9*(17), 15942–15956. DOI: 10.1109/JIOT.2022.3153847

Hui, J. S., Greenberg, M. D., & Gerber, E. M. (2014). Understanding the role of community in crowdfunding work. In *Proceedings of the 17th ACM conference on Computer supported cooperative work & social computing* (pp. 62-74). DOI: 10.1145/2531602.2531715

Hu, J. (2014). Sustainable upcycling practices in art and design. *Journal of Cleaner Production.*

Ibitoye, O., Abou-Khamis, R., ElShehaby, M., Matrawy, A., & Shafiq, M. (2025). The threat of adversarial attacks against machine learning in network security: A survey. Journal of Electronics and Elec-trical Engineering. DOI: 10.37256/jeee.4120255738

Igarashi, A., & Nagayoshi, K. (2021). Norms to be prejudiced: List experiments on attitudes towards immigrants in Japan. *Social Science Research, 102*, 102647. Advance online publication. DOI: 10.1016/j.ssresearch.2021.102647 PMID: 35094765

Izzo, Z., Smart, M. A., Chaudhuri, K., & Zou, J. (2021, March). Approximate data deletion from machine learning models. In *International conference on artificial intelligence and statistics* (pp. 2008-2016). PMLR.

Jain, A. K., & Gupta, B. B. (2018). Two-level authentication approach to protect from phishing attacks in real-time. *Journal of Ambient Intelligence and Humanized Computing*, *9*(6), 1783–1796. DOI: 10.1007/s12652-017-0616-z

Jain, A., Bagoria, R., & Arora, P. (2025). An intelligent zero-day attack detection system using unsupervised machine learning for enhanc-ing cyber security. *Knowledge-Based Systems*, *324*, 113833. Advance online publication. DOI: 10.1016/j.knosys.2025.113833

Jana, S., Biswas, R., Banerjee, C., Patra, T., Pal, M., & Pal, K. (2024). Leveraging artificial intelligence for enhancing cybersecurity: A comprehensive review and analysis. International Journal of Advanced Research in Science. *Tongxin Jishu*, •••, 173–183. DOI: 10.48175/IJARSCT-19030

Jeong, J., & Lee, J. (2024). A federated reinforcement learning framework via a committee mechanism for resource management in 5g networks. *Sensors (Basel)*, *24*(21), 7031. DOI: 10.3390/s24217031 PMID: 39517924

Jyothsna, V., Sandhya, E., Baseer, K. K., & Pydala, B. (2024). AI-driven threat detection in cloud environments. In *Advances in Information Security*. Privacy, and Ethics.

Kairouz, P., McMahan, H. B., Avent, B., Bellet, A., Bennis, M., Bhagoji, A. N., Bonawitz, K., Charles, Z., Cormode, G., Cummings, R., D'Orazio, M., Eichner, H., El Rouayheb, S., Evans, D., Gardner, J., Garrett, Z., Gascón, A., Ghazi, B., Gibbons, P. B., & Zhao, S. (2021). Advances and open problems in federated learning. *Foundations and Trends in Machine Learning*, *14*(1–2), 1–210. DOI: 10.1561/2200000083

Kaminskas, M., & Bridge, D. (2016). Diversity, serendipity, novelty, and coverage: A survey and empirical analysis of beyond-accuracy objectives in recommender systems. [TiiS]. *ACM Transactions on Interactive Intelligent Systems*, *7*(1), 1–42. DOI: 10.1145/2926720

Kannan, A., Subramanian, K. S. R., & Suresh, S. (2025). RESIGN: Alzheimer's Disease Detection Using Hybrid Deep Learning based Res-Inception Seg Network. *Current Alzheimer Research*, *22*(9), 698–709. DOI: 10.2174/011567205037382125 0612053943 PMID: 40539353

Karim, A., Shahroz, M., Mustofa, K., Belhaouari, S. B., & Joga, S. R. K. (2023). Phishing detection system through hybrid machine learning based on URL. *IEEE Access : Practical Innovations, Open Solutions*, *11*, 36805–36822. DOI: 10.1109/ACCESS.2023.3252366

Kasula, V. K., Yadulla, A. R., Konda, B., & Yenugula, M. (2023). *Fortifying cloud environments against data breaches: A novel AI-driven security framework*. World Journal of Advanced Research and Reviews.

Khalid, O., Ullah, S., Ahmad, T., Saeed, S., Alabbad, D. A., Aslam, M., Buriro, A., & Ahmad, R. (2023). An insight into the machine-learning-based fileless malware detection. *Sensors (Basel)*, *23*(2), 612. DOI: 10.3390/s23020612 PMID: 36679406

Khanh, T. T., Hai, T. H., Hossain, M. D., & Huh, E. N. (2022). Fuzzy-assisted mobile edge orchestrator and SARSA learning for flexible offloading in heterogeneous IoT environment. *Sensors (Basel)*, *22*(13), 4727. DOI: 10.3390/s22134727 PMID: 35808224

Kleemann, R., Zadelaar, S., & Kooistra, T. (2008). Cytokines and atherosclerosis: A comprehensive review of studies in mice. *Cardiovascular Research*, *79*(3), 360–376. DOI: 10.1093/cvr/cvn120 PMID: 18487233

Kleftodimos, A., Lappas, G. and Vrigkas, M. (2022). An augmented reality workflow for creating 'live' wine labels. International Journal of Entertainment Technology and Management, [online] 1(4), p.311. .DOI: 10.1504/IJENTTM.2022.10054762

Koedinger, K. R., & Aleven, V. (2016). (Classic reference) Works on ITS, scaffolding, cognitive tutors. (Mention in background)

Koh, P. W., & Liang, P. (2017, July). Understanding black-box predictions via influence functions. In *International conference on machine learning* (pp. 1885-1894). PMLR.

Kortemeyer, G. (2023). Could an artificial-intelligence agent pass an introductory physics course? *Physical Review. Physics Education Research*, *19*(1), 010132. DOI: 10.1103/PhysRevPhysEducRes.19.010132

Labadze, L., Saatloo, S., & Daumiller, F. (2023). Role of AI Chatbots in Education: Systematic Literature Review. Education and Information Technologies.

LeCun, Y., Bottou, L., Bengio, Y., & Haffner, P. (1998). Gradient-based learning applied to document recognition. *Proceedings of the IEEE*, *86*(11), 2278–2324. DOI: 10.1109/5.726791

Lee, E. S. (1966). A theory of migration. *Demography*, *3*(1), 47–57. DOI: 10.2307/2060063

Leoni, J., Gelmini, S., Panzani, G., & Tanelli, M. (2024). A derivative, integral, and proportional features extractor for fault detection in dynamic processes. *Engineering Applications of Artificial Intelligence*, *128*, 107510. Advance online publication. DOI: 10.1016/j.engappai.2023.107510

Le, T. M., Tran, H. M., Wang, K., Pham, H. V., & Dao, S. V. T. (2025). An Internet-of-Things-integrated deep learning model for fault diagnosis in industrial rotating machines. *IEEE Access : Practical Innovations, Open Solutions*, *13*, 57266–57286. DOI: 10.1109/ACCESS.2025.3553155

Létourneau, S., Turcotte, M., & Nkambou, R. (2025). *A Systematic Review of AI-Driven Intelligent Tutoring Systems in K-12*. PMC article.

Li, Y., & Ma, G. (2024). Research on cloud computing network security framework based on machine learning. Journal of Higher Education Research.

Li, Y., Sun, X., & Zhang, J. (2025). Reinforcement learning based dynamic firewall policy optimization (arXiv:2505.01423). arXiv. DOI: 10.22541/au.174949149.94911054/v

Li, H., Zhang, Y., Cao, Y., Zhao, J., & Zhao, Z. (2025). Applications of artificial intelligence in the AEC industry: A review and future outlook. *Journal of Asian Architecture and Building Engineering*, *24*(3), 1672–1688. DOI: 10.1080/13467581.2024.2343800

Li, J., Yang, Z., Chen, K., Ming, Z., Li, X., Fan, Q., Hao, J., & Cheng, L. (2024). Dependency-aware task offloading based on deep reinforcement learning in mobile edge computing networks. *Wireless Networks*, *30*(6), 5519–5531. DOI: 10.1007/s11276-023-03283-y

Lin, Y., Li, H., Wu, A., Wang, Y., & Qu, H. (2023). DMiner: Dashboard design mining and recommendation. *IEEE Transactions on Visualization and Computer Graphics*, *30*(7), 4108–4121. DOI: 10.1109/TVCG.2023.3251344 PMID: 37028006

Li, T., Sahu, A. K., Zaheer, M., Sanjabi, M., Talwalkar, A., & Smith, V. (2020). Federated optimization in heterogeneous networks. *Proceedings of Machine Learning and Systems*, *2*, 429–450.

Li, X., & Li, F. (2017). Adversarial examples detection in deep networks with convolutional filter statistics. In *Proceedings of the IEEE international conference on computer vision* (pp. 5764-5772). DOI: 10.1109/ICCV.2017.615

Li, Z., Liu, F., Chen, L., Huang, Y., & Fan, M. (2023). Exploring the Opportunities of AR for Enriching Storytelling with Family Photos between Grandparents and Grandchildren. *Proceedings of the ACM on Interactive, Mobile, Wearable and Ubiquitous Technologies*, *7*(3), 1–26. DOI: 10.1145/3610903

Long, D., & Magerko, B. (2020, April). What is AI literacy? Competencies and design considerations. In *Proceedings of the 2020 CHI conference on human factors in computing systems* (pp. 1-16). DOI: 10.1145/3313831.3376727

Luo, J., Liu, Y., Cui, Q., Zhong, J., & Zhang, L. (2024). Single-ended time domain fault location based on transient signal measurements of transmission lines. *Protection and Control of Modern Power Systems*, *9*(2), 61–74. DOI: 10.23919/PCMP.2023.000503

Luong, T. T. H. (2019). Crowd Funding - A New Funding Trend for SMEs in Vietnam Today. Corporate Finance Review, October <https://taichinhdoanhnghiep.net.vn/crowd-funding--xu-huong-tai-tro-moi-cho-cac-doanh-nghiep-nho-va-vua-o-viet-nam-hien-nay-d9422.html>

Mach, P., & Becvar, Z. (2017). Mobile edge computing: A survey on architecture and computation offloading. *IEEE Communications Surveys and Tutorials*, *19*(3), 1628–1656. DOI: 10.1109/COMST.2017.2682318

Madry, A., Makelov, A., Schmidt, L., Tsipras, D., & Vladu, A. (2017). Towards deep learning models resistant to adversarial attacks. *arXiv preprint arXiv:1706.06083*.

Maei, H. R., & Sutton, R. S. (2010, June). GQ (lambda): A general gradient algorithm for temporal-difference prediction learning with eligibility traces. In *3d Conference on Artificial General Intelligence (AGI-2010)* (pp. 100-105). Atlantis Press.

Mamodiya, U., Kishor, I., Naz, R., Almaiah, M., & Alqutaish, A. (2026). A hybrid blockchain-based framework for adaptive cyber-risk prediction and multi-layer threat mitigation in enterprise networks. *Jour-nal of Cybersecurity and Privacy*, *6*(3), 85. Advance online publication. DOI: 10.3390/jcp6030085

Mantelero, A. (2018). AI and Big Data: A blueprint for a human rights, social and ethical impact assessment. *Computer Law & Security Review*, *34*(4), 754–772. DOI: 10.1016/j.clsr.2018.05.017

Manzini, A. (2015). *Design, when everybody designs: An introduction to design for social innovation*. MIT Press. DOI: 10.7551/mitpress/9873.001.0001

Massey, D. S., Arango, J., Hugo, G., Kouaouci, A., Pellegrino, A., & Taylor, J. E. (1993). Theories of international migration: A review and appraisal. *Population and Development Review*, *19*(3), 431–466. DOI: 10.2307/2938462

McCormack, J., Gifford, T., & Hutchings, P. (2019). Autonomy, authenticity, authorship and intention in computer generated art. In *The philosophy of computer art* (pp. 81–106). Springer. DOI: 10.1007/978-3-030-16667-0_3

McMahan, H. B., Moore, E., Ramage, D., Hampson, S., & Aguera y Arcas, B. (2017). Communication-efficient learning of deep networks from decentralized data. *Proceedings of the 20th International Conference on Artificial Intelligence and Statistics (AISTATS)*, 1273–1282.

McSherry, F., & Mironov, I. (2009, June). Differentially private recommender systems: Building privacy into the netflix prize contenders. In *Proceedings of the 15th ACM SIGKDD international conference on Knowledge discovery and data mining* (pp. 627-636). DOI: 10.1145/1557019.1557090

Mihanjo, C., & Mongi, A. F. (2022). Isolation of DDoS attacks and flash events in internet traffic using deep learning techniques. Tanzania Journal of Engineering and Technology.

Mitra, S., Chakraborty, B., & Mitra, P. (2024). Smart meter data analytics applications for secure, reliable and robust grid system: Survey and future directions. *Energy*, *289*, 129920. DOI: 10.1016/j.energy.2023.129920

Miyato, T., Kataoka, T., Koyama, M., & Yoshida, Y. (2018). Spectral normalization for generative adversarial networks. *arXiv preprint arXiv:1802.05957.*

Mnih, V., Kavukcuoglu, K., Silver, D., Rusu, A. A., Veness, J., Bellemare, M. G., ... & Hassabis, D. (2015). Human-level control through deep reinforcement learning. nature, 518(7540), 529-533.

Mollick, E. (2014). The dynamics of crowdfunding: An exploratory study. *Journal of Business Venturing*, *29*(1), 1–16. DOI: 10.1016/j.jbusvent.2013.06.005

Moon, J., He, Y., Tian, X., & Hargis, J. (2024). Teaching and learning creative coding with conversational AI. *GLOKALde Journal.*

Moosavi-Dezfooli, S. M., Fawzi, A., & Frossard, P. (2016). Deepfool: a simple and accurate method to fool deep neural networks. In *Proceedings of the IEEE conference on computer vision and pattern recognition* (pp. 2574-2582). DOI: 10.1109/CVPR.2016.282

Mothukuri, V., Parizi, R. M., Pouriyeh, S., Huang, Y., Dehghantanha, A., & Srivastava, G. (2021). A survey on security and privacy of federated learning. *Future Generation Computer Systems*, *115*, 619–640. DOI: 10.1016/j.future.2020.10.007

Mourby, M., Mackey, E., Elliot, M., Gowans, H., Wallace, S. E., Bell, J., Smith, H., Aidinlis, S., & Kaye, J. (2018). Are 'pseudonymised'data always personal data? Implications of the GDPR for administrative data research in the UK. *Computer Law & Security Review*, *34*(2), 222–233. DOI: 10.1016/j.clsr.2018.01.002

Moustafa, N., & Slay, J. (2015). UNSW-NB15: A comprehensive dataset for network intrusion detection systems (UNSW-NB15 network dataset). In *2015 Military Communications and Information Systems Conference (MilCIS)* (pp. 1–6). IEEE. DOI: 10.1109/MilCIS.2015.7348942

Muratovski, G. (2021). *Research for Designers: A Guide to Methods and Practice* (2nd ed.). SAGE Publications.

Na, S., Heo, S., Han, S., Shin, Y., & Roh, Y. (2022). Acceptance model of artificial intelligence (AI)-based technologies in construction firms: Applying the Technology Acceptance Model (TAM) in combination with the Technology–Organisation–Environment (TOE) framework. *Buildings (Basel, Switzerland)*, *12*(2), 90. DOI: 10.3390/buildings12020090

Nguyen, N. H. T., Nguyen, P. N. H., Le, T. M., & Truong, Q. N., TonThat, L., & Dao, S. V. T. (2024a). Fault diagnosis for rotating machine utilizing machine learning with FFT-based feature extraction. In *2024 Tenth International Conference on Communications and Electronics (ICCE)* (pp. 619–624). IEEE. DOI: 10.1109/ICCE62051.2024.10634660

Nguyen, D. C., Pham, Q. V., Pathirana, P. N., Ding, M., Seneviratne, A., Li, J., Niyato, D., & Hwang, W. J. (2022). Federated learning for cybersecurity: A comprehensive survey. *IEEE Communications Surveys and Tutorials*, *24*(2), 1192–1237. DOI: 10.1109/COMST.2021.3122606

Nguyen, N. H. T., Le, T. M., Tran, H. G., Tran, T. D., Do, A. T., & Dao, S. V. T. (2024b). Fault diagnosis of rotating machine based on machine learning with feature selection. In *2024 International Conference on Advanced Technologies for Communications (ATC)* (pp. 684–689). IEEE. DOI: 10.1109/ATC63255.2024.10908195

Ninet, A. (2025). *Using ChatGPT to create a collection of illustrations for graphic design*. Instagram Reels.

Nisar, S., Khan, O. U., & Tariq, M. (2016). An efficient adaptive window size selection method for improving spectrogram visualization. *Computational Intelligence and Neuroscience*, *2016*, 6172453. Advance online publication. DOI: 10.1155/2016/6172453 PMID: 27642291

Nwachukwu, C., Durodola-Tunde, K., & Akwiwu-Uzoma, C. (2024). AI-driven anomaly detection in cloud computing environments. International Journal of Scientific Research Archive.

Nye, B. D., Graesser, A. C., & Hu, X. (2014). AutoTutor and family: A review of 17 years of natural language tutoring. *International Journal of Artificial Intelligence in Education, 24*(4), 427–469. DOI: 10.1007/s40593-014-0029-5

Obini, U. C., Jeremiah, C., & Igwe, S. A. (2024). Development of a machine learning based fileless malware filter system for cyber-security. *Journal of the Nigerian Society of Physical Sciences, 6*(4), 2192. Advance online publication. DOI: 10.46481/jnsps.2024.2192

Oduri, S. (2021). AI-powered threat detection in cloud environments. *International Journal on Recent and Innovation Trends in Computing and Communication, 9*(12), 57–62. DOI: 10.17762/ijritcc.v9i12.10999

Ofusori, L., Makaba, T., & Mhlongo, S. (2025). Explainability and interpretability of artificial intelligence use in cybersecurity. *Discover Computing, 28*. Advance online publication. DOI: 10.1007/s10791-025-09760-6

Onwuegbuzie, A. J., & Daniel, L. G. (2002). Uses and misuses of the independent t-test.

Open, A. I. (2023). *GPT-4 technical report.* OpenAI.

Oppenlaender, J. (2022). *The creativity of text-to-image generation* (arXiv:2206.02904). arXiv. https://arxiv.org/abs/2206.02904 DOI: 10.1145/3569219.3569352

Oppenlaender, J., Linder, R., & Silvennoinen, J. (2024). Prompting AI art: An investigation into the creative skill of prompt engineering. *International Journal of Human-Computer Interaction.*

Paladines, J., & Ramírez, J. (2020). A Systematic Literature Review of Intelligent Tutoring Systems With Dialogue in Natural Language. [cited in arXiv review]. *IEEE Access : Practical Innovations, Open Solutions, 8*, 164246–164267. DOI: 10.1109/ACCESS.2020.3021383

Pali, P., Sharma, P., & Paranjape, V. (2023). Optimizing cloud security with an intelligent framework: Combining machine learning and probabilistic techniques. International Journal of Innovative Research in Computer and Communication Engineering.

Patil, D., Rane, N. L., Desai, P., & Rane, J. (Eds.). (2024). *Trustworthy artificial intelligence in industry and society*. Deep Science Publishing. DOI: 10.70593/978-81-981367-4-9

Peng, H., & Shen, X. (2020). Deep reinforcement learning based resource management for multi-access edge computing in vehicular networks. *IEEE Transactions on Network Science and Engineering*, *7*(4), 2416–2428. DOI: 10.1109/TNSE.2020.2978856

Peng, Y., & Li, Y. (2021). Study of the windowing and overlap-add operation for a super-Gaussian random vibration test. *Shock and Vibration*, *2021*(1), 6644957. Advance online publication. DOI: 10.1155/2021/6644957

Pham, A. D., Bui, M. T., & Hoang, D. P. (2020). Employees' Motivation for Creating International Business Venture: The Role of Perceived Competence and Job Satisfaction. *Asian Academy of Management Journal*, *25*(2), 109–134. DOI: 10.21315/aamj2020.25.2.5

Pham, H. M., Pham, A. D., Tran, D. V., Vuong, N. L., & Nguyen, H. D. (2024). Integrating CSR with environmental consciousness and commitment: Pathways to employee stewardship in Vietnam's mechanical manufacturing sector. *Business Strategy & Development*, *7*(4), e70036. DOI: 10.1002/bsd2.70036

Pham, H. T., Nong, D., Simshauser, P., Nguyen, G. H., & Duong, K. T. (2024). Artificial intelligence (AI) development in the Vietnam's energy and economic systems: A critical review. *Journal of Cleaner Production*, *438*, 140692. DOI: 10.1016/j.jclepro.2024.140692

Phan, Q. N. X., Le, T. M., Tran, H. M., Tran, L. V., & Dao, S. V. T. (2024). Novel machine learning techniques for classification of rolling bearings. *IEEE Access : Practical Innovations, Open Solutions*, *12*, 176863–176879. Advance online publication. DOI: 10.1109/ACCESS.2024.3431040

PhishTank. (n.d.). *Developer information*. https://www.phishtank.com/developer_info.php

Pinchover, S. (2017). The relation between teachers' and children's playfulness: A pilot study. *Frontiers in Psychology*, *8*, 2214. Advance online publication. DOI: 10.3389/fpsyg.2017.02214 PMID: 29312069

PTC. (2023). Tactic Studio Creates Engaging Coke Augmented Reality Experience. [online] Ptc. com. Available at: https://www.ptc.com/en/case-studies/tactic-coke-augmented-reality-interactive- applications.

PTC. (n.d.). 19 Crimes Uses Augmented Reality to Differentiate Their Brand Through Innovation | PTC. [online] Available at: https://www.ptc.com/en/case-studies/augmented-reality-19-crimes-wine

Qi, T., Wu, F., Wu, C., Huang, Y., & Xie, X. (2020, November). Privacy-preserving news recommendation model learning. In *Findings of the Association for Computational Linguistics* (pp. 1423–1432). EMNLP.

Raghunathan, A., Steinhardt, J., & Liang, P. (2018). Certified defenses against adversarial examples. *arXiv preprint arXiv:1801.09344*.

Ragin, C. C. (2014). *The comparative method: Moving beyond qualitative and quantitative strategies*. Univ of California Press. DOI: 10.1525/9780520957350

Rahman, M. A., Hossain, M. S., Islam, M. S., Alrajeh, N. A., & Muhammad, G. (2020). *Privacy-preserving intrusion detection systems using federated learning: Concepts and challenges*.

Rai, K., & Kamble, M. (2023). Reviewing a new optimized ANFIS-based framework for detecting intrusion detection system with machine learning algorithms (deep learning algorithm). *International Journal of Engineering and Advanced Technology*, *12*(3), 35–42. DOI: 10.35940/ijeat.B3916.0212323

Rajarao, B. V., & Sreenivasulu, M. (2023). A hybridized logistic regression and deep learning-based approach for precise anomaly detection in cloud. *International Journal on Recent and Innovation Trends in Computing and Communication*, *11*(9s), 378–385. DOI: 10.17762/ijritcc.v11i9s.7433

Ramadhan, M., Warnars, H. L. H. S., & Razak, F. H. A. (2023). "Combining Intelligent Tutoring Systems and Gamification: A Systematic Literature Review." (cited in arXiv review)

Rane, N., Paramesha, M., Choudhary, S., & Rane, J. (2024). Business intelligence and business analytics with artificial intelligence and machine learning: Trends, techniques, and opportunities. Techniques, and Opportunities (May 17, 2024).

Raval, K. J., Jadav, N. K., Rathod, T., Tanwar, S., Vimal, V., & Yamsani, N. (2024). A survey on safeguarding critical infrastructures: Attacks, AI security, and future directions. *International Journal of Critical Infrastructure Protection*, *44*, 100647. DOI: 10.1016/j.ijcip.2023.100647

Regona, M., Yigitcanlar, T., Xia, B., & Li, R. Y. M. (2022a). Artificial intelligent technologies for the construction industry: How are they perceived and utilized in Australia? *Journal of Open Innovation*, *8*(1), 16. DOI: 10.3390/joitmc8010016

Regona, M., Yigitcanlar, T., Xia, B., & Li, R. Y. M. (2022b). Opportunities and adoption challenges of AI in the construction industry: A PRISMA review. *Journal of Open Innovation*, *8*(1), 45. DOI: 10.3390/joitmc8010045

Rezwana, J., & Maher, M. L. (2022). Understanding user perceptions, collaborative experience and user engagement in different human–AI interaction designs for co-creative systems DOI: 10.1145/3527927.3532789

Roberts, E. J., Kumar, A., & Taylor, L. (2024). Continuous learning in AI-driven cybersecurity: Challenges and solutions for model adaptation. *ACM Transactions on Privacy and Security*, *27*(2), 1–26.

Roy Choudhury, N., Paul, S., & Ghosh, S. (2024). Comparative analysis of traditional vs. AI-driven network security. In *Advances in cy-bersecurity and network defense*. IGI Global., DOI: 10.4018/979-8-3693-6552-6.ch006

Rummery, G. A., & Niranjan, M. (1994). *On-line Q-learning using connectionist systems* (Vol. 37). University of Cambridge, Department of Engineering.

Rupanetti, D., & Kaabouch, N. (2024). Combining edge compu-ting-assisted Internet of Things security with artificial intelligence: Appli-cations, challenges, and opportunities. *Applied Sciences (Basel, Switzerland)*, *14*. Advance online publication. DOI: 10.3390/app14167104

Rustemi, A., & Dalipi, F. (2025). Synergizing IoT, AI, and blockchain for smart agriculture: Challenges, opportunities, and future directions. *Internet of Things : Engineering Cyber Physical Human Systems*, *101778*, 101778. Advance online publication. DOI: 10.1016/j.iot.2025.101778

Sahingoz, O. K., Buber, E., & Kugu, E. (2024). DEPHIDES: Deep learning based phishing detection system. *IEEE Access : Practical Innovations, Open Solutions*, *12*, 8052–8070. DOI: 10.1109/ACCESS.2024.3352629

Saito, T., & Rehmsmeier, M. (2015). The precision-recall plot is more informative than the ROC plot when evaluating binary classifiers on imbalanced datasets. *PLoS One*, *10*(3), e0118432. DOI: 10.1371/journal.pone.0118432 PMID: 25738806

Saldana, J. (2009). The Coding Manual for Qualitative Researchers. *Sage (Atlanta, Ga.)*.

Salem, A., Zhang, Y., Humbert, M., Berrang, P., Fritz, M., & Backes, M. (2018). Ml-leaks: Model and data independent membership inference attacks and defenses on machine learning models. *arXiv preprint arXiv:1806.01246*.

Salem, A. H., Khan, M. T., & Li, Y. (2024). Advancing cybersecu-rity: A comprehensive review of AI-driven methodologies. *Journal of Big Data*, *11*(2), 78–96. DOI: 10.1186/s40537-024-00957-y

Sameen, M., Han, K., & Hwang, S. O. (2020). PhishHaven—An efficient real-time AI phishing URLs detection system. *IEEE Access : Practical Innovations, Open Solutions*, *8*, 83425–83443. DOI: 10.1109/ACCESS.2020.2991403

Sánchez Sánchez, P. M., Huertas Celdrán, A., Xie, N., Bovet, G., Martínez Pérez, G., & Stiller, B. (2024). FederatedTrust: A solution for trustworthy federated learning. *Future Generation Computer Systems*, *152*, 83–98. DOI: 10.1016/j.future.2023.10.013

Santos, J. V., & Jocson, J. (2024). Adoption of artificial intelligence technologies in the Philippine construction industry: A review of literature. *Journal of Interdisciplinary Perspectives*, *2*(8), 461–471. DOI: 10.69569/jip.2024.0304

Sarikaya, A., Correll, M., Bartram, L., Tory, M., & Fisher, D. (2018). What do we talk about when we talk about dashboards? *IEEE Transactions on Visualization and Computer Graphics*, *25*(1), 682–692. DOI: 10.1109/TVCG.2018.2864903 PMID: 30136958

Scheela, W., Nguyen, T. T. T., & Nguyen, T. K. A. (2018). Business angel investing in Vietnam: An exploratory study. *Journal of Private Equity*, *21*(2), 96–106. DOI: 10.3905/jpe.2018.1.067

Schmitt, M. (2023). Automated machine learning: AI-driven decision making in business analytics. *Intelligent Systems with Applications*, *18*, 200188. DOI: 10.1016/j.iswa.2023.200188

Schmitt, M. (2023). Deep learning in business analytics: A clash of expectations and reality. *International Journal of Information Management Data Insights*, *3*(1), 100146. DOI: 10.1016/j.jjimei.2022.100146

Schulman, J., Wolski, F., Dhariwal, P., Radford, A., & Klimov, O. (2017). Proximal policy optimization algorithms. arXiv preprint arXiv:1707.06347.

Schwab, P.-N. (2021). 19 Crimes: a stroke of marketing genius ... often poorly imitated. [online] Market research consulting. Available at: https://www.intotheminds .com/blog/en/19-crimes-wine- augmented-reality/

Schwienbacher, A., & Larralde, B. (2010). *Crowdfunding of small entrepreneurial ventures. Handbook of Entrepreneurial Finance*. Oxford University Press.

Sekar, P., Suba Raja, K. S., & Krishnaraj, R. (2025). DRCNN-Lesion Proxy: A hybrid CNN with lesion-inspired feature simulation for diabetic retinopathy severity classification. *Scientific Reports*, *15*(1), 37954. DOI: 10.1038/s41598-025-21337-3 PMID: 41168219

Setlur, V., Correll, M., Satyanarayan, A., & Tory, M. (2023). Heuristics for supporting cooperative dashboard design. *IEEE Transactions on Visualization and Computer Graphics*, *30*(1), 370–380. PMID: 37883257

Shaik, S. (2021). *Enhancing cloud computing security through deep learning: An artificial neural network approach*. International Research Journal of Engineering and Applied Sciences.

Shannon, C. E. (1948). A mathematical theory of communication. *The Bell System Technical Journal*, *27*(3), 379–423. DOI: 10.1002/j.1538-7305.1948.tb01338.x

Sharafaldin, I., Lashkari, A. H., & Ghorbani, A. A. (2018). Toward generating a new intrusion detection dataset and intrusion traffic charac-terization. In *Proceedings of the 4th International Conference on Infor-mation Systems Security and Privacy (ICISSP)* (pp. 108–116). DOI: 10.5220/0006639801080116

Sharma, V. K., & Kumar, H. (2023). Enablers driving success of artificial intelligence in business performance: A TISM-MICMAC approach. *IEEE Transactions on Engineering Management*, *71*, 4665–4675. DOI: 10.1109/TEM.2023.3236768

Shibli, D., & West, R. (2018). Cognitive Load Theory and Its Application in the Classroom. [online] Impact. Available at: https://my.chartered.college/impact_article/cognitive-load-theory-and-its- application-in-the-classroom/.

Shneiderman, B. (1996, September). The eyes have it: A task by data type taxonomy for information visualizations. In *Proceedings 1996 IEEE symposium on visual languages* (pp. 336-343). IEEE. DOI: 10.1109/VL.1996.545307

Shokri, R., Stronati, M., Song, C., & Shmatikov, V. (2017, May). Membership inference attacks against machine learning models. In *2017 IEEE symposium on security and privacy (SP)* (pp. 3-18). IEEE. DOI: 10.1109/SP.2017.41

Shukla, A. K., Janmaijaya, M., Abraham, A., & Muhuri, P. K. (2019). Engineering applications of artificial intelligence: A bibliometric analysis of 30 years (1988–2018). *Engineering Applications of Artificial Intelligence*, *85*, 517–532. DOI: 10.1016/j.engappai.2019.06.010

Sicular, S. (2013). Gartner's big data definition consists of three parts, not to be confused with three" v" s. Gartner, Inc, 27.

Simonyan, K., & Zisserman, A. (2014). Very deep convolutional networks for large-scale image recognition. *arXiv preprint arXiv:1409.1556.*

Şimşek, E. E. (2024). The effect of augmented reality storybooks on the story comprehension and retelling of preschool children. *Frontiers in Psychology*, *15*, 1459264. Advance online publication. DOI: 10.3389/fpsyg.2024.1459264 PMID: 39439758

Singh, S. P., & Sutton, R. S. (1996). Reinforcement learning with replacing eligibility traces. *Machine Learning*, *22*(1), 123–158. DOI: 10.1023/A:1018012322525

Siraparapu, S. R., & Azad, S. M. A. K. (2024). Securing the IoT landscape: A comprehensive review of secure systems in the digital era. *e-Prime-Advances in Electrical Engineering. Electronics and Energy*, *10*, 100798. DOI: 10.1016/j.prime.2024.100798

Slater, M., & Wilbur, S. (1997). A Framework for Immersive Virtual Environments (FIVE): Speculations on the Role of Presence in Virtual Environments. *Presence (Cambridge, Mass.)*, *6*(6), 603–616. DOI: 10.1162/pres.1997.6.6.603

Smalenberger, M., & Smalenberger, K. (2022). "Toward Accessible Intelligent Tutoring Systems: Integrating Cognitive Tutors and Conversational Agents." (cited in arXiv review)

Srivastava, A., & Thomson, S. B. (2009). *Framework Analysis: A Qualitative Methodology for Applied Policy Research (*SSRN *Scholarly Paper No. ID 2760705). Rochester, NY: Social Science Research Network.*

Sudhakar, & Kumar, S. (2020). An emerging threat fileless malware: A survey and research challenges. Cybersecurity, 3(1), Article 1–12. DOI: 10.1186/s42400-019-0043-x

Sue, D. W., Capodilupo, C. M., Torino, G. C., Bucceri, J. M., Holder, A. M. B., Nadal, K. L., & Esquilin, M. (2007). Racial microaggressions in everyday life: Implications for clinical practice. *The American Psychologist*, *62*(4), 271–286. DOI: 10.1037/0003-066X.62.4.271 PMID: 17516773

Sung, K. (2023). Understanding upcycling and circular economy and their interrelationships for design education. In *Proceedings of the Design Society*.

Sun, X., Yu, X., Huang, Q., Wang, Z., Guo, J., Huang, Z., & Xie, F. (2025). Reliability techniques and architectures for blockchain-enabled internet of things: Current applications, systematic review, and future trends. *Computer Standards & Interfaces*, *104068*. Advance online publication. DOI: 10.1016/j.csi.2025.104068

Sun, Y., Liu, J., Wang, J., Cao, Y., & Kato, N. (2020). When machine learning meets privacy in 6G: A survey. *IEEE Communications Surveys and Tutorials*, *22*(4), 2694–2724. DOI: 10.1109/COMST.2020.3011561

Sun, Z., Strang, K., & Firmin, S. (2017). Business analytics-based enterprise information systems. *Journal of Computer Information Systems*, *57*(2), 169–178. DOI: 10.1080/08874417.2016.1183977

Sutton, R. S., & Barto, A. G. (1998). *No. 1* (Vol. 1). Reinforcement learning: An introduction. MIT press.

Sweller, J. (1988). Cognitive Load during Problem Solving: Effects on Learning. *Cognitive Science*, *12*(2), 257–285. DOI: 10.1207/s15516709cog1202_4

Szegedy, C., Zaremba, W., Sutskever, I., Bruna, J., Erhan, D., Goodfellow, I., & Fergus, R. (2013). Intriguing properties of neural networks. *arXiv preprint arXiv:1312.6199.*

Tabachnick, B. G., & Fidell, L. S. (2019). *Using Multivariate Statistics* (7th ed.). Pearson.

Takeda, S. (2023). Factors related to turnover and intention to leave the care working profession in Japan: A review. *Yonago Acta Medica*, *66*(2), 196–201. DOI: 10.33160/yam.2023.05.001 PMID: 37229365

Taleb, T., Samdanis, K., Mada, B., Flinck, H., Dutta, S., & Sabella, D. (2017). On multi-access edge computing: A survey of the emerging 5G network edge cloud architecture and orchestration. *IEEE Communications Surveys and Tutorials*, *19*(3), 1657–1681. DOI: 10.1109/COMST.2017.2705720

Thakur, K., Ali, M. L., Schmeelk, S., Debello, J., & Rahman, M. M. (2026). Real-time ransomware detection using reinforcement learning agents. *Information (Basel)*, *17*(2), 194. Advance online publication. DOI: 10.3390/info17020194

The Interaction Design Foundation. (2016). What is Visual Hierarchy? [online] Available at: https:// www.interaction-design.org/literature/topics/visual-hierarchy ?srsltid=AfmBOorYhiwlcCJmCDepYaEywm8_cWnXImGjHnSEZaffssEu1LsQdkjH [Accessed 16 Feb. 2025].

Tia Sang. (2021). Angel Investors - An Essential Link in the Venture Capital Market. *VnExpress* <https://tiasang.com.vn/-doi-moi-sang-tao/nha-dau-tu-thien-than-mat-xich-thiet-yeu-cua-thi-truong-dau-tu-mao-hiem-7400>

Tolpegin, V., Truex, S., Gursoy, M. E., & Liu, L. (2020). Data poisoning attacks against federated learning systems. *Computer Security – ESORICS 2020*, 480–501. https://doi.org/DOI: 10.1007/978-3-030-58951-6_24

Tramèr, F., Kurakin, A., Papernot, N., Goodfellow, I., Boneh, D., & McDaniel, P. (2017). Ensemble adversarial training: Attacks and defenses. *arXiv preprint arXiv:1705.07204.*

Tran, V. B., Le, V. T., & Nguyen, V. K. (2019). Angel Investors and Their Role in Innovative Startup Projects. *Vietnam Journal of Science and Technology*, •••, 17–20.

Truex, S., Liu, L., Gursoy, M. E., Yu, L., & Wei, W. (2019). Demystifying membership inference attacks in machine learning as a service. *IEEE Transactions on Services Computing*, *14*(6), 2073–2089. DOI: 10.1109/TSC.2019.2897554

Tsouplaki, A., Fung, C., & Kalloniatis, C. (2025). Enhancing IoT privacy with artificial intelligence: Recent advances and future directions. *Internet of Things : Engineering Cyber Physical Human Systems*, *101752*, 101752. Advance online publication. DOI: 10.1016/j.iot.2025.101752

U.S. Copyright Office. (2023). *Works containing material generated by artificial intelligence—Policy statement.* U.S. Copyright Office.

Urano, Y., Kurosu, A., & Todorov, A. (2019). *Good visual hierarchy is good design.* Princeton University.

Urmeneta, M., & Romero, M. (2025). A PRISMA review of AI's role in supporting creativity in education. *Frontiers in Education*, *10*, 1602151. DOI: 10.3389/feduc.2025.1602151

Van Tam, N., Toan, N. Q., & Van Phong, V. (2024). Investigating potential barriers to construction digitalization in emerging economies: A study in Vietnam. *International Journal of Information Management Data Insights*, *4*(1), 100226. DOI: 10.1016/j.jjimei.2024.100226

Varma, S. A., & Reddy, K. G. (2023). An AI-based IDS framework for detecting DDoS attacks in cloud environment. Information Security Journal: A Global Perspective.

Veletsianos, G., & Russell, G. S. (2013). Pedagogical agents. In *Handbook of research on educational communications and technology* (pp. 759–769). Springer New York.

Villaronga, E. F., Kieseberg, P., & Li, T. (2018). Humans forget, machines remember: Artificial intelligence and the right to be forgotten. *Computer Law & Security Review*, *34*(2), 304–313. DOI: 10.1016/j.clsr.2017.08.007

Wang, J., Li, S., Xin, Y., & An, Z. (2019). Gear fault intelligent diagnosis based on frequency-domain feature extraction. *Journal of Vibration Engineering & Technologies*, *7*(2), 159–166. DOI: 10.1007/s42417-019-00089-1

Wang, K. (2024). Leveraging deep learning for enhanced information security: A comprehensive approach to threat detection and mitigation. *International Journal of Advanced Computer Science and Applications*, *15*(12). Advance online publication. DOI: 10.14569/IJACSA.2024.0151296

Wang, S., Wang, F., Zhu, Z., Wang, J., Tran, T., & Du, Z. (2024). Artificial intelligence in education: A systematic literature review. *Expert Systems with Applications*, *252*, 124167. DOI: 10.1016/j.eswa.2024.124167

Wang, Z., Schaul, T., Hessel, M., Hasselt, H., Lanctot, M., & Freitas, N. (2016, June). Dueling network architectures for deep reinforcement learning. In *International conference on machine learning* (pp. 1995-2003). PMLR.

Warnecke, A., Pirch, L., Wressnegger, C., & Rieck, K. (2021). Machine unlearning of features and labels. *arXiv preprint arXiv:2108.11577.*

Wei, Y., & Sekiya, Y. (2022). Sufficiency of ensemble machine learning methods for phishing websites detection. *IEEE Access : Practical Innovations, Open Solutions*, *10*, 124103–124113. DOI: 10.1109/ACCESS.2022.3224781

Welch, P. (1967). The use of fast Fourier transform for the estimation of power spectra: A method based on time averaging over short, modified periodograms. *IEEE Transactions on Audio and Electroacoustics*, *15*(2), 70–73. DOI: 10.1109/TAU.1967.1161901

Wiafe, I., Koranteng, F. N., Obeng, E. N., Assyne, N., Wiafe, A., & Gulliver, S. R. (2020). Artificial intelligence for cybersecurity: A system-atic mapping of literature. *IEEE Access : Practical Innovations, Open Solutions*, *8*, 146598–146612. DOI: 10.1109/ACCESS.2020.3013145

Wollny, S., Schneider, J., Di Mitri, D., Weidlich, J., Rittberger, M., & Drachsler, H. (2021). Are we there yet?-a systematic literature review on chatbots in education. *Frontiers in Artificial Intelligence*, *4*, 654924. DOI: 10.3389/frai.2021.654924 PMID: 34337392

Wong, S. Y., Choe, C. W. C., Goh, H. H., Low, Y. W., Cheah, D. Y. S., & Pang, C. (2021). Power transmission line fault detection and diagnosis based on artificial intelligence approach and its development in UAV: A review. *Arabian Journal for Science and Engineering*, *46*(10), 9305–9331. DOI: 10.1007/s13369-021-05522-w

Wu, M. H., Hsu, F. H., Huang, J. H., Wang, K., Hwang, Y. L., Wang, H. J., Chen, J.-X., Hsiao, T.-C., & Yang, H. T. (2024). Enhancing linux system security: A kernel-based approach to fileless malware detection and mitigation. *Electronics (Basel)*, *13*(17), 3569. DOI: 10.3390/electronics13173569

Yamamoto, R. (2010). Migrants as a crime problem: The construction of foreign criminality discourse in contemporary Japan. *International Journal of Comparative and Applied Criminal Justice*, *34*(2), 301–330. DOI: 10.1080/01924036.2010.9678831

Yan, P., & Khoei, T. T. (2025). Securing the internet of things: A comprehensive review of ransomware attacks, detection, countermeasures, and future prospects. *Franklin Open*, *11*, 100256. DOI: 10.1016/j.fraope.2025.100256

Yaseen, Z. M., Ali, Z. H., Salih, S. Q., & Al-Ansari, N. (2020). Prediction of risk delay in construction projects using a hybrid artificial intelligence model. *Sustainability (Basel)*, *12*(4), 1514. DOI: 10.3390/su12041514

Yashchenko, V. (2014, August). *Artificial intelligence theory (Basic concepts). In 2014 Science and Information Conference*. IEEE.

Yin, R. K. (2018). *Case study research and applications* (Vol. 6). Sage.

Yoshida, Y., & Miyato, T. (2017). Spectral norm regularization for improving the generalizability of deep learning. *arXiv preprint arXiv:1705.10941.*

Zaki, A., & Mohammed, S. (2024). Artificial intelligence for web application firewall (WAF): A comprehensive review. *International Research Journal of Innovations in Engineering and Technology*, *8*, 219–224. DOI: 10.47001/IRJIET/2024.811027

Zhang, S., Hu, C., Wang, L., Mihaljevic, M. J., Xu, S., & Lan, T. (2023). A malware detection approach based on deep learning and memory forensics. *Symmetry*, *15*(3), 758. Advance online publication. DOI: 10.3390/sym15030758

Zhang, X., Zhao, Z., Wang, Z., & Wang, X. (2021). Fault detection and identification method for quadcopter based on airframe vibration signals. *Sensors (Basel)*, *21*(2), 581. Advance online publication. DOI: 10.3390/s21020581 PMID: 33467463

Zhou, J., Li, R., Tang, J., & Wang, H. (2024). *Understanding nonlinear collaboration between human and AI agents: A co-design framework for creative design* (arXiv:2403.XXXX). arXiv. DOI: 10.1145/3613904.3642812

Zhu, T., Li, G., Zhou, W., & Yu, P. S. (2017). Differentially private data publishing and analysis: A survey. *IEEE Transactions on Knowledge and Data Engineering*, *29*(8), 1619–1638. DOI: 10.1109/TKDE.2017.2697856

Zou, R. (2024, September). Research on ethical issues, data privacy protection, algorithmic bias, and regulatory policy of Artificial Intelligence Technology in Digital Transformation. In *International Conference on Economic Management and Green Development* (pp. 234-242). Singapore: Springer Nature Singapore. https://doi.org/DOI: 10.1007/978-981-96-3236-7_21

About the Contributors

Anchit Bijalwan is the IEEE R10 ACEI lead ambassador for Vietnam and distinguished academician, researcher, consultant, and trainer specializing in advanced technologies and pedagogy for a diverse range of researchers and educators in higher education. Currently, he serves as an Associate Professor and leads the Computing, Games, and Data Science Academic Research Cluster (CGDS ARC) at British University Vietnam. With a robust track record in research and community service, Dr. Anchit has successfully completed numerous projects that bridge the gap between academia and real-world application. He is the author of two influential books, "Advanced Java Programming" and "Network Forensics: The Privacy and Security", and has edited eight additional volumes. His prolific contributions to his field include patents, over fifty research papers published in esteemed, high-index international journals and conferences, with a keen focus on Privacy and Security in cyber forensics. Dr. Anchit is a dynamic presence at international workshops and conferences, where he has delivered keynote addresses in countries such as the UK, Germany, Spain, El Salvador, Vietnam, and India. He is a proud member of the Association for Computing Machinery (ACM) and a senior member of IEEE (Vietnam) and involved in AICTE sponsored STTP's, ATAL FDP's. In recognition of his outstanding contributions, Dr. Anchit received the 'International Research Award' from the International Research Association in London, UK, in 2021 for his book and research project endeavors. He was also honored with the Vice-Chancellor Best Researcher Award by British University in 2023 and Dean's Research Excellence Award in 2024.

Ali Al-Dulaimi received the B.Eng. and M.Sc. degrees in electronics and communications engineering from the University of Baghdad, Iraq, in 1998 and 2001, respectively, and the M.Eng. degree in quality systems engineering and the Ph.D. degree in electrical and computer engineering from Concordia University, Montreal, Canada, in 2016 and 2020, respectively. He contributed as a Postdoctoral Research Fellow to a collaborative project between Concordia and York Universities and was the Lead Instructor for the Community Initiatives Support Program in Montreal, from 2018 to 2020. From 2022 to 2024, he was the Quality and Project Manager of the Berries Group, London, ON, Canada, after leading a field research team with Tiger Produce Canada. He is currently an Associate Professor and the Head of the School of Computing and Innovative Technologies, British University Vietnam (BUV). His research focuses on developing innovative prognostic health management (PHM) methodologies and algorithms, leveraging advanced deep learning techniques to enhance fault diagnosis and failure prognosis in dynamic systems. By integrating deep learning with stochastic filtering and forecasting, he significantly contributes to maintenance engineering and quality assurance, improving system reliability, and operational efficiency across various industries. His scholarly pursuits foster a deeper understanding of predictive analytics in engineering and health management systems, including developing general-purpose theories in machine learning through advanced deep learning techniques.

***

Abubakr Abdulrahman is a Teaching Assistant at the Faculty of Engineering and Information Technology, Taiz University, Yemen. His research interests include cybersecurity, cloud computing, malware analysis, intrusion detection systems, artificial intelligence, and machine learning applications in information security. He has participated in several academic and research projects related to network security and cyber threat detection.

Manohari B is currently pursuing PhD in the School of Computer Science Engineering and Information Systems, Vellore Institute of Technology, Vellore Campus. Her educational background includes M.E Computer Science Engineering(2020-2022) at Sri Krishna College of Engineering and Technology, Coimbatore and B.E Computer Science Engineering(2013-2017) at Government College of Technology, Coimbatore. Her research interests include machine learning and deep learning techniques, as well as cyber security.

Dibya Bora is currently serving as Dean at the School of Computing Sciences, The Assam Kaziranga University, Jorhat, India. He is also the Vice President of the Institution's Innovation Council (IIC) Cell and an active member of the IPR Cell. He earned his PhD in Computer Science from Barkatullah University (formerly Bhopal University) with specialization in Artificial Intelligence(Computer Vision and Machine Learning). He got the university 1st rank in PG and distinction in UG. With more than a decade of academic and research experience, Dr. Bora has held several leadership roles, including Associate Dean and Head of the Department of IT. He is a prolific researcher with over 1,457 citations, an h-index of 15, and an i10-index of 21. His research interests include Machine Learning, Computer

Vision, Soft Computing, and emerging AI technologies. Dr. Bora has over 70 scholarly contributions, including 43 journal and conference research papers, 6 book chapters, 5 preprints, and 22 conference presentations. He is an editor of research books published by reputed publishers such as IGI Global, Bentham Science, CRC Press etc.

Alex Gaja Ortega completed his master's degree in Architecture. After various academic, professional, and research experiences in Spain, Poland, France, and Japan, in 2015, Alex arrived in Vietnam where he developed his career both in design consultancy for international offices and higher education. He was also awarded with a second master's degree in Education Management. In universities, he is appreciated for his devotion to students and organizing engaging activities such as professional collaborations, field trips, events, etc. Currently developing a doctorate about ethnic minorities in Vietnam, he is an outgoing Spaniard who enjoys trips to nature and embraces cultural appreciation.

Tuan Minh Le graduated from the International University, Vietnam National University, Vietnam with a bachelor's degree in electrical engineering, in 2018 and M.SC in Electronics Engineering, in 2022. His research concentrates on Embedded systems the Internet of Things (IoT) and machine learning (ML), and optimization.

Hamza Mutaher is a lecturer in cyber security at Birmingham City University where he also serves as the deputy course lead of the Cyber Security course. Dr Hamza holds a Ph.D. in Computer Science with specialization in security of Software Defined Networks (SDNs). Dr Hamza has an M.Sc. in Computer Networking and a Bachelor of Computer Applications. Previously, Dr Hamza was a lecturer and Program lead of cyber security and cloud computing program at British University Vietnam. He also had decent experience working in industry as a system administrator. With a research focus on SDN security, Dr. Hamza has made notable contributions to the development of secure SDN architectures. He proposed various security authentication techniques using different cryptography methods. Dr Hamza's research publications have been published in reputed journals, international conferences and book chapters.

Nguyen Van Tang serves as a senior lecturer, vice-dean of the Faculty of Technology and Data Science, Hanoi Foreign Trade University. He received a Ph.D. degree in Computer Science (in March, 2009) from Japan Advanced Institute of Science and Technology (JAIST). The main areas of his research are formal methods: mathematics, computer algorithms, software engineering, model checking, theorem proving, optimization, data analysis, blockchain technology. He has published about 30 papers in international conferences and journals, 02 professional books. He has been joined practical R&D projects on new topics such as AI, deep learning, data analysis and blockchain with software Industry. He is also a Digital Transformation consultant for international organizations such as The World Bank, Food and Agriculture Organization of the United Nations (FAO), Swiss Import Promotion Programme (SIPPO), Agence Française de Développement (AFD), etc.

Thi Thu Cuc Nguyen is a Teaching Assistant at Foreign Trade University, Hanoi. She graduated from Master of Science in Marketing Analytics from the University of Southampton, UK. Her research focuses on consumer behaviour, multicultural markets, and data-driven marketing.

Tien Nguyen is a software engineer based in Vietnam, where he applies his technical expertise to build and develop innovative software solutions. With a strong foundation in engineering principles, he is committed to delivering high-quality work and continuously expanding his skills in an ever-evolving industry. Passionate about technology and its potential to create meaningful impact, Tien brings both precision and dedication to every project he takes on.

Senthil P is currently pursuing the National Post-Doctoral Fellow Program in the Department of Computer Science and Engineering at Srinivas University, Mukka, and Mangalore, India and the International Post-Doctoral Fellow Program in the Department of Computer Science and Engineering at the National Kaohsiung University of Science and Technology, Taiwan. He also serves as a faculty member at the SRM Institute of Science and Technology (Formerly SRM University), Tiruchirappalli, India. Dr. Senthilkumar earned his Ph.D. in Computer Science and Engineering (CSE) from Anna University, Chennai, and holds both bachelor's and master's degrees in CSE, obtained in 2005 and 2008, respectively. He has made significant contributions to academia with over 18 years of teaching experience, authoring 30 research papers in international journals, and presenting 25 papers at conferences. He was contacted at psenthilexcel@gmail.com.

Thuy Pham is currently a Teaching Fellow and Acting Programme Lead at British University Vietnam, where she delivers the BA (Hons) Contemporary Creative Practice programme. Her background spans both industry practice and education. After completing her BA in Multimedia Design, she worked as a Junior Art Director at DSTNCT, a youth marketing agency, developing visual concepts for clients including Shangri-La Hotel Singapore, Maserati Philippines, and Grab Singapore. Following her Master of Research in Communication, she returned to Vietnam and joined Shopee, gaining experience in livestream production and content creation. Her passion for agency work eventually led her to VaynerMedia, where she collaborated with Disney Studios on social content for film releases. This breadth of professional experience now informs her teaching, allowing her to bridge the gap between academic theory and industry expectations. Her research interests centre on creative education management and creative practice, with a particular focus on the ever changing context of design, trends and technology.

Dharun S is a B.Tech student in Information Technology at KGiSL Institute of Technology, Coimbatore. His interests lie in Artificial Intelligence, Machine Learning, Deep Learning, and Generative AI. He has worked on projects involving computer vision and natural language processing, including coral health monitoring using deep learning and intelligent document-based chatbot systems. Dharun has completed internships through programs such as Summer of AI (IIIT Hyderabad, Swecha & Meta) and the Infosys Springboard Virtual Internship, gaining practical exposure to AI technologies. He also served as a Campus Ambassador for E-Cell, IIT Bombay, an institution ranked 2nd in India and 129th globally in the QS World University Rankings 2026. He was recognized in the Top 10 Tier of Summer Analytics 2025 at IIT Guwahati, secured second place at Hackspora 2K25 at Karpagam Academy of Higher Education, Coimbatore, and received the Special Achiever Award at the Annual Day of KGiSL Institute of Technology in 2026.

Kanaga Suba Raja S is a Professor and Associate Dean of the School of Computing at SRM Institute of Science and Technology, Tiruchirappalli, with 18 years of teaching and research experience. He obtained his Ph.D. in Computer Science and Engineering in 2013 and has successfully guided numerous graduate and undergraduate students. Under his supervision, nine Ph.D. scholars have completed their degrees, and six scholars are currently pursuing their Ph.D. at Anna University, Chennai. He has received Rs. 17,00,000 in research funding from the MSME Idea Hackathon 1.0 and has published 91 research papers in international journals and conferences. His research contributions include 20 SCI / WoS indexed journal papers, 78 Scopus-indexed documents, 453 Scopus citations (h-index: 13), and 1013 Google Scholar citations (h-index: 16). He has also been actively involved in organizing three international conferences (e-ICICN 2020, e-ICICN 2021, IVCMASM 2022) and three national conferences (NCICN 2012, NCICN 2013, NCICN 2014), along with attending 40 national-level workshops. His contributions to innovation include two granted patents and five published patents, along with six book chapters and one online book. His research interests focus on Wireless Body Area Networks, Data Science, and Machine Learning, and he possesses expertise in Computer Science, Machine Learning, Administration, Leadership, and Research Skills.

Nashwan Saif is a Computer Network and Distributed Systems student at Taiz University, Yemen. He is interested in cloud computing, cybersecurity, artificial intelligence, and deep learning technologies. He is actively involved in academic projects and research activities related to information technology.

Siddharth Selvakumar is a B.Tech student in Computer Science and Engineering at SRM Institute of Science and Technology, Tiruchirappalli, India. His areas of interest include Artificial Intelligence, Machine Learning, Deep Learning, Edge AI, Computer Vision, and Intelligent Embedded Systems. He has worked on research and development projects involving OCR, healthcare AI, predictive systems, and edge computing. He has participated in multiple national-level hackathons, research initiatives, and technical internships, with a focus on developing practical and impactful AI-driven solutions.

Duc-Tan Tran is an Associate professor and Vice Dean of Faculty of Electrical and Electronic Engineering, Phenikaa University. He has published over 150 research papers. His publications received the "Best Paper Award" at the 9th International Conference on Multimedia and Ubiquitous Engineering (MUE-15), and International Conference on Green and Human Information Technology (ICGHIT-2015). He was the recipient of the award for the excellent young researcher from Vietnam National University in 2008, Hanoi and the third prize in the contest "Vietnamese Talents" in 2008. His main research interests include the representation, processing, analysis, and communication of information embedded in signals and datasets. He serves as a TP Co-chair, technical committee program member, track chair, session chair and reviewer of many international conferences and journals.

Dao Ha Trang is a final-year student at British University Vietnam pursuing a Bachelor's degree in Contemporary Creative Practice. Her academic interests focus on creative production, visual culture, and interdisciplinary artistic practices. She is particularly interested in exploring how contemporary art and design intersect with digital media and storytelling.

Asha V is a Ph.D. Research scholar in the Department of Computer Science and Engineering at SRM Institute of Science and Technology, Tiruchirappalli, Tamil Nadu, India. She received the B.Tech. Degree in Information Technology and the M.E. degree in Computer Science and Engineering from Anna University, Chennai, in 2012 and 2015, respectively. She has over 10 years of teaching experience and has authored 10 research articles in international journals, 10 papers in international conferences and 2-book chapter. Her current research focuses on computer networks, network security, and intelligent firewall systems. Her broader research interests include network security, image and data processing, biometrics, medical image analysis, and pattern recognition. She can be contacted at ashamephd@gmail.com and ashamecse@gmail.com

Index

www.ingramcontent.com/pod-product-compliance
Lightning Source LLC
LaVergne TN
LVHW080906300926
849162LV00003B/572
* 9 7 9 8 3 3 7 3 9 5 4 9 4 *